COLLOQUIA MATHEMATICA
SOCIETATIS JÁNOS BOLYAI, 22.

NUMERICAL METHODS

Edited by:

P. RŐZSA

NORTH-HOLLAND PUBLISHING COMPANY
AMSTERDAM — OXFORD — NEW YORK

Budapest, Hungary, 1980

ISBN North-Holland: 0444 85407 X
ISBN Bolyai: 963 8021 38 1
ISSN Bolyai: 0139 3383

Joint edition published by

JÁNOS BOLYAI MATHEMATICAL SOCIETY

and

NORTH-HOLLAND PUBLISHING COMPANY

Amsterdam — Oxford — New York

In the U.S.A. and Canada:

NORTH-HOLLAND PUBLISHING COMPANY

52 Vanderbilt Avenue

New York, N.Y. 10017

Printed in Hungary
ÁFÉSZ, VÁC
Sokszorosító üzeme

FOREWORD

The Colloquium on Numerical Methods held in Keszthely from 5th to 10th September, 1977 was the third in
this series organized by the Bolyai János Mathematical
Society. The first colloquium on the same subject was
held in Tihany in 1968, the second in Keszthely in 1973.
At the colloquium dealt with here, there were 117 participants from 20 countries.

The main purpose of the Colloquium was to provide
participants and, indeed, readers of this volume, with
an overview on the up-to-date results and problems of the
two major fields dealing with numerical methods, namely
on those of numerical algebra and on those of numerical
solutions of partial differential equations. It was for
this reason that the short communications of the
Colloquium were given in two parallel sections. As the
organizing committee (that consisted of P. Rózsa as
chairman, Katalin Balla as secretary, M. Hosszú, Anna
Lee, J. Gergely and L. Veidinger as members) had no wish
to restrict the topics of the lectures to the main fields
mentioned above, a number of lectures on related topics
were accepted as well. A total of 69 lectures was deli-
vered: of these, there were fourteen talks of 40 minutes,
the others were short communications of fifteen minutes.
Even though certain papers whose topics seemed to be far
from the centre of interest were welcomed at the
Colloquium, the Organizing Committee was, however, obliged
to omit them from this volume; papers already published
elsewhere have similarly not been included. Thus this
volume contains 41 lectures only.

The scientific program of the Colloquium was complemented by social events which greatly contributed to the overall success of the Colloquium. Thus, participants were invited to a concert in the Music Hall of Festetich Castle and went on a half-day's excursion to Badacsony. At Badacsony the József Egry Museum could be visited and a dinner was arranged in Kisfaludy House where the participants had the opputrunity to spend the time in each other's company in an informal atmosphere and in a unique place with its unforgettable panorama.

Finally, grateful acknowledgements are due to Csaba J. Hegedüs for his valuable assistance in editing the manuscript before sending it to press.

PÁL RÓZSA

CONTENTS

SCIENTIFIC PROGRAM

Monday, 5. September

 9.00 - Opening of the Colloquium
Plenary Session Morning, Chairman: H. Heinrich
10.00 - 10.40   A.A.Samarskii: New iterative methods for
                solving difference equations (in Russian)
10.45 - 11.25   O. Taussky Todd: Simultaneously triangu-
                larizable matrices

Plenary Session Afternoon, Chairman: J. Gergely
15.30 - 16.10   A.A. Abramov-I.B. Konyukhova - V.I. Ulya-
                nova - E.S. Birger: On methods of nume-
                rical solution of boundary value problems
                for systems of linear ordinary diffe-
                rential equations (in Russian, presented
                by K. Balla)

Section A, Chairman: E. Schechter
16.30 - 16.45   K. Balla: On error estimates of the sub-
                stitution of the boundedness condition on
                solutions of systems of ordinary diffe-
                rential equations with regular singularity
16.50 - 17.05   E.L. Ortiz: The Lancos' tau method
17.10 - 17.25   A. Galántai: New stability property
                concerning stiff methods
17.30 - 17.45   L. Beretta: A package for the numerical
                solution of ordinary differential initial
                value problem

Monday, 5. September:

Section B, Chairman: E. Vincze

16.30 - 16.45 L. Elsner: Some remarks on iterative
               methods for eigenvectors

16.50 - 17.05 A. Hadjidimos: Accelerated overrelaxation
               method

17.10 - 17.25 W.R. Chmielowiec: On algorithm of the
               computation of the discriminations

17.30 - 17.45 F. Fazekas: Méthodes numériques matricielles
               pour l'analyse des systemes et processus
               aléatoires

Tuesday, 6. September:

Plenary Session Morning, Chairman: M. Fiedler

 9.00 - 9.40   H.J. Stetter: Defect correction and dis-
               cretization methods

 9.45 - 10.25 R.T. Gregory: Computing eigenvalues of
               non-hermitian matrices using a norm-
               -reducing Jacobi-like method

Section A, Chairman: H.J. Stetter

10.45 - 11.00 Gy. Sonnevend: On the optimization of
               adaptive lagorithms

11.05 - 11.20 A. Kokoszkiewicz: An explicit difference
               scheme of solving the Chauchy problem
               for a degenerate hyperbolic equation

11.25 - 11.40 Phan vǎn Hap: On a form of the represen-
               tation of nonlinear operators in normed
               spaces and on a class of k-order iterative
               methods (in Russian)

11.45 - 12.00 M. Dryja: Alternating-direction Galerkin
               methods for parabolic problems

Tuesday, 6. September:

Section B, Chairman: R.T. Gregory
10.45 - 11.00   G.N. de Oliveira: The characteristic values
                of the matrix $A+X^{-1}BX$
11.05 - 11.20   Gy. Varga: Nonzero eigenvalues of singular
                matrices, Jordanian normal form of
                matrices
11.25 - 11.40   M. Petkov: On a solution of the equation
                $(A\lambda^2+B\lambda+C)x=0$
11.45 - 12.00   L. Gerencser: On a stabilization method
                of A. Miele (Convergence and applications)

Plenary Session Afternoon, Chairman: E.M. Bruins
16.00 - 16.40   D. Greenspan: Arithmetic applied mathematics

Section A, Chairman: D. Greenspan
17.00 - 17.15   R. Gorenflo: Mass conserving discretization
                of diffusion equations
17.20 - 17.35   E. Schechter: On the numerical solution
                of the equation of filtration
17.40 - 17.55   T. Desperat: Free boundary problems in
                the theory of fluid flow through porous
                media
18.00 - 18.15   R. Kersner: On some properties of dege-
                nerate parabolic equations (in Russian)
18.20 - 18.35   A.V. Gulin: Stability of difference
                schemes (in Russian)

Section B, Chairman: R.J. Plemmons
17.00 - 17.15   J. Gergely: Methods for inversion of
                matrices
17.20 - 17.35   O. Pokorná: Pseudoinverse matrices
17.40 - 17.55   G. Maeß: Iterative Lösung rechteckiger
                linearer Gleichingssysteme und verallge-
                meinerte Matrixinverse

Tuesday, 6. September:

18.00 - 18.15   J. Mikloško: A fast algorithm for repeated
                computation of linear recurrence relation

Wednesday, 7. September:

Plenary Session Morning, Chairman: O. Taussky Todd
 9.00 - 9.40   V.N. Faddeeva - D.K. Faddeev: Parallel
                computations in linear algebra
 9.45 - 10.25   M. Fiedler: Combinatorial aspects in the
                theory of matrices and numerical analysis

Section A, Chairman: A.A. Samarskii
10.45 - 11.00   K. Barra - M. Kovács: Computational
                aspects in the problem of system identi-
                fication
11.05 - 11.20   É. Gyurkovics - T. Vörös: On a compu-
                tational method of the time-optimal
                control problem (in Russian)
11.25 - 11.40   L. Adamczyk: Monotone spline interpo-
                lations and optimal control problem
                (in Russian)

Section B, Chairman: A. Hadjidimos
10.45 - 11.00   D. Richter: About handling great linear
                systems with respect to using external
                storage
11.05 - 11.20   L. Malina: On carry over methods for
                special systems of linear equations
11.25 - 11.40   J. Abaffy: A class of dual matrix methods

Thursday, 8. September:

Plenary Session Marning, Chairman: J. Todd
 9.00 - 9.40   L. Collatz: Numerical method for singular
                boundary value problems

Thursday, 8. September:

9.45 - 10.25 R. Ansorge: Existence and numerical rep-
              resentation of generalized solution of
              nonlinear partial initial value problems

Section A, Chairman: M. Hosszú

10.45 - 11.00 R. Zezula: Pseudoinverses and their use
               to iterative solution of singular operator
               equations

11.05 - 11.20 F. Aleixo de Oliveira: Numerical solution
               of two point boundary value problem by
               spline function

11.25 - 11.40 J. Chábek: Exponentielle spline inter-
               polation

11.45 - 12.00 S. Sallam: On the stability of quasi-
               double step spline function approximations
               for solutions of initial value problems

Section B, Chairman: G.N. de Oliveira

10.45 - 11.00 R.J. Plemmons: Generalizations of the
               Ostrowski-Reich theorem

11.05 - 11.20 L. Halada: A recursive algorithm for the
               determinant of a t-diagonal matrix

11.25 - 11.40 A. Meskó: An optimum interpolation method
               for twodimensional data

11.45 - 12.00 L. Hibbey: Remarks on building matrices

Plenary Session Afternoon, Chairman: D.K. Faddeev

16.00 - 16.40 I.V. Fryazinov: Difference schemes for
               the Poisson equation in a polygon (in
               Russian)

Section A, Chairman: R. Ansorge

17.00 - 17.15 Ž. Mejran: The direct solution of the
               discrete Poisson equation

Thursday, 8. September:

17.20 - 17.35   G. Avdelas - A. Hadjidimos: The use of
                Wachspress parameters for the numerical
                solution of three-dimensional elliptic
                problems
17.40 - 17.55   A.S. Ilinskii: Direct inverstigation
                methods for the exterior boundary value
                problems of elliptic equations (in Russian)
18.00 - 18.15   H.R. Farzan - Gy. Molnárka: On uniform
                convergence and stability of a finite-
                -difference scheme for weakly nonlinear
                parabolic equations with cylindrical
                symmetry (in Russian)
18.20 - 18.35   Gy. Adler: Numerical problems in the
                digital simulation of thermal secondary
                oil recovery by underground combustion

Section B, Chairman: E.L. Ortiz
17.00 - 17.15   T. Janaszak: Functions in economy (in
                Russian)
17.20 - 17.35   M. Kovács: A continuous method for the
                nonlinear programming problem with
                equality constrains (in Russian)
17.40 - 17.55   E. Vincze: Über ein verbessertes Lösungs-
                verfahren der klassischen Iterations-
                gleichung
18.00 - 18.15   W. Solak: Iteration methods for inverse
                and convex operators
18.20 - 18.35   P.C.G. Vassiliou: (presented by A. Yeyios)
                Predicting the service distribution in
                manpower systems

Friday, 9. September:

Plenary Session Morning, Chairman: P. Rózsa

  9.00 - 9.40   J. Todd: The many limits of mixed means
  9.45 - 10.25   T.Frey: On the stability of spline
                approximations

Section A, Chairman: T. Frey
10.45-- 11.00   F. Schipp: Schnelle Fouriertransformation
                und bedingter Erwartungswert
11.05 - 11.20   I. Környei: Fehlerabschätzungen der
                Kubaturenformeln uneigentlicher Integrale
11.25 - 11.40   E.M. Bruins: The history of numerical
                methods

Section B, Chairman: L. Veidinger
10.45 - 11.00   F.A. Musa - H. Eleyan: The deflexion of a
                simply supported rectangular plate under
                uniform pressure
11.05 - 11.20   F. Juhász - K. Mályusz: Application of
                numerical methods in the cluster analysis
11.25 - 11.40   H. Schwetlick: A superlinearly convergent
                method for determining critical points of
                curves implicitly defined by nonlinear
                equations

Plenary Session Afternoon, Chairman: L. Collatz
16.00 - 16.45   H. Heinrich: Über Gauß' Beiträge zur
                Numerischen Mathematik
17.00   Closing of the Colloquium

ABAFFY, J., Computer and Automation Institute of the
Hungarian Academy of Sciences, 1111 Budapest,
Kende u. 13-17. Hungary

ADAMCZYK, L., Oskar Lange Academy of Economics, Institute
of Cybernetics, 53-345 Wrocław, ul. Komandorska
118/120, Poland

ADLER, Gy., Mathematical Institute of the Hungarian
Academy of Sciences, 1053 Budapest, Reáltanoda u.
13-15., Hungary

ALBRYCHT, J., 60-323 Poznan, ul. Grochowska 40/10, Poland

ANSORGE, R., Universität Hamburg, Institut für Angewandte
Mathematik, D-2000 Hamburg 13, Bundesstr. 55, GFR

ARANY, Ilona, MÜM SzÁMTI, 1089 Budapest, Reguly A. u.
57/59, Hungary

AVDELAS, G., University of Ioannina, Department of
Mathematics, Ioannina, Greece

ÁCS, G., Central Research Institute for Physics of the
Hungarian Academy of Sciences 1525 Budapest, P.O.B.
49., Hungary

BAJCSAY, P., Technical University of Budapest, 1111 Budapest, Müegyetem rkp. 3/9., Hungary

BALATONI, J., Bolyai János Mathematical Society, 1368
Budapest, pf. 240. Hungary

BALLA, Katalin, Computer and Automation Institute of the
Hungarian Academy of Sciences, 1111 Budapest, Kende
u. 13-17, Hungary

BANCSIK, Zs., Technical University of Heavy Industry,
Mathematical Institute, 3515 Miskolc-Egyetemváros
Hungary

BARRA, K., Computing Centre for Universities, 1093 Buda-
        pest, Dimitrov tér 6. Hungary

BARTA, S., Research Institute for Applied Computer Sciences,
        H-1536 Budapest, P.O.Box 227, Hungary

BERETTA, L., ENEL Centro Ricerca di Automatica, 20133
        Milano, Via V. Peroni 74/2, Italy

BOWMAN, Bonita, University of Wisconsin, Department of
        Mathematics, Madison, Wisconsin 53706, USA

BREDENDIEK, Elisabeth, Rechenzentrum der Universität,
        2000 Hamburg 13, Rothenbaumchaussee 81, GFR

BRUINS, E.M., Amsterdam-Z 1, Joh. Verhulststraat 185,
        The Netherlands

CHÁBEK, J., 602 00 Brno, Barvičova 23, Czechoslovakia

RONKA-CHMIELOWIEC, W., Oskar Lange Akademy of Economics,
        Institute of Cybernetics, 53-345 Wroclaw, ul.
        Komandorska 118/120, Poland

COATMÉLEC, C., Institut National des Sciences Appliquées,
        20 Avenue des Buttes de Coesmes, BP 14 A, 35031
        Rennes Cedex, France

COLLATZ, L., Universität Hamburg, Institut für Angewandte
        Mathematik, D-2000 Hamburg 13, Bundesstrasse 55, GFR

DESPERAT, T., 00-132 Warszawa, Grzybowska 9 m 702, Poland

DIÓSI, L., Central Research Institute for Physics of the
        Hungarian Academy of Sciences, 1525 Budapest, P.O.
        Box 49., Hungary

DRASNY, Ágota, Computer and Automation Institute of the
        Hungarian Academy of Sciences, 1111 Budapest,
        Kende u. 13-17. Hungary

DRYJA, M., University of Warsaw, Department of Mathematics,
        00-901 Warsaw, PKiN 8p. 850, Poland

EGELI, Gy., Central Research Institute for Physics of the
        Hungarian Academy of Sciences, 1525 Budapest, P.O.
        Box 49., Hungary

ELSNER, L., Universität Bielefeld, Fakultät für Mathematik,
     4800 Bielefeld, Postfach 8640, GFR

FADDEEV, D.K., Institute of Mathematics, Fontanka 25,
     191011 Leningrad, USSR

FADDEEVA,Vera N., Institute of Mathematics, Fontanka 25,
     191011 Leningrad, USSR

FARAGÓ, I.,   MÜM SzÁMTI   1089 Budapest, Reguly A. u. 57/59.
     Hungary

FARKAS, Éva Central Research Institute for Physics of the
     Hungarian Academy of Sciences, H-1525 Budapest,
     P.O.Box 49. Hungary

FARZAN, H.R., Eötvös Lorånd University, Department of
     Mathematics, 1088 Budapest, Múzeum krt 6/8. Hungary

FAZEKAS, F., Technical University of Budapest, 1111 Buda-
     pest, Műegyetem rkp. 3/9. Hungary

FIEDLER, Matematicky Ustav ČSAV, 115 67 Praha 1, Žitna 25,
     Czechoslovakia

FÓNYAD, Z., Technical University of Heavy Insdustry,
     Mathematical Institute, 3515 Miskolc-Egyetemváros,
     Hungary

FÓTI, Galina, Computing Centre for Universities, 1093
     Budapest, Dimitrov tér 8. Hungary

FRYAZINOV, I.V., Institute for Applied Mathematics,
     Moscow A-47, Miusskaja pl. 4. USSR

FREY, T., Technical University of Budapest, 1111 Budapest,
     Műegyetem rkp. 3/9. Hungary

GALÁNTAI, A., University of Agricultural Sciences, Mathe-
     matical Institute, 2103 Gödöllő, Hungary

GERENCSÉR, L., Computer and Automation Institute of the
     Hungarian Academy of Sciences, 1111 Budapest,
     Kende u. 13-17. Hungary

GERGELY, J., Computer and Automation Institute of the
     Hungarian Academy of Sciences, 1111 Budapest,
     Kende u. 13-17. Hungary

GORENFLO, R., Freie Universität, FB Mathematik, D-1000
     Berlin /West/ 33, Arnimallee 2-6, West Berlin

GREENSPAN, D., University of Texas, Mathematics Department,
     Arlington, Texas 76019, USA

GREGORY, R.T., University of Tennessee, Department of
     Computer Science, Knoxwille, Tennessee 37916, USA

GULIN, A.V., Institute for Applied Mathematics,
     Moscow A-47, Miusskaja pl. 4. USSR

GYURKOVICS, Éva, Computing Centre for Universities, 1093 Bu-
     dapest, Dimitriv tér 8, Hungary

HADJIDIMOS, A., University of Ioannina, Department of
     Mathematics, Ioannina, Greece

HALADA, L., Institute of Technical Cybernetics, Slovak
     Academy of Sciences, 809 31 Bratislava, Dúbravska 1,
     Czechoslovakia

vǎn HAP, P., University of Hanio, Hanoi, Vietnam

HEINRICH, H., Technische Universität Dresden, Sektion
     Mathematik,   8027 Dresden, Zellerschen Weg
     12/14, GDR

HIBBEY, L., Institute of Energetics, 1027 Budapest, Bem
     rkp. 33., Hungary

HORAČEK, J., Faculty of Mathematics and Physics of the
     Charles University, Department of Theoretical Physics,
     121 16 Prague 2. Ke Karlovu 3, Czechoslovakia

HOSSZÚ, M., University of Agricultural Sciences, Mathe-
     matical Institute, 2103 Gödöllő, Hungary

ILINSKII, A.S., Computer Centre, Moscow State University,
     117234 Moscow, USSR

JANASZAK, T., Oskar Lange Academy of Economics, Institute
     of Cybernetics, 53-345 Wrocław, ul Komandorska
     118/120, Poland

JANOSY, J.S., Central Research Institute for Physics of
     the Hungarian Academy of Sciences, 1525 Budapest,
     P.O.Box 49. Hungary

JUHÁSZ, F., Computer and Automation Institute of the
     Hungarian Academy of Sciences, 1111 Budapest, Kende
     u. 13-17. Hungary

KERSNER, R., Computer and Automation Institute of the
     Hungarian Academy of Sciences, 1111 Budapest,
     Kende u. 13-17., Hungary

KOKOSZKIEWICZ, A., 00-132 Warszawa, Grzybowska 9 m. 712,
     Poland

KOLLÁR, A., Technical University of Heavy Industry,
     Mathematical Institute, 3515 Miskolc-Egyetemváros
     Hungary

KOVÁCS, Margit, Computing Centre for Universities, 1093
     Budapest, Dimitrov tér 8, Hungary

KÖRNYEI, I., Computing Centre for Universities, 1093 Bu-
     dapest, Dimitrov tér 8., Hungary

KRISZTINKOVICS, F., University of Agricultural Sciences,
     Mathematical Institute, 2103 Gödöllő, Hungary

LEE, Anna, Mathematical Institute of the Hungarian
     Academy of Sciences, 1053 Budapest, Reáltanoda u.
     13-15. Hungary

LUKÁCS, Márta, Computer and Automation Institute of the
     Hungarian Academy of Sciences, 1111 Budapest, Kende
     u. 13-17. Hungary

MAEß, G., W.-Pieck-Universität, Sektion Mathematik,
     25 Rostock, Universitätsplatz 1, GDR

MALINA, Ľ., UAM VT PF-UK, 816 31 Bratislava, Mlynská dolina,
     Czechoslovakia

MÁLYUSZ, K., Research Institute for Applied Computer
     Sciences, H-1536 Budapest, P.O.Box 227, Hungary

MEJRAN, Z., University of Warsaw, Palace Kulturi i Nauki,
     Warsaw, Poland

MELKES, F., VÚES, 657 65 Brno, Mostecká 26, Czechoslovakia

MESKÓ, Andrea, Compuer and Automation Institute of the
     Hungarian Academy of Sciences, 1111 Budapest,
     Kende u. 13-17. Hungary

MIKLOŠKO, J., Institute of Technical Cybernetics, Slovak
    Academy of Sciences, 809 31 Bratislava, Dubravska 3,
    Czechoslovakia
MIKULÍK, M., VÚES, 657 65 Brno, Mostecká 26,
    Czechoslovakia
MOLNÁRKA, Gy., Eötvös Loránd University, Department of
    Mathematics, 1088 Budapest, Muzeum krt. 6/8.
    Hungary
MÓRICZ, F., József Attila University, Bolyai Institute
    6720 Szeged, Aradi vértanúk tere 1., Hungary
MUSA, F., University of Kuwait, Department of Mathematics,
    Kuwait
OBÁDOVICS, Gy. J., MÜM SzÁMTI 1089 Budapest, Reguly A.
    u. 57/59, Hungary
OLIVEIRA, Fernanda Aleixo, University of Coimbra,
    Department of Mathematics, Portugal
de OLIVEIRA, G.N., University of Coimbra, Instituto de
    Matematica, Coimbra, Portugal
ORTIZ, E.L., University of London, Imperial College,
    London SW7 2BZ, England
ÓHEGYI, E., Research Institute for Applied Computer
    Sciences, H-1536 Budapest, P.O.Box 227, Hungary
ÖRDÖG, L., Bolyai János Matematical Society, 1368 Budapest,
    P.O.Box 240., Hungary
PERNECZKY, L., Central Research Institute for Physics of
    the Hungarian Academy of Sciences, H-1525 Budapest,
    P.O.Box 49., Hungary
PETKOV, M., ul. "Prof. G. Boncev" BL.8. 1113 Sofia 13,
    Bulgaria
PLEMMONS, R.J., University of Tennessee, Computer Science,
    Department, Knoxville, Tennessee 37916, USA
POKORNÁ, Olga, MFF UK, 118 00 Praha 1, Malostranské
    Nam. 25, Czechoslovakia

POVILAITIS, S., National Planing Office, Computer Centre,
    1149 Budapest, Angol u. 27., Hungary

RENNER, G., Computer and Automation Institute of the
    Hungarian Academy of Sciences, 1111 Budapest, Kende
    u. 13-17. Hungary

RICHTER, D., Zentrum für Rechentechnik der Akademie,
    1199 Berlin, Rudower Chaussee 5, GDR

RÓZSA, P., Technical University of Budapest, 1111 Budapest,
    Műegyetem rkp. 3-9. Hungary

RUPP, Mårta, Institute for Electrical Power Research,
    1051 Budapest, Zrinyi u. 1., Hungary

SALLAM, S.M. University of Assiut, Dept. of Mathematics,
    Assiut, Egypt.

SAMARSKII, A.A., Institute for Applied Mathematics, Moscow
    A-47, Miusskaja pl. 4. USSR

ŠANTAVÁ, S., 602 00 Brno, Barvicova 85, Czechoslovakia

SÁNDOR, I., Technical University of Budapest, 1111 Buda-
    pest, Műegyetem rkp. 3-9. Hungary

SCHECHTER, E., University of Cluj, Department of Mathematics,
    3400 Cluj, str. Kogalniceanu 1, Rumania

SCHIPP, F., Eötvös Loránd University, Department of
    Mathematics, 1088 Budapest, Múzeum krt. 6-8. Hungary

SCHMIDT, Rita, Hahn Meitner Institut GmbH, D-1000 Berlin
    39, Glienicker Str. 1000, West Berlin

SCHNEIDER, A., Technische Hochschule, Sektion Mathematik,
    Karl-Marx-Stadt, Reichenhainer Str. 41, GDR

SOLAK, W., Kraków, ul. Dluga 17/2, Poland

SONNEVEND, Gy., Eötvös Loránd University, Department of
    Mathematics, 1088 Budapest, Múzeum krt 6/8. Hungary

STETTER, H.J., Technischer Universität Wien, A-1040 Wien,
    Gußhaus   Str. 27, Austria

SCHWETLICK, H., Technische Universität Dresden, Sektion
    Mathematik, 8027 Dresden, Mommsenstr. 13, GDR

SZILÁGYI, P., Research Institute for Applied Computer
     Sciences, H-1536 Budapest, P.O.Box 227, Hungary
TAUSSKY TODD, Olga, California Institute of Technology,
     Department of Mathematics, 253-37, Pasadena,
     CA 91125, USA
TODD, J., California Institute of Technology, Department
     of Mathematics, 253-37, Pasadena, CA 91125, USA
VARGA, Gy., Computer and Automation Institute of the
     Hungarian Academy of Sciences, 1111 Budapest,
     Kende u. 13-17. Hungary
VASS, Sz., Central Research Institute for Physics of the
     Hungarian Academy of Sciences, H-1525 Budapest, P.O.
     Box 49. Hungary
VEIDINGER, L., Mathematical Institute of the Hungarian
     Academy of Sciences, 1053 Budapest, Reáltanoda u.
     13-15., Hungary
VÉRTES, A., Central Research Institute for Physics of the
     Hungarian Academy of . Sciences, H-1525 Budapest,
     P.O.Box 49., Hungary
VICSEK, Mária, Computer and Automation Institute of the
     Hungarian Academy of Sciences, 1111 Budapest,
     Kende u. 13-17., Hungary
VINCZE, E., Technical University of Heavy Industry,
     Mathematical Institute, 3515 Miskolc-Egyetemváros,
     Hungary
VÖRÖS, T., Computing Centre for Universities, 1093 Buda-
     pest, Dimitrov-tér 8., Hungary
YEYIOS, A., University of Ioannina, Department of
     Mathematics, Ioannina, Greece
ZEÖLD, L., National Planing Office, Computer Centre,
     1149 Budapest, Angol u. 27., Hungary
ZEZULA, R., Matem. ústav KU, 186 00 Praha 8, Sokolovská
     83, Czechoslovakia

COLLOQUIA MATHEMATICA SOCIETATIS JÁNOS BOLYAI
22. NUMERICAL METHODS, KESZTHELY (HUNGARY), 1977.

A CLASS OF DUAL-MATRIX METHODS

J. ABAFFY

Methods not using linear minimization and enabling
the minimum of a quadratic form with a positive semidefinite Hesse matrix in finite steps have been intensely
investigated in recent years. H u a n g passed two
important remarks in his paper [1].

1. If $f(x)$ is a quadratic function with positive semidefinite Hessian A, then conjugate directions can
be determined without linear minimization.

2. Let $f(x)$ be a quadratic function with an $n \times n$
positive semidefinite Hessian A, let g_i be the
gradient of $f(x)$ in the point x_i and finally
$y_i = g_{i+1} - g_i$. If g_k of the k-th step is a linear
combination of the previous y_i directions, and if

$$(1) \qquad s_k = -\sum_{i=1}^{k-1} \frac{s_i s_i^T}{s_i^T y_i} g_k \, ,$$

then $x_{k+1} = x_k + s_k$ is a minimum point of $f(x)$.

Here we intend to construct a class of dual-matrix
methods. We note that this class of methods contains non-

symmetric matrix sequences, too.

The class of dual-matrix methods is as follows: Let H_1 be an $n \times n$ positive definite matrix, $x_1 \in R^n$ so that $g_1 \neq 0$ and let B_1 be an arbitrary matrix. For $i = 1, 2, \ldots$ define the following steps of iteration:

$$(2) \qquad s_i = -\alpha_i H_i^T g_i, \qquad\qquad \alpha_i \neq 0$$

with α_i arbitrary or for a non quadratic function it is such that $f(x_{i+1}) < f(x_i)$, where

$$(3) \qquad x_{i+1} = x_i + s_i$$

$$(4) \qquad y_i \equiv g_{i+1} - g_i$$

$$(5) \qquad H_{i+1} = H_i + \gamma_i H_i y_i s_i^T - \frac{\gamma_i \, s_i^T y_i + 1}{y_i^T H_i y_i} H_i y_i y_i^T H_i$$

$$(6) \qquad B_{i+1} = B_i + \frac{1 - \beta_i y_i^T B_i y_i}{s_i^T y_i} s_i s_i^T + \beta_i s_i y_i^T B_i -$$

$$- \frac{\delta_i y_i^T B_i y_i + 1}{s_i^T y_i} B_i y_i s_i^T + \delta_i B_i y_i y_i^T B_i \, .$$

Here γ_i, β_i and δ_i are arbitrary parameters. If a direction g_k is a linear combination of the previous y_i-s then

$$(7) \qquad s_k = -B_k g_k .$$

Naturally this class has a quadratic termination property.

Note that both sequences H_i, B_i can be non-symmetric, too (all methods of Huang are symmetric ones).

Further note that if

$$\beta_i = -\frac{\delta_i y_i^T B_i y_i + 1}{s_i^T y_i}$$

that is, the matrix sequence B_i is symmetric, then we obtain the Broyden class of quasi-Newton methods. The termination criterion can be determined according to Huang as

$$\left| \frac{g_k^T s_k}{g_k^T g_k} \right| < \varepsilon,$$

where ε is given. After n steps, the subsequent step is determined as

$$H_{k+1} = B_k$$
$$\quad\quad \text{if } B_i \text{ symmetric, and } \delta_i \le 0$$
$$B_{k+1} = B_k$$

(8)   or

$$H_k = H_1$$

$$B_k = B_1 \quad\quad \text{otherwise.}$$

In his diplom work László Vavra has tested some members of the class for the following test functions

		min.p.	f.v.
(A):	$f(x) = 100(x_2 - x_1^2)^2 + (1-x_1)^2$	(1,1)	0
(B):	$f(x) = (x_2 - x_1^2)^2 + (1-x_1)^2$	(1,1)	0

		min.p.	f.v.

$$\text{(C):}\quad f(x)=(x_1+10x_2)^2+5(x_3-x_4)^2+ \qquad (0,0,0,0) \qquad 0$$

$$+(x_2-2x_3)^4+10(x_1-x_4)^4$$

$$\text{(D):}\quad f(x) = (e^{x_1}-x_2)^4+100(x_2-x_3)^6+ \qquad (0,1,1,1\pm n\pi) \qquad 0$$

$$+\mathrm{tg}^4(x_3-x_4)+x_1^8+(x_4-1)^2$$

$$\text{(E):}\quad f(x) = x_1(2x_1-x_3-1)+x_2(x_2-3)+ \qquad (2/13,3/2, \qquad -2.865.$$

$$+x_3(2x_3+x_4+1)+x_4(x_4-1) \qquad\qquad -5/13,9/13)$$

Note that function (D) is a little more complicated than the others.

If we determine a cost function as

$$K_M = \sum_{i=1}^{m} (\ell_i(\varepsilon)+n\cdot d_i(\varepsilon)),$$

where

$$m \qquad\qquad - \text{ number of iterations,}$$

$$\ell_i(\varepsilon) \quad - \text{ number of function evaluations in one step of iteration,}$$

$$d_i(\varepsilon) \quad - \text{ number of gradient evaluations in one step of iteration,}$$

then K_M is suitable for measuring the efficiency of the methods. By choosing appropriately the parameters β_i, γ_i, δ_i, three methods of Huang (H-1, H-2, H-3) were chosen and four other methods.

The results are in the following tables:

Table 1.

test func.	method	min. points				n.of it.	cost.f.	n.of m.
		x_1	x_2	x_3	x_4			
A (−1,2,1)	1	$0.332.10^{-12}$	$0.99...$	$0.99...$		46	156	*H*−2
	2	$0.192.10^{-13}$	$1.00...$	$1.00...$		38	134	*H*−3
	3	$0.137.10^{-15}$	$1.00...$	$1.00...$		57	205	unsym
	4	$0.112.10^{-16}$	$1.00...$	$1.00...$		56	213	unsym
	5	$0.154.10^{-14}$	$1.00...$	$0.99...$		42	151	
	6	$0.626.10^{-19}$	$1.00...$	$1.00...$		54	205	
	7	$0.225.10^{-16}$	$1.00...$	$1.00...$		71	435	*H*−1
B (−1,2,1)	1	$0.00...$	$1.00...$	$1.00...$		13	40	*H*−2
	2	$0.760.10^{-15}$	$1.00...$	$1.00...$		11	34	*H*−3
	3	$0.105.10^{-20}$	$1.00...$	$1.00...$		17	56	unsym
	4	$0.00...$	$1.00...$	$1.00...$		16	55	unsym
	5	$0.718.10^{-16}$	$1.00...$	$1.00...$		12	38	
	6	$0.997.10^{-16}$	$1.00...$	$1.00...$		12	41	unsym
	7	$0.939.10^{-14}$	$1.00...$	$1.00...$		17	64	*H*−1
C (−3,−1,0,1)	1	$0.513.10^{-16}$	$-0.58.10^{-4}$	$0.58.10^{-5}$	$0.10.10^{-4}$	53	270	*H*−2
	2	$0.363.10^{-15}$	$-0.10.10^{-3}$	$0.10.10^{-4}$	$-0.24.10^{-4}$	49	250	*H*−3
	3	$0.100.10^{-15}$	$0.83.10^{-4}$	$-0.83.10^{-5}$	$0.30.10^{-4}$	73	580	unsym
	4	$0.246.10^{-12}$	$0.55.10^{-3}$	$-0.55.10^{-4}$	$0.17.10^{-3}$	59	481	unsym

Cont. of Table 1.

test func.	method	x_1	x_2	x_3	x_4	n.of it.	cost.f.	n.of m.
			min. points					
C	5	$0.125.10^{-13}$	$-0.25.10^{-3}$	$-0.25.10^{-4}$	$0.15.10^{-3}$	$0.15.10^{-3}$	40	205	
$(-3,-1,0,1)$	6	$0.354.10^{-13}$	$0.35.10^{-3}$	$-0.35.10^{-4}$	$0.12.10^{-3}$	$0.12.10^{-3}$	68	549	unsym
	7	$0.886.10^{-12}$	$-0.10.10^{-3}$	$0.10.10^{-4}$	$0.38.10^{-3}$	$0.38.10^{-3}$	60	351	$H-1$
	1	$0.131.10^{-7}$	0.0334	1.026	1.009	1.000...	51	259	$H-2$
	2	$0.237.10^{-8}$	0.0129	1.007	1.006	1.000...	58	321	$H-3$
D	3	$0.229.10^{-7}$	-0.0390	0.9684	0.989	1.000...	37	196	unsym
$(1,2,2,2)$	4	$0.171.10^{-8}$	-0.0215	0.980	0.994	1.000...	34	180	unsym
	5	$0.922.10^{-8}$	0.0278	1.023	1.009	1.000...	51	280	
	6	$0.122.10^{-7}$	-0.0312	0.971	0.990	1.000...	31	165	unsym
	7	$0.125.10^{-9}$	0.0136	1.011	1.001	0.999...	62	343	$H-1$
	1	-2.8654	0.1538	1.500...	-0.3846	0.6923	6	31	$H-2$
	2	-2.8654	0.1538	1.500...	-0.3846	0.6923	6	31	$H-3$
E	3	-2.8654	0.1538	1.500...	-0.3846	0.6923	20	247	unsym
$(20,20,20,20)$	4	-2.8654	0.1538	1.500...	-0.3846	0.6923	22	288	unsym
	5	-2.8654	0.1538	1.500...	-0.3846	0.6923	6	31	
	6	-2.8654	0.1538	1.500...	-0.3846	0.6923	21	252	unsym
	7	-2.8654	0.1538	1.500...	-0.3846	0.6923	6	31	$H-1$

Comparing these with the Davidon-Fletcher-Powell
method and the conjugate direction method of Powell, the
following results have been obtained:

test f.	PO		D-F-P		worst			best		
	n.of it.	cost. f.	n.of it.	cost. f.	n.of it.	cost. f.	n.of m.	n.of it.	cost. f.	n.of m.
A	25	532	12	146	71	435	H-1	38	134	H-3
B	3	148	7	84	17	64	H-1	11	34	H-3
C	107	2669	391	3900	73	580	unsym 1	40	205	5
D	76	2065	552	4878	62	343	H-1	31	165	unsym 3
E	25	825	6	139	22	288	unsym 2	6	31	

Table 2.

Note that even the worst dual matrix method is more
efficient than the D-F-P or Powell method for function
(D).

REFERENCES

[1] H.Y. Huang, Method of dual matrices for function
     Minimisation, *JOTA* 13(1974), 519-537.

[2] J. Abaffy, On a class of dual matrix methods (in
     Linear Algebra Appl. (1979) (to appear)

J. Abaffy
Computer and Automation Institute
of the Hungarian Academy of Sciences
Budapest, Kende u.13-17.
H-1111

COLLOQUIA MATHEMATICA SOCIETATIS JÁNOS BOLYAI
22. NUMERICAL METHODS, KESZTHELY (HUNGARY), 1977.

ON METHODS OF NUMERICAL SOLUTION OF BOUNDARY VALUE PROBLEMS FOR SYSTEMS OF LINEAR ORDINARY DIFFERENTIAL EQUATIONS

A.A. ABRAMOV — E.S. BIRGER —
NADYEZHDA B. KONYUKHOVA — VALENTYINA I. ULYANOVA

INTRODUCTION

As far as is known, the boundary value problems for systems of linear ordinary differential equations are often encountered when solving problems of mathematical physics. For this reason many thoughts were contributed to the development of the methods for solving such problems and a vast amount of literature is available in this topic.

The following general considerations are of great importance. Let us assume that it is necessary to solve some linear multipoint problem (it may be with conditions that connect the values of unknown function at different points) for a linear system. This problem could be treated as follows: We shall determine some particular solution of a starting nonhomogeneous linear system by solving some initial value problem (completely arbitrary, not related to the boundary value problem at hand). We shall also determine some fundamental

33

system of solutions, which corresponds to a homogeneous
system, also by solving the initial value problem for
each function of this system. The general solution of
the starting system of equations is presented in the
form of an expression which contains these calculated
functions and arbitrary constants that enter linearly.
These constants are determined from a system of linear
algebraic equations which are obtained by taking into
account the boundary conditions. It is of interest that
such a method, despite its apparent universality, is
sometimes practically quite unsuitable, as it leads to
catastrophic losses of accuracy. But this is an important
method, since the solution of the problem at hand can be
reduced to the solution of several initial value problems. This fact suggests a general idea of developing
methods for solving these problems as to express the
solution in terms of some auxiliary functions, so that
1) the initial problem is reduced to several initial-
   value problems for these auxiliary functions;
2) the solution of these auxiliary initial value problems and other problems, which occur when calculating
   the unknown function in terms of these auxiliary
   problems, will not result in large losses of accuracy.

It is apparent that in this case one should restrict oneself to well-conditioned initial problems. It
is difficult to know which of the boundary value problems
are stable as related to small variations in the values
that appear in the problems. Some simple considerations
on this problem are given in [1], chapter IX.

The marching method (or a factorization method) was
proposed even in the forties and the fifties for certain
types of well-conditioned boundary value problems. But,
this method sometimes called for some restrictive conditions which are to be imposed on the initial problem.

Later the method, based on similar considerations as the
marching method, but free from the above mentioned draw-
backs, were advanced in a number of papers. In a classical marching algorithm the solution of auxiliary Riccati
equations may have singularities. It is understood that
a classical marching can be prolonged by changing over
to new variables when this singularity is approached.
The emergence of different variants of marching is mainly
associated with the desire to avoid this change-over,
because, from the computational point of view this
change-over is quite cumbersome in the general case.
Some methods of this kind will be described below. We
shall also dwell on some problems associated with the
use of the above mentioned methods for those cases which
involve difficulties in practical realizations. The
classical and the most latest theories of the marching
method are described in the majority of books on compu —
tational mathematics. We recommend especially [1-3].

§.1.

We shall dwell in greater detail on a variant of
the marching method [4] which served as basis for several other methods.

Let us assume that a system of n differential
equations is given on the interval $a \leq x \leq b$

(1) $\qquad y' + P(x)y = f(x).$

Let the condition $y(\hat{x}) \in M^{(s)}$ be specified in the point
$\hat{x}$. Here $M^{(s)}$ is some s-dimensional linear manifold.
Then for every x there exists an s-dimensional linear
manifold $M^{(s)}(x)$ such that for every solution $y(x)$
of system (1)

$$y(\hat{x}) \in M^{(s)} \iff y(x) \in M^{(s)}(x)$$

holds. Condition $y(x) \in M^{(s)}(x)$ is obtained by transferring the condition $y(\hat{x}) \in M^{(s)}$ to the point x by using equation (1). For solving a boundary value problem, when the boundary conditions are given for several points, it is necessary to transfer these conditions to that point in which the solution is to be found, and then solve the obtained system of linear algebraical equations.

The manifold $M^{(s)}(x)$ is uniquely determined. It does not depend on any ways of its specification. Hence, for a numerical realization it is natural to specify it by quantities that change as smoothly as possible on varying x. One of the methods to achieve this is as follows: we shall preset the basis $T(x)$ in the conjugate space in the orthogonal complement of $M^{(s)}(x)$ and take the projection $T(x)$ on the subspace created by $T(x+dx)$ as $T(x+dx)$ (it is determined uniquely). The projection is done by choosing a scalar product associated with matrix $Q(x)$. Choosing this matrix depends on the research worker by whom the main points of the problem are taken into account. Finally, the transfer of the condition

$$(2) \qquad \psi_0^* y(\hat{x}) = \beta_0 \qquad (\det(\psi_0^* \psi_0) \neq 0)$$

is determined by the formulae ($\psi(x)$ is a matrix the columns of which form the basis in $T(x)$):

$$(3) \qquad \psi' - P^* \psi + \psi (\psi^* Q \psi)^{-1} \psi^* Q P^* \psi = 0,$$

$$(4) \qquad \psi(\hat{x}) = \psi_0,$$

$$(5) \qquad \beta' + \psi^* P Q \psi (\psi^* Q \psi)^{-1} \beta = \psi^* f,,$$

$$(6) \qquad \beta(\hat{x}) = \beta_0.$$

The condition $y(x) \in M^{(s)}(x)$ takes the form

$$\psi^*(x) y(x) = \beta(x).$$

This method has the following properties:

(1)   it is numerically stable exactly to that extent as the behaviour of $M^{(s)}(x)$ is stable on varying x;

(2)   the estimates

$$\frac{\max\limits_{[a,b]} r(x)}{\min\limits_{[a,b]} r(x)} \le e^{\int_a^b m(t)\,dt},$$

$$\frac{\max\limits_{[a,b]} s(x)}{\min\limits_{[a,b]} s(x)} \le e^{\int_a^b m(t)\,dt},$$

hold [5], where $r(x) = |\psi^*(x) Q(x) \psi(x)|$,

$$s(x) = |(\psi^*(x) Q(x) \psi(x))^{-1}|^{-1},$$

$Q(x)$ is presented in the form

$$Q(x) = G^*(x) G(x),$$

$$m(x) = |G^{*-1}(x) Q'(x) G^{-1}(x)|$$

and $\psi(x)$ is some solution of equation (3). (By a matrix norm we mean the Euclidean norm: $|U| = \sqrt{\max \lambda_{U^* U}}$). Often it is sufficient to take $Q(x) \equiv E$, where E is the identity matrix. Then the relation

(7) $\qquad \psi^*(x)\psi(x) \equiv \text{const}$

can serve as control for the numerical integration of
equation (3). This method was applied extensively in the
solution of physical problems (see, for example [5],
[6]-[9]*).

A realization of the described algorithm for transferring the boundary conditions when $Q(x) \equiv E$ has been
developed in ALGOL [10]. Two modified versions of this
method are given in [1], [11]. The variant [11], en-
suring the fulfilment of the condition

$$\psi^*(x)\psi(x)+\beta(x)\beta^*(x) \equiv \text{const}$$

instead of (7), enables even a stronger control of the
numerical stability of calculations. In the method [1],
through the use of an equation similar to (3), the
basis of $M^{(s)}(x)$ itself is transferred instead of the
basis of $T(x)$, and the concept of inverse marching
arises; this method permits stable solving boundary
value problems for system (1) in those cases where the
matrix $P(x)$ has singularities at the ends of the
finite or infinite interval.

§.2.

We do not dwell on this problem in detail, though
at present, several stable methods for solving boundary
value problems are being used. These methods are based
on different ideas.

---

* Here and further on when a reference is made to the
  review report [6], we shall mean the papers cited
  therein.

2.1. Concurrently with the method in [4], the following
method of transfer of boundary conditions for the system
(1) was proposed in [12]. The integration inverval is
divided into a number of small subintervals (orthogonali-
zation steps). The basis of the manifold $M^{(s)}(x)$ is
transferred in each subinterval by the use of the initial
system (1). At the end points of the subintervals, the
basis system of vectors is subjected to orthogonaliza-
tion and normalization, and this protects the basis from
degeneration. The closure of this algorithm (when ortho-
gonalization steps tend to zero) has properties similar
to those of the variant [1] of the method in [4] (see
[1], chapter IX, [13]). The method [12] has found extensive application in a series of problems (see, for
example, [14]-[16]). The method [17] is close to the
described variant of the marching method. It is based
on the idea of orthogonal transformation of the variables.

In practice, the variants of the general classical
marching method (see [18-21]) are also used in solving
boundary value problems for system (1). Here, we shall
not dwell on these variants.

2.2. Several variants of the marching method have been
proposed for second order equations [22]-[26] and for
systems of such equations [27]-[29]. Let us consider a
boundary value problem on the interval $a \le x \le b$ for
a system of n equations of second order:

(8) $\qquad y'' = P(x)y'+Q(x)y+f(x),$

(9) $\qquad A_1 y'(a)+A_2 y(a) = \gamma_1,$

$$(10) \qquad B_1 y'(b) + B_2 y(b) = \gamma_2,$$

where $P(x)$ and $Q(x)$ are arbitrary square matrices of order n with continuous elements, A_i, B_i are square matrices of order n; these matrices are such that the $n \times 2n$ rectangular matrices $[A_1 \ A_2]$ and $[B_1 \ B_2]$ have rank n. In the general case, a variant for the transfer of boundary conditions has been proposed in [29] for non-selfadjoint problems. The variant is based on the idea of orthogonal transformation of variables. But this method is quite cumbersome.

2.2.1. To solve (8)-(10) by the use of a "bilateral" marching variant, which is based on the idea of [4], proceed as follows: the linear manifold $M^{(n)}(x)$ of that solutions of (8), which fulfil condition (9), is given in the form

$$(11) \qquad S(x)y' + C(x)y = \gamma(x)$$

in the (y', y)-space. The marching equations for S, C and γ are obtained from (8), (11) and from the condition

$$S(x)S^*(x) + C(x)C^*(x) = \text{const}$$

for all $x \in [a, b]$. On transferring the boundary conditions (9) and (10) over the whole interval $[a, b]$ through the use of these equations, we find $y(x)$ as a solution in the point x of the corresponding linear algebraic problem.

2.2.2. For $n = 1$ the equations for S, C and γ are of the form:

$$S' = K(x,S,C)S - SP(x) - C,$$

$$C' = K(x,S,C)C - SQ(x),$$

$$\gamma' = K(x,S,C)\gamma + Sf(x),$$

where

$$K(x,S,C) = \frac{\mathrm{Re}\{CS^* + Q(x)SC^* + SS^*P(x)\}}{SS^* + CC^*}$$

is a real-valued function [23]. For inverse marching, the function

$$\delta^*(x) = C^*(x)y' - S^*(x)y$$

is taken and a differential equation is obtained for this function. The solution of problem (8)-(10) is obtained from the following formulae

$$y(x) = \frac{\gamma C^* - \delta^* S}{SS^* + CC^*} \, ,$$

$$y'(x) = \frac{C\delta^* + \gamma S^*}{SS^* + CC^*} \, .$$

The stability of calculations is guaranteed, if the initial problem is well-conditioned.

2.2.3. The selfadjoint problems (8)-(10) are of interest. For these problems the number of parameters, which define $M^{(n)}(x)$, can be reduced. Now we shall consider a case when $n = 1$ in (8)-(10), and all the quantities that determine the problem assume real values, but they are

arbitrary otherwise. Then the following method of solving
problems (8)-(10) is as stable as this problem is well-
-conditioned [22].

We shall normalize the boundary conditions (9) and
(10) so that $A_1^2+A_2^2 = 1$, $B_1^2+B_2^2 = 1$. And we shall write
the equation in the Hesse normal form for $M^{(1)}(x)$ that
is a straight line, which corresponds to the left-hand
boundary condition;

$$(12) \qquad y \sin \Theta(x)+y'\cos \Theta(x) = u(x).$$

We shall transfer the relationship (12) over the whole
interval $[a, b]$ by solving the initial value problem:

$$(13) \qquad \begin{aligned} &\Theta'-\sin^2\Theta-P(x)\sin \Theta \cos \Theta+Q(x)\cos^2\Theta = 0, \\ &\sin \Theta(a) = A_2, \quad \cos \Theta(a) = A_1; \end{aligned}$$

$$(14) \qquad \begin{aligned} &u'-[Q(x)\sin \Theta \cos \Theta+\sin \Theta \cos \Theta+ \\ &+P(x)\cos^2\Theta]u = f(x)\cos \Theta, \\ &u(a) = \gamma_1. \end{aligned}$$

At the right end we shall find $y(b)$ and $y'(b)$ by
taking the boundary conditions (10) into account. For
inverse marching the function

$$(15) \qquad v(x) = y \cos \Theta(x)-y'\sin \Theta(x)$$

is taken and a differential equation is obtained for
this function. Then the solution of the problem (8)-
-(10) is determined from the following formulae

$$y(x) = u(x)\sin \Theta(x)+v(x)\cos \Theta(x),$$

$$y'(x) = u(x)\cos\Theta(x) - v(x)\sin\Theta(x).$$

If the end points a and b lie close to the singularities of equation (10), then the equations of direct marching (13) and (14) should be integrated from the points a and b to some middle point $\hat{x}\in(a, b)$ and the equation of inverse marching for the function (15) should be integrated in the directions towards a and b from the point $\hat{x}$. If a metric is introduced into the (y', y)-space for every x by the quadratic form

$$y^2 + m^2 y'^2,$$

where $m^2(x)$ is some positive function, the choice of which is at our disposal, then (12) is written as

$$y\sin\Theta + m^2 y'\cos\Theta = u(x).$$

The rest of the formulae becomes somewhat complicated.

2.2.4. The method [22] has been generalized in [27] for the selfadjoint problem (8)-(10) for an arbitrary value of n. Let us assume that $P(x) \equiv 0$ in the equation (8), $Q(x)$ is a Hermitian matrix of order n and

$$A_1 A_2^* = A_2 A_1^*, \qquad B_1 B_2^* = B_2 B_1^*.$$

The Hermitian matrix $\hat{\Theta}(x)$ serves as an analogue to the function $\Theta(x)$ (it is symmetrical in the real case). The marching equations are written for the unitary matrix

$$u(x) = -\exp 2i\hat{\Theta}(x).$$

As a result, the problem (8)-(10) is solved in the following manner: Over the whole interval $[a, b]$, we shall transfer the relationship

$$\frac{1}{2}(E-u(x))y' - \frac{i}{2}(E+u(x))y = \xi(x),$$

where $u(x)$ and $\xi(x)$ satisfy appropriate differential equations. This time the condition $u^*(x)u(x) = E$ is satisfied for all values of x. For inverse marching, we shall introduce the function

$$\eta(x) = \frac{i}{2}(E+u(x))y' + \frac{1}{2}(E-u(x))y.$$

A differential equation is also obtained for this function. After solving the corresponding initial value problems for $u(x)$, $\xi(x)$ and $\eta(x)$, we find the solution of the problem (8)-(10) from the following formulae

$$y(x) = \frac{i}{2}(E+u^*)\xi + \frac{1}{2}(E-u^*)\eta,$$

$$y'(x) = \frac{1}{2}(E-u^*)\xi - \frac{i}{2}(E+u^*)\eta.$$

The method is generalized to more general selfadjoint systems of second order and to Hamiltonian system of n equations of first order. The stability of calculations is guaranteed, if the initial problem is well-conditioned.

2.3. We see that in the methods considered, the way of defining (pre-assigning) $M^{(s)}(x)$ is the first problem.

In particular, it is essential, by how many parameters this subspace is given. In a classical marching method it is given by $s\times(n-s)$ values. In the variant described in §.1, this number equals $(n-s)\times n$. The dimension of the Grossman space G (space of all

s-dimensional subspaces of the n-dimensional space)
equals $s\times(n-s)$. For this reason, those marching methods
are the most economical in which the auxiliary initial
value problems are raised for a system of dimension
$s\times(n-s)$ (as in the classical marching method). But
apparently, (as in the case of the classical marching
method) equations without feasible singularities cannot
be introduced this time. Geometrically, it indicates that
it is not possible to introduce a standard system of
coordinates in G without singularities. In order to
form an absolutely universal marching method that is
most economical with respect to the number of equations,
it is necessary to imbed a manifold $\widetilde{G}$ in the linear
space of the lowest dimensionality q such that $\widetilde{G}$ could
be mapped onto G so that this mapping is locally a
diffeomorphism. The value of q is not known to us if
the values n and s are given.

§.3.

The following problems are very interesting: find
the solution of a system of n equations

(16) $\qquad y'+P(x)y = f(x), \quad a \leq x \leq b,$

which satisfies the condition

(17) $\qquad R_1 y(x_1)+R_2 y(x_2)+\ldots+R_N y(x_N) = g.$

Here, $x_1,\ldots,x_N$ are given points in the interval
$[a, b]$ and $R_1,\ldots,R_N,\ g$ are given matrices (the so-
-called problem with undecomposed conditions).

It has been mentioned earlier that the solution of
this problem can be reduced to the solution of several
initial value problems. But, two basic points still
remained

1) which of these problems are well-conditioned;

2) which quantities must be introduced in the case when the problem is well-conditioned in order to obtain a well-conditioned method for solving the initial problem. The last problem becomes non-trivial even if condition (17) turns into the condition $y(a) = y(b)$ i.e. a problem of finding periodic solutions of the system (16) is raised. This problem is widely encountered in practice (as regards this problem, see [30], [31]: some general considerations concerning the general problem (16), (17) are also given in [30]). For the case, if (17) is of the form

(18) $Ay(a)+By(b) = C,$

an excellent, general method is proposed in [17]. This method is the following:

Assuming

$$\hat{y}(x) = \begin{bmatrix} y(x) \\ y(a+b-x) \end{bmatrix} , \quad \hat{f}(x) = \begin{bmatrix} f(x) \\ -f(a+b-x) \end{bmatrix} ,$$

$$\hat{P}(x) = \begin{bmatrix} P(x) & 0 \\ 0 & -P(a+b-x) \end{bmatrix} ,$$

we get for $\hat{y}(x)$ a problem on the interval $[a,(a+b)/2]$ with decomposed boundary conditions of the form:

(19) $\hat{y}'+\hat{P}(x)\hat{y} = \hat{f}(x), \quad a \le x \le (a+b)/2,$

(20) $\psi_1^* \hat{y}(a) = C, \quad \psi_1^* = [A \ B],$

(21) $\qquad \psi_2^* \hat{y}((a+b)/2) = 0, \qquad \psi_2^* = [E, -E].$

Apparently (this has not been studied), the auxiliary
boundary value problem (19)-(21) is stable, if the starting boundary value problem (16)-(18) is stable.

Another method for solving problem (16), (18) has
been proposed and realized in [32], [33]. We would like
to emphasize the fact that a general consideration is
still missing in the general case, which would enable
to solve problem (16)-(17) without requiring additional
analysis, and on the only assumption that this problem
would be well-conditioned.

§.4.

Some other problems also call for their solution by
one or other variant of the marching method, for example
it is the case in eigenvalue problems or nonlinear
problems (when solving the latter by iterative methods
which reduce the initial problem to a sequence of linear
problems). Additional calculational difficulties arise
in solving problems with singularities at the end points
or inside the interval. Ref. [6] gives a review of works,
in which the problems of local stable transfer of boundary conditions out of, or through points of singularity
are solved for some classes of linear and nonlinear
problems. We shall illustrate below (in §.4 and §.5) the
developed methods by examples of some typical problems
of physics by restricting ourselves only to a small
group of problems which have been touched upon in [6].
Here, we cannot cover the whole diversity of applications
either, even which were solved both by the authors of
this paper and by other authors.

4.1. The marching methods are particularly convenient in solving linear homogeneous eigenvalue problems as they do not call for the storage of "marching functions" in computer memory. Now we shall discuss the following problem in more detail, which arises, for example, when solving Schrödinger equation by variational methods [6], [7]:

$$(22) \qquad y''-A(\lambda,\ t)y = 0, \qquad 0 < t < \infty,$$

$$(23) \qquad \lim_{t \to 0} y(t) = 0,$$

$$y = \begin{bmatrix} y_1 \\ \vdots \\ y_n \end{bmatrix}.$$

$$(24) \qquad \lim_{t \to \infty} y(t) = 0,$$

Here, λ is the unknown parameter and the matrix $A(\lambda,\ t)$, when $t \to 0$ and $t \to \infty$, has asymptotic representations of the form:

$$(25) \qquad A(\lambda,\ t) = t^{-2}(A_0+A_1 t+A_2 t^2+\ldots), \qquad t \to 0,$$

$$(26) \qquad A(\lambda,\ t) = A_\infty+\widetilde{A}_1/t+\widetilde{A}_2/t^2+\ldots, \qquad t \to \infty.$$

Let us assume that $\mathrm{Re}\ \lambda_{A_0}+(\mathrm{Im}\ \lambda_{A_0})^2 > 0$ and the eigenvalues of matrix A_∞ do not lie on the nonpositive real semi-axis, then the dimensions of subspaces of the solutions of (22), which are bounded at zero and at infinity, are equal to n. The problem is to determine those values of the parameter λ for which the problem (22)--(24) has nontrivial solutions.

The transfer of boundary conditions (23), (24) out of points of singularity, can be accomplished in the following manner [6], [38], [53]. The solutions of (22), which satisfy condition (23), form an n-dimensional

48

linear subspace $M^{(n)}(t)$ in the (y',y)-space. This subspace, for sufficiently small values of t, can be specified in the form

$$(27) \qquad ty' = \beta(\lambda, t)y, \qquad t \leq t_0.$$

Here, the $n \times n$ matrix $\beta(x)$ is the solution of the initial value problem

$$(28) \qquad t\beta' + \beta^2 - \beta - t^2 A(\lambda, t) = 0, \qquad 0 < t \leq t_0,$$

$$(29) \qquad \beta(\lambda, 0) = \frac{1}{2}E + \frac{1}{2}\sqrt{E + 4A_0},$$

$$(30) \qquad \beta(\lambda, t) \sim \sum_{k=0}^{\infty} \beta_k(\lambda)t^k, \qquad t \to 0,$$

where the coefficients β_k are determined by formal substitution of expansions (25), (30) into (28). (In formulae (29) and later in (33), by the root of matrix D we mean a matrix C such that $C^2 = D$ and the eigenvalues of the matrix C lie in the right-hand half-plane).

   Similarly, the n-dimensional subspace of solutions of (22), which satisfy the condition (24), can be presented in the form

$$(31) \qquad y' = \alpha(\lambda, t)y, \qquad t \geq t_\infty$$

for sufficiently large values of t. Here, the $n \times n$ matrix α is the solution of the initial value problem

$$(32) \qquad \alpha' + \alpha^2 - A(\lambda, t) = 0, \qquad t_\infty \leq t < \infty,$$

$$(33) \qquad \lim_{t \to \infty} \alpha(\lambda, t) = -\sqrt{A_\infty},$$

$$(34) \qquad \alpha(\lambda, t) \sim \sum_{k=0}^{\infty} \alpha_k(\lambda)/t^k, \qquad t \to \infty,$$

where $\alpha_k(\lambda)$ are determined by formal substitution of the expansions (26), (34) into (32).

As a result, problem (22)-(24) is reduced to an equivalent problem on the finite interval $[t_0, t_\infty]$, where, for the boundary condition in the point t_0, relationship (27) is entered, and the relationship (31) is taken in the point t_∞ (with the use of the expansions (30), (34)). The problem is then solved on the finite interval by "shooting" for the parameter λ: for a fixed value of λ, we transfer the boundary conditions (27), (31) from the points t_0 and t_∞ to some point $\hat{t} \in (t_0, t_\infty)$ (after having applied some stable variant of the marching method); we find the eigenvalues of the problem (22)-(24) from the condition of nontrivial compatibility of the system of linear algebraic equations obtained in the point $\hat{t}$.

4.2. In the case of selfadjoint eigenvalue problems, some methods, like, for example, the method of phase functions [34] for one equation of second order, or a variant [27] of the marching method for systems of such equations, enable us to find eigenvalues of given serial numbers.

As an example, we shall consider the problem (22)--(24) with real-valued coefficients for

$$(35) \qquad n = 1, \quad A(\lambda, t) = \lambda^2 + \hat{A}(t), \quad \hat{A}(\infty) = 0,$$

so that in the expansions (25) and (26), $A_\infty = \lambda^2 > 0$, $A_0 > 0$. Such problems for the radial Schrödinger equation occur in many fields (see, for example, [6], [7], [34], [35]). From the general theory of differential equations, it is known that for the problem (22)-(24) there exists, on the assumptions (35), a finite or infinite sequence of eigenvalues

$$\lambda_0^2 > \lambda_1^2 > \ldots > \lambda_k^2 > \ldots \Rightarrow 0$$

and the eigenfunction $y_k(t)$ has exactly k zeros on the interval $(0, \infty)$.

By the Prüfer transformation $y' = y \, \mathrm{ctg}\, \Theta$, the corresponding problem (22), (27), (31) on the finite interval is reduced to the boundary value problem for the function $\Theta(t)$:

$$(36) \qquad \Theta' = \cos^2\Theta - (\lambda^2 + \hat{A}(t))\sin^2\Theta, \qquad t_0 \leq t \leq t_\infty,$$

$$(37) \qquad \Theta(t_0) = \mathrm{arcctg}(\beta(\lambda, t_0)/t_0),$$

$$(38) \qquad \Theta(t_\infty) = \mathrm{arcctg}(\alpha(\lambda, t_\infty)) + \kappa\pi,$$

where $\kappa = 0, 1, 2, \ldots$ is the serial number of eigenvalues and

$$\beta(\lambda, t) = \beta_0 + \frac{A_1}{2\beta_0} t + o(t), \qquad t \to 0,$$

$$\beta_0 = \frac{1}{2} + \frac{1}{2}\sqrt{1 + 4A_0},$$

$$\alpha(\lambda, t) = -\lambda - \frac{\widetilde{A}_1}{2\lambda t} + o(\tfrac{1}{t}), \qquad t \to \infty.$$

Equation (36) can be stably integrated in the directions away from the points of singularity. The monotonic dependence of Θ on λ for a fixed value of t, makes the "shooting" for the parameter λ especially convenient in the problem (36)-(38). The generalization of this

method for selfadjoint problems of systems of equations
of second order is given in [36].

4.3. The calculation of complex eigenvalues of non-selfadjoint differential operators is a very cumbersome
problem. In [37] a method has been described, which
enables to calculate approximately the eigenvalues that
lie inside a given contour in the complex plane. For
determining the quantities that are used in the method,
and which are expressed in terms of solutions of auxiliary non-homogeneous linear problems, the method of
transfer of boundary conditions is applied.

Some methods for solving particular non-selfadjoint
problems have been described in [6]. One of these methods
is the method of "continuous motion along a parameter"
(which is formally introduced). The eigenvalues of the
problem are known or can be easily calculated for particular values of this parameter.

As an example, let us consider the problem (22)-
-(24) for

$$n = 1, \quad A(\lambda, t) = \mu + c/t^2 + \lambda \hat{A}(t),$$

$$\hat{A}(0) \neq 0, \quad \lim_{t \to \infty} t^2 \hat{A}(t) = 0,$$

where $c > 0$ is fixed, $\hat{A}(t)$ is a real-valued function
of t, the given parameter μ may assume both positive
and negative values; if $\mu < 0$, the radiation condition

$$(24') \quad \lim_{t \to \infty} y'(t)/y(t) = i\sqrt{-\mu}$$

is given at infinity instead of condition (24). The problem is to determine the eigenvalues of problem (22),
(23), (24') for $\mu = -\mu_0 < 0$. (See [6] for examples of

such problems in nuclear physics).

For $\mu = \mu_0 > 0$, we solve the selfadjoint problem (22)-(24), which was already mentioned in 4.2., by the phase method.

Based on the known eigenvalue $\lambda(\mu_0)$, we move with a small step of μ along the semicircle of radius μ_0 in the upper half-plane of the complex variable μ after replacing (24') by an approximated condition at the end point

$$y'(t_\infty) = i\sqrt{-\mu-\lambda\hat{A}(t_\infty)-c/t_\infty^2}\,y(t_\infty).$$

(Here, the root is taken with positive real part). When $\mu = -\mu_0 < 0$, the marching equation should be integrated not along the real axis, but along the ray in the complex plane

$$t = t_0+\tau e^{i\gamma}, \quad 0 < \gamma < \pi, \quad 0 \le \tau \le \tau_\infty$$

for the reason of numerical stability.

4.4. The problem of finding eigenfunctions of linear problems, even in the case if the respective eigenvalues are at hand, is a complicated computational problem. Some recommendations regarding the construction of eigenfunctions, together with their practical realization, can be found in [11], [23], [36], [44].

Major difficulties arise if the interval $[a, b]$ is large or the matrix of the linear system has singularities at the ends of the interval (see problem (22)--(24)). For some singular problems, the eigenfunctions can be stably found, for example, by the use of a variant

[1] of the method [4].*

It is of interest that in many physical problems (see [6], [43]) the calculation of homogenous functionals from eigenfunctions is needed rather than the calculation of eigenfunctions themselves. Such functionals can be found stably without calculating actual eigenfunctions by solving some auxiliary initial value problems which can be stably integrated in the needed direction [52]. For example, in one of the papers, which have been listed in [6], one had to calculate a functional of the form

$$(39) \qquad J(y_k(t)) = \int_0^\infty \frac{y_k^2(t)}{t^3}\, dt \Big/ \int_0^\infty y_k^2(t)\, dt,$$

where $y_k(t)$ was the k-th eigenfunction of the problem (22)-(24) on the basis of assumptions given in 4.2. For calculating J, the auxiliary functions

$$\varphi_\nu(t) = \Big(\int_\nu^t \xi^{-3} y_k^2(\xi)\, d\xi\Big) \Big/ (y_k^2(t) + y_k'^2(t)),$$

$$\psi_\nu(t) = \Big(\int_\nu^t y_k^2(\xi)\, d\xi\Big) \Big/ (y_k^2(t) + y_k'^2(t)),$$

were introduced. Here, $\nu = 0, \infty$.

The functions $\varphi_\nu(t)$ satisfy the equation

$$(40) \qquad \varphi_\nu' = -\sin 2\theta (1 + \lambda_k^2 + \hat{A}(t))\varphi_\nu + \frac{1}{t^3} \sin^2\theta, \qquad 0 < t < \infty,$$

(see problem (36)-(38)) and the boundary conditions

---

* These problems will be discussed in detail in another paper.

$$\text{(41)} \qquad \lim_{t \to 0} \varphi_0(t) = \frac{1}{A_0(1+\sqrt{1+4A_0}\,)} \quad , \qquad \lim_{t \to \infty} \varphi_\infty(t) = 0.$$

For the functions $\psi_\nu(t)$, we have

$$\text{(42)} \qquad \psi_\nu' = -\sin 2\theta(1+\lambda_k^2+\hat{A}(t))\psi_\nu+\sin^2\theta, \qquad 0 < t < \infty,$$

$$\text{(43)} \qquad \lim_{t \to 0} \psi_0(t) = 0, \quad \lim_{t \to \infty} \psi_\infty(t) = -\frac{1}{2\lambda_k(1+\lambda_k^2)} \; .$$

Having replaced the boundary conditions (41), (43) by asymptotic formulae in the points t_0 and t_∞, and having integrated the problems (40)-(41), (42)-(43) in combination with (36)-(38) from these points to the point $\hat{t} \in (t_0, t_\infty)$, we determine the unknown functional (39) from the formula

$$J(y_k(t)) = \frac{\varphi_0(\hat{t})-\varphi_\infty(\hat{t})}{\psi_0(\hat{t})-\psi_\infty(\hat{t})} \; .$$

Equations (40), (42), just as equation (36), are stably integrated from singular points.

4.5. The marching methods proved to be very convenient when solving some problems of radiophysics in which the unknown physical quantities (surface impedances, wave reflection and transfer coefficients etc.) can be found from the solutions of boundary condition transfer equations without finding non-trivial solutions of the original homogeneous boundary value problems (see [6], [20], [21], [39]). Thus, in [39], an equation of the form

$$\text{(44)} \qquad f_\lambda'' + \frac{1}{\xi} f_\lambda' - \frac{\lambda^2}{\xi^2} f_\lambda = \eta e^{-\xi^2} f_\lambda, \qquad 0 < \xi < \infty,$$

is considered. In this equation $\lambda = 0,1$; η is a complex parameter. It is necessary to determine complex constants β_0, β_1 in the asymptotics

$$f_0(\xi) \sim \alpha_0(\ln \xi + \beta_0), \qquad \xi \to \infty,$$

$$f_1(\xi) \sim \alpha_1(\xi + \beta_1/\xi), \qquad \xi \to \infty,$$

for the solutions of the equation (44) which satisfy the condition of boundedness for $\xi = 0$.

After presenting β_0, β_1 in the form

$$\beta_0 = \lim_{\xi \to \infty} [f_0(\xi)/(\xi f_0'(\xi)) - \ln \xi],$$

$$\beta_1 = \lim_{\xi \to \infty} \xi^2 [2f_1(\xi)/(\xi f_1(\xi))' - 1],$$

we notice that for their calculation it is sufficient to find the limit of ratios $f_\lambda'(\xi)/f_\lambda(\xi)$ for large ξ, $\lambda = 0,1$. In the finite point ξ_0, the condition of boundedness at zero of the solutions of (44) is equivalent to the condition

$$(45) \qquad \xi_0 f_\lambda'(\xi_0) = \gamma_\lambda(\xi_0) f_\lambda(\xi_0),$$

where $\gamma_\lambda(\xi) = \lambda + \dfrac{\eta}{2(1+\lambda)} \xi^2 + o(\xi^2)$, $\xi \to 0$. For determining β_0 and β_1, it is now sufficient to transfer the boundary condition (45) by the use of the variant 2.2.2. of the marching method from the point ξ_0 to a sufficiently remote point ξ_∞.

§.5.

At present, much interest is shown in the solution of nonlinear boundary value problems.

The iterative methods are widely spread. They reduce the initial problem to a sequence of linear boundary value problems (see [1], [40]-[42].

5.1. Let us assume that a two-point boundary value problem of the form

$$(46) \qquad y' = f(t, y), \qquad a \leq t \leq b,$$

$$(47) \qquad g(y(a), y(b)) = 0, \qquad y = \begin{bmatrix} y_1 \\ \vdots \\ y_n \end{bmatrix}, \qquad f = \begin{bmatrix} f_1 \\ \vdots \\ f_n \end{bmatrix},$$

$$g = \begin{bmatrix} g_1 \\ \vdots \\ g_n \end{bmatrix}$$

is considered on the interval $a \leq t \leq b$. Here, the vector functions $f(t, y)$ and $g(v, w)$ are assumed to be sufficiently smooth. The function $\hat{y}(t)$ is termed as the isolated solution of the problem (46)-(47) if the appropriate linearized problem (close to this solution) has a unique solution, i.e. the problem of the form

$$(48) \qquad z' - A(t)z = 0, \qquad a \leq t \leq b,$$

$$(49) \qquad U_a z(a) + U_b z(b) = 0,$$

where $A(t) = \dfrac{\partial f}{\partial y}(t, \hat{y}(t))$, $U_x = \dfrac{\partial g}{\partial y(x)}(\hat{y}(a), \hat{y}(b))$, $x = a, b$, has only a trivial solution [45]. For deter\-mining the isolated solution of the problem (46)-(47), we apply Newton's iterative method. The linear boundary value problem, which occurs in every step of the itera\-tive procedure and has a unique solution under the assumption of problem (48)-(49), can be solved by one of the methods described under §§. 1-3.

5.2. The nonlinear singular problems are often encountered in practice (see, for example, [46]-[50]). For some classes of such problems, the transfer of boundary conditions from points of singularity can be accomplished (by analogy with linear problems) by constructing appropriate nonlinear manifolds [6] in the neighbourhood of singular points. We shall restrict ourselves to considering an example.

The following type of boundary value problem occurs in many problems of physics [46], [48]-[50], [54]):

$$(50) \qquad g'' - g + g^3/r^2 = 0, \quad 0 < r < \infty,$$

$$(51) \qquad \lim_{r \to 0} g(r) = 0,$$

$$(52) \qquad \lim_{r \to \infty} g(r) = 0.$$

On the basis of the results of [46], the problem (50)-(52) has a countable set of nontrivial solutions $\{g_n(r)\}$, $n = 0,1,2,\ldots$, which are continuously differentiable on the interval $[0,\infty)$, and $g_n(r)$ has exactly n zeros on the interval $(0,\infty)$. In addition, there exists a limit $\lim_{r \to 0} g_n(r)/r$. The transfer of boundary conditions (51), (52) from singular points can be carried out in the following manner. The solutions of (50), which satisfy condition (51), form a one-dimensional nonlinear manifold in the (g', g)-space. This manifold for sufficiently small values of r can be presented in the form

$$rg' = \beta(r, g), \quad r \le r_0.$$

Here, $\beta(r, g)$ is the solution of the problem

$$r\frac{\partial \beta}{\partial r} + \frac{\partial \beta}{\partial g}\beta - \beta + g^3 - r^2 g = 0, \quad r \leq r_0, \quad |g| \leq \delta,$$

$$\beta(0, g) = g + \gamma(g).$$

This solution can be expanded into a convergent power series of r and g, while $\gamma(g) = [g]_3$ is the only holomorphic solution of the problem

$$\frac{d\gamma}{dg}(g+\gamma) + g^3 = 0, \quad |g| \leq \delta, \quad \gamma(0) = 0$$

in the neighbourhood of the point zero.

With an accuracy up to higher order terms, we have

$$\beta(r, g) = \beta_1(r)g + \beta_3(r)g^3,$$

where

$$\beta_1(r) = 1 + (1/3)r^2 - (1/45)r^4 + o(r^4), \quad r \to 0,$$

$$\beta_3(r) = -1/3 + (4/45)r^2 + o(r^2), \quad r \to 0.$$

Similarly, the condition (52) segregates a one-dimensional manifold in the space of solutions of (50). This manifold can be given for large values of r in the form

$$g' = \alpha(r, g), \quad r \geq r_\infty,$$

where $\alpha(r, g)$ is the solution of problem

$$\frac{\partial \alpha}{\partial r} + \frac{\partial \alpha}{\partial g}\alpha - g + g^3/r^2 = 0, \quad r \geq r_\infty, \quad |g| \leq \delta,$$

$$\lim_{r \to \infty} \alpha(r, g) = -g.$$

This solution can be expanded into a convergent series of integer powers in g with coefficients which can be expanded into an asymptotic series of inverse integer powers of r.

With an accuracy up to higher order terms, we have

$$\alpha(r, g) = -g + \alpha_3(r)g^3,$$

where $\alpha_3(r) = \dfrac{1}{4r^2}(1 - \dfrac{1}{2r} + \dfrac{3}{8r^2} + o(\dfrac{1}{r^2})), \quad r \to \infty.$

As a result, on the finite interval $[r_0, r_\infty]$, we get a boundary value problem in the form

$$g'' - g + g^3/r^2 = 0, \qquad r_0 \leq r \leq r_\infty,$$

$$r_0 g'(r_0) = \beta_1(r_0)g(r_0) + \beta_3(r_0)g^3(r_0),$$

$$g'(r_\infty) = -g(r_\infty) + \alpha_3(r_\infty)g^3(r_\infty).$$

Using the initial approximation $g^{(0)}(r) = Cre^{-r}$, the fundamental solution of this problem (positive on the interval $[r_0, r_\infty]$), can be found by the quasilinearization method. For solving a linear boundary value problem, it is convenient to apply the variant [22] of the marching method in each step of iteration. This method for solving problem (50)-(52) was applied in [54].

An iterative method of solving some nonlinear boundary value problems with a singularity at one end, is described in [51].

REFERENCES

[1] N.S. Bakhvalov, *Numerical Methods* (in Russian),
Nauka, Moscow, 1973.

[2] I. Babuška - E. Vitasek - M. Prager, *Numerical
Processes of Solving Differential Equations*, Mir,
Moscow, 1969.

[3] J. Taufer, *Lösung der Randwert - Probleme für
System von Linearen Differentialgleichungen*,
Akademia, Praha, 1973.

[4] A.A. Abramov, On transfer of boundary conditions
for systems of linear ordinary differential equations (a variant of the marching method) (in
Russian), *ZhVM i MF*, 1, No.3, (1961), 542-545.

[5] A.A. Abramov, Methods of solving some linear problems, *Thesis for the Doctor's degree*, (in Russian),
Moscow, VTs AN SSSR, 1974.

[6] A.A. Abramov - E.S. Birger - N.B. Konyukhova -
- V.I. Ulyanova, Numerical segregation of the
bounded solutions for systems of ordinary differen-
tial equations, in Colloquia Mathematica Societatis
János Bolyai, 15. *Differential equations* (ed. M.
Farkas), Keszthely (Hungary), 1975.

[7] K. Balla, On calculation of nuclear models by
hyperspherical function method, *MTA Számitástechni-
kai és Automatizálási Kutató Intézete (Budapest),
Közlemények.* 17/1976, 27-39.

[8] A.A. Abramov - Yu.A. Ivanov - V.I. Ulyanova,
Instability of two-layer geostrophic flow in North
Atlantic conditions (in Russian), *Journal "Okeano-
logiya"*, 15, fasc. 4, (1975), 565-573.

[9] L.S. Klabukova - N.A. Stadnikova, Calculation of a
     conical body subjected to a load concentrated at
     its vertex (in Russian), *ZhVM i MF*, 14, No.4, (1974),
     955-969.

[10] E.S. Birger, Algorithm of transfer of boundary
     conditions for systems of linear ordinary differen-
     tial equations, - In: *"Algorithms and algorithmic
     languages"*, Moscow, VTs AN SSSR, 1973, fasc. 6,
     3-14.

[11] V.I. Ulyanova, On transfer of non-homogeneous
     boundary conditions and on calculations of eigen-
     functions (in Russian), *ZhVM i MF*, 16, No.4, (1976),
     1057-1059.

[12] S.K. Godunov, On numerical calculation of boundary
     value problems for systems of linear ordinary dif-
     ferential equations (in Russian), *UMN*, XVI,
     3(99), (1961), 171-174.

[13] A.F. Shapkin, Closure of two computational algo-
     rithms based on the idea of orthogonalization (in
     Russian), *ZhVM i MF*, 7, No.2, (1967), 411-416.

[14] V.M. Lyubimov - G.I. Pshenichnov, Perturbation
     method in the theory of flat shells (in Russian),
     Proceedings of AN SSSR, MTT, 1976, No.4, 143-147.

[15] N.V. Valishvili, *Methods of Calculating Shells of
     Rotation Through the Use of Digital Computers*,
     (in Russian), "Mashinostroenie", Moscow, 1976.

[16] B.A. Orlov, On numerical method of calculating off-
     -axially symmetrical vibrations of a cylindrical
     shell filled with a fluid (in Russian), *Proceedings
     of AN SSSR, MTT*, No.4, 1974, 90-94.

[17] K. Moszynski, A method of solving the boundary
     value problem for a system of linear ordinary dif-
     ferential equations, *Algorytmy*, Vol.2. No.3, (1964),
     25-43.

[18] V.L. Biderman, Application of the marching method
     for numerical solution of structural mechanics
     problems, *MTT*, 1967, No. 5, 62-66.

[19] D.G. Baratov - V.M. Valov - R.A. Rzaev - V.L. Rykov
     - I.M. Shalashov, Stress-deformed state of parabo-
     lic shell of rotation subjected to external magne-
     tic pressure, *PMTF*, 1974, No.3, 133-142.

[20] P.E. Krasnushkin, A method of solving general boun-
     dary value problem of propagation of long and
     very-long waves around the Earth, *DAN SSSR*, 171,
     No.1, (1966), 61-64.

[21] P.E. Krasnushkin - R.B. Baibulatov, Calculation of
     reflection and electromagnetic wave propagation
     coefficients for a spherically laminated anisotropic
     layer by impedance and field scaling method, *DAN
     SSSR*, vol. 188, No.2, (1969), 300-303.

[22] A.A. Abramov, A variant of the marching method (in
     Russian), *ZhVM i MF*, 1, No.2, (1961), 349-351.

[23] E.S. Birger - N.B. Konyukhova, Numerical computation
     of propagation of radiowaves in vertically-inho-
     mogeneous troposphere (in Russian), *Radiotekhnika
     i elektronika*, 14, No.7. (1969), 1147-1156.

[24] L.M. Degtyarev - A.P. Favorskii, Flow variant of
     the marching method (in Russian), *ZhVM i MF*, 8,
     No.3, (1968), 679-684.

[25] L.M. Degtyarev - A.P. Favorskii, Flow variant of the marching method for difference problems with vigorously changing coefficients (in Russian), *ZhVM i MF*, 9, No.1. (1969), 211-218.

[26] N.N. Kalitkin, Marching for an infinite thermal conductivity coefficient (in Russian), *ZhVM i MF*, 8, No.3, (1968), 684-686.

[27] V.B. Lidskii - M.G. Neigauz, A marching method for a selfadjoint system of the second order (in Russian), *ZhVM i MF*, 2, No.1, (1962), 161-165.

[28] T.H. Barrett, A Prüfer transformation for matrix differential equations, *Proc. Amer. Math. Soc.*, 8(1957), 510-518.

[29] P.I. Monastyrny, On marching method for systems of the second order (in Russian), *ZhVM i MF*, 11, No.4, (1971), 925-933.

[30] A.A. Abramov - V.B. Andreev, On the application of marching method for the determination of periodic solutions of differential and difference equations (in Russian), *ZhVM i MF*, 3, No.2, (1963), 376-381.

[31] V.M. Goloviznin - A.P. Favorsky, Cyclic flow marching, *Preprint IPM AN SSSR*, No.76, 1976.

[32] A. Kalnins, Free vibrations of rotationally symmetric shells, *JASA*, v. 36, No.7, (1964).

[33] A. Kalnins, Study of rotation shells when symmetrical and asymmetrical loads act on them (in Russian), *Prikladnaya mekhanika (Proc. of the American Society of Mechanical Engineers)*, No.3, 1964.

[34] V.V. Babikov, *A Method of Phase Functions in Quantum Mechanics*, "Nauka", Moscow, 1976.

[35] M.S. Marinov - V.S. Popov, A phase method for the
Dirac equation and calculation of critical charge
(in Russian), *ZhTF*, 65, fasc. 6(12), (1973), 2141-
-2154.

[36] M.G. Neigauz - G.V. Shkadinskaya, A method of calculating Rayleigh surface waves in a vertically
inhomogeneous half-space (in Russian), *Vyčisl.
Seism.* No.2, Moscow, "Nauka", 1966, 121-128.

[37] V.V. Ditkin, Approximate calculation of eigenvalues
lying in the defined part of a complex plane (in
Russian), *ZhVM i MF*, 16, No.5, (1976), 1102-1109.

[38] K. Balla, On error estimation of replecement of
solutions boundedness condition for systems of
linear ordinary differential equations with a regular singularity (in Russian), *ZhVM i MF*, 18, No.2,
(1978), 370-378.

[39] S.M. Dikman - N.B. Konyukhova, Numerical research
of skin-effect in a plasma pinch (in Russian),
ZhVM i MF, 18, No.2, (1978), 394-405.

[40] R. Bellman - R. Kalaba, *Quasi-Linearization and
Nonlinear Boundary Value Problems*, "Mir", Moscow,
1968.

[41] M.A. Krasnoselskii - G.M. Vainikko - P.P. Zabreiko -
- Ya.B. Rutitskii - V.Ya. Stetsenko, *An Approximate
Solution of Operator Equations*, "Nauka", Moscow,
1969.

[42] E.P. Zhidkov - G.I. Makarenko - I.V. Puzynin,
Continuous analog of Newton method in nonlinear
problems of physics. *Fizika elementarnykh chastits
i atomnogo yadra.* (*EChAYa*), vol.4, fasc.1, Moscow,
Atomizdat, 1973, 127-166.

[43] A.M. Badalyan - Yu.A. Simonov, Three-body problem.
     An equation for partial waves. *YaF*, 1966, 3, fasc.6,
     1032-1047.

[44] A.A. Abramov - Yu.A. Ivanov - V.I. Ulyanova, Eigen-
     functions of the zonal flow stability problem (in
     Russian), Ž. *"Okeanologiya"*, XVI, fasc.1, (1976),
     41-46.

[45] H.B. Keller - A.B. White, Difference methods for
     boundary value problems in ordinary differential
     equations, *SIAM J. Numer. Anal.*, 12, No.5, (1975),
     791-802.

[46] G.H. Ryder, Boundary value problems for a class of
     nonlinear differential equations, *Pacific J. Math.*,
     22, No.3, (1967), 477-503.

[47] E.V. Bogomolny - M.S. Marinov, Calculation of mono-
     pole mass in calibrating theory. *YaF*, 1976, vol.23,
     No.3, 676-680.

[48] T.I. Belova - N.A. Voronov - I.Yu. Kobzarev - N.B.
     Konyukhova, Particle-like solutions of scalar Higgs
     equation (in Russian), *ZhTF*, 73, fasc. 5(11) (1977),
     1611-1622.

[49] I.L. Bogolyubskii, Oscillating particle-like solu-
     tions of nonlinear Klein-Gordon equation (in Rus-
     sian), *Pis'ma v ZhTF*, 24, fasc.10, (1976), 579-583.

[50] V.B. Glasko - F. Leryust - Ya.P. Terletskii - S.F.
     Shushurin, Study of particle-like solutions of
     nonlinear scalar field equation (in Russian), *ZhTF*,
     35, fasc. 2(8), (1958), 452-457.

[51] N.B. Konyukhova, On iterative solution of nonlinear
     boundary value problems segregating small solutions
     of some systems of ordinary differential equations

[51] with a singularity (in Russian), *ZhVM i MF*, 14,
     No.5, (1974), 1221-1231.

[52] E.S. Birger, Computation of functionals of eigen-
     functions of boundary value problems for systems of
     linear ordinary differential equations (in Russian),
     ZhVM i MF, 13, No.1. (1973), 227-233.

[53] K. Balla, On the replacement of the condition of
     boundedness for certain systems of linear ODE-s
     with regular singularity (in this book).

[54] N.A. Voronov - N.B. Konyukhova, Stability of par-
     ticle-like solutions of some nonlinear wave equa-
     tions, *Preprint ITEP*-26, Moscow, 1978.

A.A. Abramov - E.S. Birger
Nadyezhda B. Konyukhova - Valentyina I. Ulyanova
117333 Moscow
ul. Vavilova 40
Computer Centre
Acad. Sci. USSR

MONOTONE SPLINE INTERPOLATIONS AND OPTIMAL CONTROL PROBLEM

L. ADAMCZYK

The aim of the paper is to formulate and prove theorems on monotone spline interpolations and to examine, in what sense obtained estimates are the best. It is shown that the optimal property of the obtained interpolant allows to solve explicitely a certain optimal control problem.

1. NOTATION AND BACKGROUND

Let $\Delta : = \{0 = x_0 < x_1 < \ldots < x_k = 1\}$ and let $y_0, y_1, \ldots, y_k$ be real numbers such that $y_{i-1} \neq y_i$ $i = 1, 2, \ldots, k$. It was shown independently by WOLIBNER [8] and YOUNG [9] that there exists a polynomial $p(x)$ such that $p(x_i) = y_i$, $i = 0, 1, \ldots, k$, and p is monotone in each of the intervals $I_i : = [x_{i-1}, x_i]$, $i = 1, 2, \ldots, k$. Both proofs, however, fail to give any information about the degree of the polynomial needed. In this note we consider the problem of piecewise monotone interpolation (PMI), but use splines as interpolating functions.

DEFINITION 1.1. Let $S_n^j(\Delta) = S_n^j(x_0, x_1, \ldots, x_k)$, $0 \le j \le n-1$, be the set of all functions $s \in C^j[0,1]$ and such that s agrees with a polynomial of degree $\le n$ on I_i, $i=1,2,\ldots,k$. $s \in S_n^j$ is said to be a spline of order n with deficiency $n-j$. (cf., AHLBERG, NIELSON, WALSH [1]).

PASSOW in [5] proved the following result:

THEOREM 1.1. *Let* $0 = x_0 < x_1 < \ldots < x_k = 1$ *and let* y_i, $i=0,i,\ldots,k$, *be arbitrary. Then, for each* n, *there exists a unique* $s \in S_{2n-1}^{n-1}$ *such that* $s(x_i)=y_i$, $i=0,1,\ldots,k$, *and* s *is monotone on each of the intervals* I_i, $i=1,2,\ldots,k$.

REMARK 1.1. The theorem is the best possible in the sense that S_{2n-1}^{n-1} are splines of minimal deficiency for which PMI is always possible.

The method of the Theorem 1.1. can be used to show that for each n there exists $s \in S_{2n}^n$ which solves the PMI problem. In this case, however, the interpolating spline is not unique.

Let P_n be the set of all algebraic polynomials of degree $\le n$, and $E_n(f) := \inf\{\|f-p\|_\infty, p \in P_n\}$. For f monotonically increasing, let $E_n^*(f) := \inf\{\|f-p\|_\infty, p \in P_n, p'(x) \ge 0$ on $[0,1]\}$.

LORENTZ in [4] proved the following result:

THEOREM 1.2. [4, p.209] *If* $f \in C^1[0,1]$ *is increasing then* $E_n^*(f) \le (c/n)\omega(f'; 1/n)$, *where* ω *is the modulus of continuity of* f' *and* c *is an absolute constant.*

We apply the method of this theorem and the methods of KORNEICHUK [2] to monotone approximation of splines with specified deficiency which are given by theorem 1.1.

2. BEST PMI AND APPROXIMATION IN THE L_1 AND CHEBYSHEV NORMS

Let X be a family of real-valued functions on $[0,1]$. For $f \in X$ we write $\|\cdot\|_\infty$ and $\|\cdot\|_1$ to denote the Chebyshev and the L_1 norm respectively. Let $g = (g_1, g_2, \ldots, g_{n-1})$ be a vector function, where $g_i \in X$, $i = 1, 2, \ldots, n-1$.

DEFINITION 2.1. A real-valued function $s(\Delta; f; g) \in S_{2n-1}^{n-1}$ is said to be a Hermite interpolation spline provided the following conditions are satisfied

$$(2.1) \qquad s(\Delta; f; g)(x_i) = f(x_i), \qquad i = 0, 1, \ldots, k,$$

$$(2.2) \qquad s^{(j)}(\Delta; f; g)(x_i) = g_j(x_i), \qquad j = 1, 2, \ldots, n-1.$$

REMARK 2.1. The unique spline $s \in S_{2n-1}^{n-1}$ by Theorem 1.1. is the spline $s(\Delta; f; 0)$, where $f(x_i) = y_i$ $i = 0, 1, \ldots$ $\ldots, k$.

We need to recall the definition of the modulus of continuity.

DEFINITION 2.2. Modulus of continuity is defined for $f \in C_B(R)$, $t \geq 0$ by

$$(2.3) \qquad \omega(f;t) := \sup_{|h| \leq t} |f(x+h) - f(x)|,$$

having the properties that $\omega(f;t)$ is a continuous, monotonely decreasing function of t, with $\omega(f;t) \to 0$ for $t \to 0_+$.

For a fixed modulus of continuity $\omega(t) \neq 0$, let

(2.4) $H^\omega := \{f \in X : \omega(f;t) \le \omega(t), \quad 0 \le t \le 1\}.$

LEMMA 2.1. *Equivalently, we can write* $H^\omega = \{f \in X :$
$|f(x') - f(x'')| \le \omega(|x'-x''|)$ *for each* $x', x'' \in [0,1]\}.$ H^ω
in a convex and closed subset in $C[0,1]$ *(and* $L_1[0,1]$*)*
and satisfying the condition $f \in H^\omega \iff -f \in H^\omega.$

Let $x \in I_i$. If the $2n$ real numbers

(2.5) $f(x_{i-1}), \ f(x_i), \ g_j(x_{i-1}), \ g_j(x_i) \quad j=1,2,\ldots,n-1,$

are preassigned, then the 2-point Hermite interpolation
problem (2.1) and (2.2) has a unique solution. This
solution is given by the formula

$$s(\Delta; \ f; \ g)(x) =$$

$$(2.6) \quad = \sum_{j=0}^{n-1} g_j(x_{i-1}) L_j(h_i; \ x-x_{i-1}) +$$

$$+ \ (-1)^j g_j(x_i) L_j(h_i; \ x_i-x),$$

where

$$(2.7) \quad L_j(h; \ x) := \begin{cases} \dfrac{(h-x)^n}{j!(n-1)!} \displaystyle\sum_{r=0}^{n-1-j} \dfrac{(n-i+r)!}{r!h^{n+r}} x^{j+r}, & x \in [0,h] \\[2ex] 0 & x \notin [0,h] \end{cases}$$

$$j=0,1,\ldots,n-1, \quad g_0(x):=f(x), \ h_i=x_i-x_{i-1}.$$

For the polynomials $L_j(x)$ of degree $\le 2n-1$ the
following conditions are satisfied

$$(2.8) \qquad L_j^m(h;\,0) = \delta_{jm}, \quad L_j^m(h;\,h) = 0, \qquad m=0,1,\ldots,n-1,$$

where $\delta_{jm} = 1$ if $j=m$ and $\delta_{jm} = 0$ for $j \neq m$. The formula (2.6) is immediately obtained from K r y l o v [3, p.51].

From (2.6) we have

$$(2.9) \qquad s(\Delta;f;g;)(x) = s(\Delta;f;0)(x)+R(g)(x),$$

where the polynomial $R(g)(x)$ does not depend on f and

$$(2.10) \qquad s(\Delta;f;0)(x) = f(x_{i-1})L_0(h_i;u)+f(x_i)L_0(h_i;h_i-u) ,$$

where $u := x-x_{i-1}$, $x_{i-1} \leq x \leq x_i$, $0 \leq u \leq h_i$.

REMARK 2.2. For $f \in X$ and $0 \leq x \leq 1$

$$(2.11) \qquad s(\Delta;-f;g) = -s(\Delta;f;0)+R(g).$$

Let $e(f)(x) := f(x)-s(\Delta;f;0)(x)$, and let $\|\cdot\|$ be one of the norms $\|\cdot\|_\infty$ and $\|\cdot\|_1$. We have the following

THEOREM 2.1. If n is a natural number ≥ 2, then

$$(2.12) \qquad \inf_{g \in X^{n-1}} \sup_{f \in H^\omega} \|f-s(\Delta;f;g)\| =$$

$$= \sup_{f \in H^\omega} \|f-s(\Delta;f;0)\|.$$

PROOF. Polynomial $R(g)(x)$ does not depend on f, thus, for any $g \in X^{n-1}$

$$(2.13) \quad \|e(f)\| \leq \max\{\|e(f)-R(g)\|, \|e(f)+R(g)\|\}$$

holds. From (2.11) and from the definition

$$\|e(f)-R(g)\| = \|f-s(\Delta;f;g)\|$$

and

$$\|e(f)+R(g)\| = \|(-f)-s(\Delta;-f;g)\|.$$

From the Lemma 2.1 and (2.13) it follows that

$$\sup_{f \in H^\omega} \|e(f)\| \leq \sup_{f \in H^\omega} \|e(f)-R(g)\|.$$

Hence

$$(2.14) \quad \sup_{f \in H^\omega} \|e(f)\| \leq \inf_g \sup_{f \in H^\omega} \|e(f)-R(g)\|.$$

Thus, we have the relation (2.12), since the opposite inequality is trivial.

REMARK 2.3. Equality (2.12) will be true if set H is a bounded and closed subset of metric space $C[0,1]$ (or $L_1[0,1]$) satisfying the condition $f \in H \Leftrightarrow -f \in H$. We obtain now

THEOREM 2.2. *For a concave modulus of continuity* $\omega(t) \neq 0$ *and natural number* $n \geq 2$ *hold*

$$(2.15) \quad \sup_{f \in H^\omega} \| f - s(\Delta; f; 0) \|_\infty = \omega(\|\Delta\|/2),$$

where $\quad \|\Delta\| := \max_{1 \leq i \leq k} h_i.$

REMARK 2.4. Equality in (2.15) is fulfilled by the function

$$f(x) := \begin{cases} \omega(x - x_{i-1}) & \text{for} \quad x_{i-1} \leq x \leq x_{i-1} + h_i/2 \\[2ex] \omega(x_i - x) & \text{for} \quad x_{i-1} + h_i/2 \leq x \leq x_i \end{cases}$$

THEOREM 2.3. *For a concave modulus of continuity* $\omega(t) \neq 0$ *and natural number* $n \geq 2$

$$\sup_{f \in H^\omega} \| f - s(\Delta; f; 0) \|_1 \leq 2 \sum_{i=1}^{n} \int_0^{h_i/2} \omega(t)\, dt.$$

3. THE MONOTONE PERFECT B-SPLINES

DEFINITION 3.1. For an appropriate division Δ, function $f(x)$ is said to be a perfect spline, provided that $f \in S_n^{n-1}$ and $|f^{(n)}(x)| = \text{const.}$

Equivalently, we can say that the $k-1$ polynomial components of $f(x)$ are of the form $f(x) = (-1)^i C\, x^n +$ + lower degree terms, in I_i $i = 0, 1, \ldots, k$, where the constant C does not depend on i.

$$(3.1) \quad \text{Let} \quad \Delta := \{ -1 = x_n < x_{n-1} < \ldots < x_1 < x_0 = +1 \}.$$

The corresponding divided difference of order n is

$$(3.2) \qquad f(x_0,x_1,\ldots,x_n) = \sum_{i=0}^{n} \frac{f(x_i)}{W'(x_i)} \;,$$

where $w(x) := (x-x_0)(x-x_1)\ldots(x-x_n)$.

It is well known that for the division (3.1) we may express the divided difference of a function $f \in C^n$ in the form (cf. R i c e [6])

$$(3.3) \qquad f(x_0,x_1,\ldots,x_n) = \frac{1}{n!} \int_{-1}^{1} M(x) f^{(n)}(x)\,dx \;,$$

where $M(x)$ is a certain special spline function called a B-spline.

$M(x)$ we can express in the form

$$(3.4) \qquad M(x) = M(x;x_0,x_1,\ldots,x_n) = \sum_{i=0}^{n} \frac{n(x_i-x)_+^{n-1}}{W'(x_i)} \;,$$

where

$$u_+ := \begin{cases} u & \text{if} \quad u \geq 0 \\[2ex] 0 & \text{if} \quad u < 0. \end{cases}$$

THEOREM 3.1. (S c h o e n b e r g [7]). *The B-spline $M(x) = M(x;x_0,x_1,\ldots,x_n)$ of degree $n-1$ based on the knots $x_i = \cos(\pi i/n)$ $i=0,1,\ldots,n$ is a perfect B-spline such that*

$$(3.5) \qquad |M^{(n-1)}(x)| = 2^{n-2}(n-1)!, \quad -1 < x < +1, \quad x \neq x_i.$$

From this theorem we immediately obtain the following

PROPOSITION 3.1. *If we define*

$$(3.6) \quad f_0(x) = \int_{-1}^{x} M(t)dt \quad -1 \leq x \leq +1,$$

where $M(t)$ is from Theorem 3.1, then $f_0(x)$ is a perfect spline function in $[-1,1]$ with knots $x_1, x_2, \ldots \ldots, x_{n-1}$ such that

$$(3.7) \quad |f_0^{(n)}(x)| = 2^{n-2}(n-1)! \quad in \quad (-1,1), \quad x \neq x_i$$

and satisfying the boundary conditions

$$f_0^{(k)}(-1) = 0 \qquad k=0,1,\ldots,n-1$$
$$(3.8)$$
$$f_0(1) = 1, \quad f_0^{(k)}(1) = 0, \quad k=1,2,\ldots,n-1$$

and $f_0(x)$ is strictly increasing in $[-1,1]$. $f_0(x)$ is convex in $[-1,0]$ and concave in $[0,1]$.

In additional, spline $f_0(x)$ has the optimal property (S c h o e n b e r g [7]) that allows to solve certain optimal control problem.

EXAMPLE. A point moves on the y-axis such that $y = F(t)$ and that the $F^{(k)}(t)$ $k=0,1,\ldots,n-1$ are all absolutely continuous, while $F^{(n)}(t)$ satisfies at all times the inequality $|F^{(n)}(t)| \leq A$ (A is preassigned). We assume that the point starts from rest at $y=0$ at the time $t=0$, i.e. $F^{(k)}(0) = 0$ $k=0,1,\ldots,n-1$ and reaches $y=1$ also at rest at the time $t = T(T > 0)$, i.e. $F(T) = 1$, $F^{(k)}(T) = 0$ $k=1,2,\ldots,n-1$.

We are to find the shortest times T_0 in which this motion can be performed and are to describe the nature of this optimal motion.

SOLUTION. Let $T_0 := 2(2^{n-2}(n-1)!\ 1/A)^{1/n}$. The motion corresponding to

$$F_0(t) = 1\ f_0(2t/T_0 - 1), \qquad 0 \le t \le T_0,$$

where f_0 is given by (3.6), is the optimal one and T_0 is the least time in which the motion can be performed.

REFERENCES

[1] J.H. Ahlberg - E.N. Nielson - J.L. Walsh, *The Theory of Splines and Their Applications* , Acad. Press. New York, 1967.

[2] N.P. Korneichuk, *Extremal Problems of the Approximation Theory*, Nauka, Moscow, 1976.(in Russian)

[3] W.I. Krylov, *Computation of Integrals*, Nauka, Moscow, 1967. (in Russian)

[4] G.G. Lorentz, Monotone Approximation, in *Inequalities III*, Acad. Press, New York 1972, pp. 201-215.

[5] E. Passow, Piecewise monotone spline interpolation, *J. Approx. Theory*, Vol. 12, No 3,(1974),240-241.

[6] J.R. Riece, *The Approximation of Functions*, Vol.2, Addison-Wesley Publ. Co. 1969.

[7] I.J. Schoenberg, The perfect B-splines and a time-optimal control problem, *Israel J. Math.*, Vol. 10. 1971.

[8] W. Wolibner, Sur un polynôme d'interpolation, *Colloq. Math.* 2 1951.

[9]   S.W. Young, Piecewise monotone polynomial interpol-
      ation, *Bull. Amer. Math. Soc.* 73. 1967.

L. Adamcyk
Oskar Lange Academy of Economics,
Institute of Cybernetics, 53-345 Wrocław,
ul. Komandorska 118/120.
Poland

COLLOQUIA MATHEMATICA SOCIETATIS JÁNOS BOLYAI
22. NUMERICAL METHODS, KESZTHELY (HUNGARY), 1977.

A TWO-DIMENSIONAL MATHEMATICAL MODEL AND A RELATED OVERSHOOTING-FREE NUMERICAL METHOD FOR OIL MINING USING UNDERGROUND COMBUSTION

GY. ADLER

INTRODUCTION

The technique of underground combustion is more than 25 years old. Nevertheless - as far as we know - there exists no commercially available numerical simulator to describe the process. The published results are limited to special aspects of the very complex physical--chemical problem. (See [2], [4].) Only two papers (see [1], [3]) are attacking the problem in some generality, i.e. without any a priori assumptions on the evolution of the process and without incorporating essential mathematical simplifications into the model itself. These papers are dealing with the one-dimensional problem. The author's paper [1] describes a very stable method but this is based on a non-massconserving scheme. The method presented here is mass-conserving.

The basic problem from the numerical point of view is the stability (i.e. the magnitude of the applicable maximum time step), since the number of unknowns is

necessarily very high, hence the computer requirements
of the simulation are considerable.

In our model we included all the basic phenomena of
the process (namely combustion, evaporation/condensation,
solution of gases in liquids, heat convection, heat
conduction, effects of gravitation, effects of the over-
and underburden) in order to study and to solve the at-
tached numerical problems. At the same time we reduced
the number of the chemical components to the essential
minimum in order to keep the computer requirements as
law as possible. The extension of the method (or better
saying of the computer program) to more chemical components and three space-dimensions would not involve new
mathematical problems.

We would like to emphasize that in the model we do
not include any a priori knowledge concerning the evolution of the combustion process (e.g. not even the existence of a combustion front will be supposed). We pos-
tulate the basic physical-chemical laws only, all other
phenomena (partly known from practice) will result from
the model.

THE MATHEMATICAL MODEL

We have two variants of the model: a radial and an
areal version. In the radial case we suppose that the
oil reservoir is the $\{R_1 \leq r \leq R_2, \ 0 \leq z \leq D\}$ cylinder
which is limited by the over- and underburden from above
and from below. The over- and underburden are impermeable
to mass flow but they are heat conductors. The injection
well is the $r = R_1$ boundary, the production takes
place on the $r = R_2$ boundary. In the areal version we
use Cartesian coordinates, and the wells can be represented by any number of the grid blocks.

We consider the following chemical components in
the system: two different oil components, water, vapours
of the oil components and of the water, oxygen, nitrogen,
combustion gas, combustion gas dissolved in the mixture
of the oils and in the water. (By components we mean the
components described by individual saturations which are
not necessarily chemically different subtances, like
water vapour.)

The components form the following three phases:

gas phase: gases and vapours,

oil phase: oils and combustion gas dissolved in
the oils,

water phase: water and combustion gas dissolved
in water.

The system will be described by the saturations of
the components, the pressure and the temperature. The
combustion gas dissolved in a particular liquid is
described by a fictitious saturation defined as the
difference of the saturation of the solution (solvent +
combustion gas) and of the pure solvent.

Hence we have 13 unknowns: 11 saturations, the
pressure and the temperature.

By definition the saturations satisfy the following
conditions:

(1) $\qquad \sum S = 1, \qquad S \geq 0.$

(Summations always refer to all 11 components.)

By D a r c y 's law the saturations satisfy the
following flow- (or continuity) equations:

(2) $\qquad \phi \dfrac{\partial(\rho S)}{\partial t} = F + \phi G + \mathrm{div}\,[k\psi\rho S \ \mathrm{grad}(p-\rho z)].$

The pressure equation can be derived from equations (2):

$$\Phi\,\frac{\partial p}{\partial t}\,\sum\left(\frac{1}{\rho}\,\frac{\partial \rho}{\partial p}\,S\right) = \sum\frac{1}{\rho}(F+\Phi G)-$$

$$(3)$$

$$-\,\Phi\,\frac{\partial T}{\partial t}\,\sum\left(\frac{1}{\rho}\,\frac{\partial \rho}{\partial T}\,S\right)+\sum\frac{1}{\rho}\mathrm{div}\,[k\psi\rho S\ \mathrm{grad}\,(p-\rho z)]\,.$$

Finally the temperature satisfies the equation

$$(4)\qquad \frac{\partial(TQ)}{\partial t} = \Psi+L_{\mathrm{cond}}+L_{\mathrm{conv}},$$

where

$$\Psi = \Phi G_{\mathrm{heat}}+T_{\mathrm{inj}}\sum(FC),\qquad Q = \rho_r C_r+\Phi\sum(\rho SC)$$

$$L_{\mathrm{cond}} = \mathrm{div}(K\ \mathrm{grad}\ T)$$

$$L_{\mathrm{conv}} = \mathrm{div}\,[T\kappa\psi\rho SC\ \mathrm{grad}\,(p-\rho z)]\,.$$

In these equations F is the injection or production rate, G is the rate of the chemical processes, G_{heat} is the reaction rate generated by the chemical processes and T_{inj} is the temperature of the injected materials. The convection term L_{conv} is not present in the over- and underburden.

The combustion is governed by the Arrhenius-law

$$(5)\qquad U_H = Z_H(\rho_H S_H)^{\vartheta_H}(\rho_A S_A)^{\vartheta_A}e^{-\frac{E_H}{RT}}$$

$$(H = \mathrm{oil},\ \mathrm{oil\ vapour})\,.$$

The equilibrium of the phases is controlled by the following conditions:

Solutions:

$$(6)\qquad p_s = P_{sH}\qquad (H = \mathrm{water},\ \mathrm{oil})$$

vapours:

$$(7) \qquad p_H^* = \sigma_H P_H^* \, ,$$

where

$$P_{sH} = \ell_H \frac{\text{mols of solved combustion gas}}{\text{mols of solvent } H} T,$$

$$P_H^* = P_H^*(T) \qquad \text{(function of temperature only)}$$

$$\sigma_H = \begin{cases} 1 & \text{for water} \\ \dfrac{M_{O_i}}{M_{O_1} + M_{O_2}} & \text{for } H = O_i \qquad (i=1,2). \end{cases}$$

The system $(1)+(2)+(3)+(4)$ consists of 14 equations (equation (2) must be considered separately for all 11 chemical components) for the 13 unknowns.

The flow equations (2) for given pressure distribution p form a hyperbolic system, while equations (3) and (4) for the pressure and the tempereature are of the parabolic type.

THE NUMERICAL METHOD

The equations are solved by a mass-conserving finite difference formula using a block-centered grid system. (In the radial case the boundary grid points for $r = R_1$ and $r = R_2$ are on the boundary.) The pressure is treated implicitely, i.e. it is calculated simultaneously for all grid points. The saturations are calculated from equations (2) and afterwards they are corrected in order to satisfy equation (1).

The temperature is computed in two steps corresponding to the decomposition

$$\text{(8)} \qquad T^{n+1} - T^n = (T^{n+1} - T^{n+1/2}) + (T^{n+1/2} - T^n),$$

where $T^{n+1/2}$ and T^{n+1} are computed from the equations

$$\text{(9)} \qquad \frac{Q^n}{\Delta t}(T^{n+1/2} - T^n) = L_{\text{cond}}^{n+1/2},$$

$$\text{(10)} \qquad \frac{1}{\Delta t}[(TQ)^{n+1} - T^{n+1/2} Q^n] = L_{\text{conv}}^{n+1/2} + \psi^{n+1/2}.$$

This decomposition has various advantages. The (implicite) computation of $T^{n+1/2}$ is unconditionally stable, while the (explicite) solution of (10) is a small part of the subroutine dealing with the flow equations and is stable as long as the solution of the flow equations is stable.

The basic problem of the whole method, as far as the stability is concerned, is two-fold:

    (i) how the calculate the time-averaged coefficients,

    (ii) how to calculate the source terms G in the difference equations (calculation of the reaction rate and of the phase transitions).

Concerning problem (i) we just mention that the coefficients of the pressure- and flow equations are computed by a double averaging method: during the iterations within one time step we use means of the old and updated values and means of the consecutive updated values. This highly improves the stability. In simple two-phase flow cases (without combustion) we were able to achieve 1.5 block/time step front velocity.

As far as space-averaging is concerned, we do not use essentially new ideas: we are using upstream weighting wherever this improves stability.

Concerning problem (ii), we proceed in the following way. In the case of the combustion rate, we determine an estimate from below of the oxygen consumption within each grid block by assuming that there is no oxygen supply to the grid block during the time step. (Under such conditions the oxygen saturation in the block satisfies an ordinary differential equation.) By assuming a constant oxygen saturation we obtain an estimate from above of the oxygen consumption. A convenient weighted average of the two estimates assures that there is no over-shooting and the error is small. In test computation it was possible to produce overshooting free cases where practically all the injected oxygen was burnt in one single block.

The phase equilibrium conditions (6) and (7) form a very complicated system. This system is decomposed into smaller systems in the following way:

(α) equations (6) (separately for the water and oil phase) are solved under the assumption that the pressure and the saturation of the gas phase is constant;

(β) equations (7) (separately for each liquid) are solved under the assumption that the saturation of the gas phase is constant.

This decompositions furnishes simple linear equations for the phase-transitions (namely for the changes of the saturations) and assures a fast convergence of the iteration. (There is no separate iteration for the system (6)+(7), they are solved only once in each iteration cycle including all the difference equations.)

Let us denote by ΔS the changes of the saturations corresponding to the reaction rate (5) and the equilibrium conditions (6) and (7). (Usually

$$S^{n+1} - S^n \neq \Delta S,$$

since the difference on the left hand side includes
changes resulting from the flow process, as well.) The
source terms will be given by

$$G = \frac{\rho \Delta S}{\Delta t}$$

and G_{heat} will be calculated correspondingly.

The calculations are performed in the following
order:

1. chemical processes (combustion, evaporation/condensation, solution)
2. heat conduction
3. updating of relative permeabilities
4. updating of densities
5. flow of materials ⎫
6. heat flow ⎬ simultaneously
7. correction of the saturations.

Here the correction of the saturations is a delicate
step to satisfy the conditions (1).

NUMERICAL RESULTS

We have performed a long series of tests with the
simulator in order to check whether it exhibits the
proper trends and characteristics of the processes simulated under various conditions. We tested separately
the heat conduction, evaporation-condensation, solution
and combustion processes in linear and two-dimensional
cases, with and without gravitational effects. Moreover,
we tested the effects of various combustion methods
(like wet combustion) and the influence of oil evapora-

tion and of the solution of combustion gas in oil and
water. All these test showed the correctness of the
model.

The computer program (written in FORTRAN IV) needs
approximately 130 Kbytes of memory when using a 10x5
grid mesh for each of the reservoir, the underburden and
overburden. The CPU time on a ROBOTRON-40 computer varied
from 0.03 sec (for the linear case) to 0.08 sec (for the
two-dimensional case with maximum grid size) for one
grid point per time- and iteration step.

NOTATIONS

t time

r radial coordinate

z vertical coordinate

$\Delta t = t_{n+1} - t_n$

S saturation

p pressure

T temperature (abs.)

ρ density

C specific heat

μ viscosity

k relative permeability

$\psi = k/\mu$

M mol

p_H partial pressure of gas H

p_H^* saturated vapour pressure of liquid H

ℓ_H bubble point pressure coefficient

Z_H Arrhenius-coefficient

ϑ Arrhenius-exponent

Φ porosity

κ absolute permeability

K heat condition coefficient

INDICES

H general index of materials

g gas phase

O oil phase

O_1, O_2 oil components

w water, water phase

s combustion gas

sH combustion gas dissolved in liquid H

$\overset{*}{H}$ vapour of the liquid H

r rock

n time step

REFERENCES

[1] Gy. Adler, A linear model and a related very stable numerical method for thermal secondary oil recovery, *The Journal of Canadian Petroleum Technology*, 14(1975), 56-65.

[2] V.J. Berry Jr. - D.R. Parrish, A theoretical analysis of heat flow in reverse combustion, *Trans. AIME*, 219(1960), 124.

[3] B.S. Gottfried, A mathematical model of thermal oil recovery in linear systems, *Soc. Petr. Eng. J.* (Sept. 1965), 196-210.

[4] P.H. Holst - P.S. Karra, The size of the steam zone in wet combustion, *Soc. Petr. Eng., Preprint* 4504 (1973).

Gy. Adler
Mathematical Institute of the Hungarian Academy
of Sciences, 1053 Budapest, Realtanoda u.13-15.
Hungary

EXISTENCE AND NUMERICAL REPRESENTATION OE GENERALIZED SOLUTIONS OF NONLINEAR PARTIAL INITIAL VALUE PROBLEMS

R. ANSORGE

1. INTRODUCTION

Many results of the modern theory of difference schemes for partial initial value problems can be interpreted as generalizations of results due to L a x and R i c h t m y e r [1]. In their join paper, the structure and the behaviour of difference methods for linear initial value problems were analysed. The main theorem was an equivalence theorem establishing the equivalence of the numerical stability of the difference method to its pointwise convergence.

It is remarkable that this result (which was partly also already given by K a n t o r o v i č h [2],) was formulated in a way not depending on the type of the given initial value problem (hyperbolic, parabolic, etc.). Only linearity was required and the occuring difference operators were not allowed to depend on the time variable t.

During the last twenty years, the results of L a x and R i c h t m y e r were generalized in several

directions; p.e.: Stability theory of linear initial
boundary value problems (see p.e. [3]), equivalence
theorems for finite difference schemes with singular
coefficients [4], abstract formulations of Lax' equivalence theorem (i.e. formulations not only connected with
initial value problems) [5], [6], convergence and equivalence theorems for linear inhomogoneous [7], semilinear
[8], [9], [10], quasilinear [11] and nonlinear [12] initial value problems etc. L a x and R i c h t m y e r
showed that the numerical stability is not only sufficient
(and necessary) for the convergence of the method in case of the
numerical computation of genuine solutions but also in
case of the computation of generalized solutions. But the
rate of this convergence could only be estimated for
genuine solutions fulfilling the consistency condition.

No earlier than ten years later, the rate of the
convergence was also estimated for certain classes of
generalized solutions of linear initial value problems
by P e e t r e and T h o m é e [13].

The numerical computability of generalized solutions
is of interest because many physical phaenomenons can
only be described by such generalized solutions. In the
linear case, the existence of generalized solutions
follows from an elementary theorem of Functional Analysis.

The existence of generalized solutions for semilinear problems, the generalization of the Lax-Richtmyer-
-theory to semilinear problems (including the generalized
solutions), and the rate of convergence (also including
generalized solutions) was presented by the author on the
Keszthely-conference in 1973 (see also [14]) under the
assumption that the differential operator (with respect
to the space variables) does not depend on t.

Hence, the existence of generalized solutions, the
computability of these solutions by difference methods

and the convergence rate with respect to this computation
is an additional direction of the above mentioned generalizations of the Lax-Richtmyer results.

In this paper, some of the results concerning this
direction (which were found since 1973) will be considered
without going into details.

For convenience, we shall restrict the results on
one-step methods although most of the results are true
for multi-step methods too.

2. THE INITIAL VALUE PROBLEM AND ITS DISCRETIZATION

Let us consider initial value problems of the following type: Assume that there is given a normed space
M of (vector valued) functions u. We omit independent
"space"-variables $x \in \mathbf{R}^d$ of these functions.

We want to find an one-parameter family $\{u(t)\} \subset M$
(the parameter t is called "time"-variable) such that
this family fulfills the problem

$$u_t = A(t)u, \quad 0 \leq t \leq T$$

(1)

$$u(0) = u_0.$$

Generally, the operators $A(t)$ $(0 \leq t \leq T)$ are assumed
to be differential operators (not necessarily linear)
with respect to the space variables having a common
domain $M_A \subset M$ and mapping M_A into M. Assume that (1)
has unique (genuine) solutions for all $u_0 \in D \subset M_A$. The
solution belonging to an initial function $u_0 \in D$ is
described by

$$u(t) = E_0(t)u_0 ,$$

where the "solution operators" $E_0(t)$ $(0 \leq t \leq T)$ are
assumed to be continuous on D (i.e. the solutions
depend continuously on the data).

In order to solve (1) numerically, we devide $[0, T]$
by an equidistant (time-) step size h. We omit the step
sizes Δx as well as we already omitted the space variables assuming that certain relations

$$\Delta x = g(h) \rightarrow 0 \qquad (\text{for} \quad h \rightarrow 0)$$

are prescribed.

Using the denotation $t_n := nh$ $(n=0,1,2,\ldots)$ and
denoting the seeked approximation to the function $u(t_n)$
by u_n, we describe the (one-step) difference scheme by

$$
\begin{aligned}
u_n &= C(t_{n-1}, h)u_{n-1} \qquad (n=1,2,\ldots) \\
&= \prod_{\nu=0}^{n-1} C(t_\nu, h)u_0.
\end{aligned}
$$

(2)

We extend the discrete functions u_n $(n=1,2,\ldots)$
(discrete with respect to the space variables) to the
inter-grid points (p.e. by interpolation or other well-
-known techniques) in order to get

(3) $\qquad u_n \in M \qquad (n=0,1,2,\ldots).$

The "difference operators" $C(t, h)$ are then assumed to
be continuous operators from M into M for all
$t \in [0, T]$ and for all sufficiently small step sizes h
(say: for all $h \in [0, h_0]$, $h_0 > 0$). We suppose that the
difference scheme (2) is consistent with the given
initial value problem (1) on D, i.e. the local truncation error is bounded by a functional $\varepsilon(h, u_0)$ (not

depending on t) with $\varepsilon(h, u_0) = o(h)$ (for $h \to 0$):

$$\|C(t, h)u(t) - u(t+h)\| \leq \varepsilon(h, u_0)$$

(4)

$$\text{for all } t \in [0, T],$$
$$\text{for all } h \in [0, h_0],$$
$$\text{for all } u_0 \in D,$$
$$u(t) = E_0(t)u_0.$$

The method is said to be of order p on D, if $\varepsilon(h, u_0) = O(h^{p+1})$ for all $u_0 \in D$.

DEFINITION. If U is a subset of M with $D \subset U$ (but $U-D \neq \emptyset$) and if $E(t)$ is a continuous dense extension of the operator $E_0(t)$ from the domain D to the domain U,

$$u(t) = E(t)u_0, \quad u_0 \in U, \qquad 0 \leq t \leq T$$

is called a "generalized solution" of problem (1).

DEFINITION. (2) is called (pointwise) *convergent* on $V \subset U$, iff for all sequences $\{n_j\} \subset \mathbb{N}$ with $\lim_{j \to \infty} n_j = \infty$, for all sequences $\{h_j\} \to 0$ with $\{n_j h_j\} \subset [0, T]$ and with $\lim_{j \to \infty} n_j h_j = t$ the relation

$$\lim_{j \to \infty} Q_j u_0 = \lim_{j \to \infty} u_{n_j} = E(t)u_0$$

holds for all $u_0 \in V$ and for all $t \in [0, T]$; here, Q_j is an abbreviation for $\prod_{\nu=0}^{n_j-1} C(\nu h_j, h_j)$.

REMARK. For $V \not\subset D$, also generalized solution are included in this definition.

3. THE CLASSICAL LINEAR CASE

In the classical case considered by L a x and
R i c h t m y e r, M was assumed to be a Banach space,
A was assumed to be a linear operator not depending on
t, and it was supposed that M_A and D are linear subspaces which are dense in M.

Following from these assumptions, the $E_0(t)$ become
linear and (individually) bounded operators. This leads
(by an elementary theorem of Functional Analysis) to the
unique existence of linear bounded extensions $E(t)$ and
hence to the existence of generalized solutions for all
$u_0 \in M$ (i.e. $U = M$).

According to the linearity and t-independence of
A, also the difference operators $C(h)$ are assumed to be
linear $(M \to M)$ and to be independent of t. (2) then
gives

$$(5) \qquad u_n = C^n(h) u_0,$$

and the method is called "(numerically) stable" if the
solution of the difference equation depends equicontin-
uously of the data, i.e. if there is a constant $\varkappa_0$ such
that

$$\| C^n(h) \| = \varkappa_0 \quad \text{for all} \quad n \in \mathbf{N}, \text{ for all} \quad h \in [0, h_0]$$
$$(6)$$
$$\text{with} \quad nh \in [0, T].$$

The equivalence theorem of L a x and R i c h t -
m y e r now gives under the assumptions of this chapter:
*If the scheme (2) is consistent with problem (1),
it converges on M if and only if it is stable.* The
proof is more or less an application of the Banach-
-Steinhaus-theorem (see p.e. [15], p. 204).

Moreover, if the method is of order p on D, it follows

$$(7) \qquad \| u_n - u(t_n) \| \leq \varkappa_0 T \frac{\varepsilon(h, u_0)}{h} = 0 \, (h^p) \quad \text{for all} \quad u_0 \in D$$

(hence: the rate of convergence on D is at least equal to p).

In their paper mentioned above, P e e t r e and T h o m é e described precisely the convergence rate $q = q(N)$ $(0 < q(N) \leq p)$ for certain subsets $N \subset M - D$.

4. SEMILINEAR PROBLEMS

In Keszthely 1973 (see also [14]), the results of chapter 3 were generalized to discretizations of semi-linear problems of the type

$$(8) \qquad u_t = Fu + G(t)u, \qquad 0 \leq t \leq T$$

$$u(0) = u_0,$$

where F again was a linear differential operator with respect to the space variables (which was again not al-lowed to depend on t); the $G(t)$ $(0 \leq t \leq T)$ were assumed to be nonlinear but locally lipschitz-continuous operators mapping M into M.

The existence of generalized solutions was shown by properties of the approximating difference scheme itself, and the generalization of P e e t r e's and T h o m è e's result concerning the rate of convergence for generalized solutions was mainly based upon the fact (shown by T h o m p s o n [15]) that the initial value problem (8) is equivalent to the integral equation

97

(9) $\qquad u(t) = E^{lin}(t)u_0 + \int_0^t E^{lin}(t-\tau)G(\tau)u(\tau)d\tau,$

where $\{u(t)\}$ is the seeked solution of (8) and where $E^{lin}(t)$ are the solution operators of the corresponding linear problem

(10) $\qquad \begin{aligned} u_t &= Fu, & 0 \le t \le T \\ u(0) &= u_0. \end{aligned}$

The restriction that F does not depend on t was necessary to ensure the semi-group property

(11) $\qquad E^{lin}(t+\tau) = E^{lin}(t)E^{lin}(\tau)$

of the family $\{E^{lin}(t)\}$ which was needed by T h o m p-s o n to show the equivalence of (8) and (9). (11) describes "Hadamard's principle of nature".

In order to show that the results we got for the t-independent case are true for the case $F = F(t)$ too, we introduce for the description of the solutions of the corresponding linear problem

$$u_t = F(t)u, \qquad 0 \le t \le T$$

$$u(0) = u_0$$

a two-parameters family $\{E_0^{lin}(t, s)\}$ of linear continuous solution operators:

$$u(t) = E_0^{lin}(t, s)u(s), \qquad 0 \le s \le t \le T$$

with

$$E_0^{\text{lin}}(s, s) = I \text{ (identity), for all } s \in [0, T],$$

$$(12) \qquad E_0^{\text{lin}}(t, s) E_0^{\text{lin}}(s, r) = E_0^{\text{lin}}(t, r), \text{ for all}$$

$$r, s, t \in [0, T], \quad 0 \le r \le s \le t \le T.$$

(12) is a weaker form of Hadamard's principle which can be expected to be true also for problems where F depends on t.

The problem of solving

$$u_t = F(t)u + G(t)u, \qquad 0 \le t \le T$$
$$(13) \qquad u(0) = u_0 \qquad\qquad .$$

is then equivalent with the problem to solve

$$(14) \qquad u(t) = E^{\text{lin}}(t, 0)u_0 + \int_0^t E^{\text{lin}}(t, \tau)G(\tau)u(\tau)d\tau$$

which generalizes T h o m p s o n's result. Here, the operators $E^{\text{lin}}(t, s)$ are the continuous extensions of the operators $E_0^{\text{lin}}(t, s)$.

This equivalence was shown by R e i c h e l t [17][1] and allows a complete transfer of the results concerning the convergence rate from the t-independent to the t- -dependent semilinear problem.

These results can briefly be expressed as follows: *Compared with the corresponding linear initial value problem, the occurence of the locally lipschitz-continuous operators $G(t)$ ($0 \le t \le T$) does neither influence the*

---

[1] In connection with certain parabolic problems, two-parameters families of solution operators were also studied by Kato [18].

*occurence of convergence on the whole nor the rate of the
convergence (and this is true for genuine solutions as
well as for generalized solutions) and also the dependence on t does not influence this rate.*

Hence: The results of P e e t r e and T h o m é e
are true also for linear problems where the differential
operator depends on *t* as well as for semilinear problems of the type considered in this chapter.

5. QUASILINEAR PROBLEMS

Let us now consider quasilinear initial value problems which are very important from the point of view of
applications (because many physical phaenomenous are
mathematically simulated by quasilinear differential equations) and which are very general because most of the
nonlinear initial value problems can be transformed into
problems with systems of quasilinear equations (which are
included here because the functions *u* were allowed to
be vector-valued).

Again three questions arise:

1) Existence of generalized solutions.

2) Convergence of difference schemes in connection not
   only with genuine solutions but also with these generalized solutions.

3) Rate of convergence with respect to generalized solutions (the rate concerning genuine solution fulfilling
   a consistency conditions can mostly be determined
   easily).

The third problem is not yet solved!

In order to answer questions 1 and 2, we have to ask
for theorems which ensure (similarly to the linear case)
the existence of unique continuous extensions of continuous

solution operators, and we have to ask for theorems which
generalize the Banach-Steinhaus-theorem (in order to
generalize the Lax-Richtmyer-theorem or at least one of
the directions of this theorem).

In this' context, let us refer to a theorem which is
(in this formulation) due to $R\ i\ n\ o\ w$ ([19], p. 78/79):
Let M be a Banach-space, $\{Q_j\}$ a sequence of continuous
operators from $U \subset M$ into M. Let D be a subset of U
which is dense in U and E_0 a continuous operator from
D into M.

    Assume: (a) The operators Q_j are equicontinuous
               on U

          (b) $\lim\limits_{j \to \infty} Q_j = E_0$ on D (pointwise).

Then there is one and only one continuous extension E
of the operator E_0 from the domain D to the domain
U and the relation

$$(15) \qquad \lim_{j \to \infty} Q_j = E \quad \text{on} \quad U \quad \text{(continuously)}$$

holds.

REMARK. Later on we will identify the operators Q_j
with the iterated difference operators, D will be identified with a set of initial functions for which genuine
solutions exist and on which a consistency conditions
holds, E_0 will be identified with the solution operator
$E_0(t)$ (for every fixed $t \in [0, T]$) and E then gives the
solution operator $E(t)$ of generalized solutions arising
from initial functions $u_0 \in U$ (see the denotations in
chapter 2).

REMARK. For the linear case, the equicontinuity of
the iterated difference operators is equivalent to the

numerical stability of the method. Condition (b) of the
theorem means the (pointwise) convergence with respect
to genuine solutions arising from $u_0 \in D$.

Hence: If we use this interpretation of R i n o w's
theorem and if it should be possible to realize the assumptions (a) (stability) and (b) (pointwise convergence
of the method with respect to genuine solutions fulfilling the consistency condition) for certain classes of
problems and corresponding classes of difference schemes,
the theorem gives the existence of generalized solutions
on U (by properties of the difference scheme) as well
as the convergence of the method with respect to these
generalized solutions (see (15)).

This possibility of interpretation of R i n o w's
theorem that fits for discretization algorithms was
already mentioned by the author in 1969 [20] but was only
realized in connection with semilinear problems [14].

We will now use this theorem for a general class of
quasilinear problems, following some ideas which are due
to K r e t h [21].

Let us consider problems of type (1) of the particular form

$$u_t = F(t, u)u, \qquad 0 \leq t \leq T$$

(16)

$$u(0) = u_0$$

given in a Banach space M.

The operators $F(t, u)$ are assumed to be linear
operators (normally: differential operators with respect
to the space variables) for each fixed $t \in [0, T]$, for
each fixed "internal" $u \in M$, having a common domain
$M_F \in M$:

$$F(t, u) : M_F \to M.$$

Linearity of these operators means linearity with respect to the "external" $u \in M_F$

$$(\text{p.e.:}\ [F(t, u)v](x) = f(t)u^2(x)v_{xx}(x), \quad M = C^0_{2\pi}$$

$$\|u\| = \max_{0 \le x \le 2\pi} |u(x)|, \quad f(t) \ge 0, \quad M_F = C^2_{2\pi}).$$

REMARK. An additional term $G(t)u$ in the differential equation (16) with locally lipschitz-continuous operators $G(t)$ mapping M into M does not influence the behaviour of the convergence of corresponding consistent difference schemes (according to the results we got from the semilinear case) and is therefore omitted (only for convenience).

Assume that (16) has unique genuine global solutions (i.e. with common $T > 0$) for all $u_0 \in D$, where D is a bounded subset of M_F.

Assume further:

$$(17) \qquad \sup_{u_0 \in D}\ \max_{0 \le t \le T}\ \|E_0(t)u_0\| =: r < \infty;$$

here, the $E_0(t)$ are again the solution operators of problem (16) mapping D into M_F which are assumed to be continuous on D.

Using explicit difference schemes only, the quasilinear structure of (16) normally leads to a quasilinear structure of the difference scheme:

$$
\begin{aligned}
(18) \qquad u_n &= \overline{C}(t_{n-1}, h, u_{n-1})u_{n-1} =: C(t_{n-1}, h)'u_{n-1} =: \\
&= \prod_{\nu=0}^{n-1} C(\nu h, h)u_0.
\end{aligned}
$$

The operators $\overline{C}(t, h, z)$ are assumed to be linear continuous operators from M into M for every fixed $t\in [0, T]$, for every fixed $h\in [0, h_0]$ (with a certain $h_0 > 0$), and for every $z\in K$ where K is a closed ball with centre 0 and a certain radius $s > r$.

Because of $E_0(0) = I$, we have $D\subset K$.

If U is a subset of M (on which we want to show the existence of generalized solutions) with $D \underset{dense}{\subset} U$, we also have $U\subset K$ (because K is closed)

Assume that N is a certain subset of M (describing the structure of the elements of the bounded set U; compare the example below) with

$$M_F\subset N, \quad U\subset N, \quad C(t, h): N \to N.$$

REMARK. $C(t, h): N \to N$ means that the difference scheme preserves the structure of the elements of N and hence also of the elements of U.

Suppose, there is a functional $\varkappa(N \to \mathbf{R})$ such that the following relation holds:

$$\| \{\overline{C}(t, h, w)-\overline{C}(t, h, v)\} u\| \leq \varkappa(u)h\|w-v\| \quad \text{for all}$$

(19) $t\in [0, T]$, for all $h\in [0, h_0]$, for all $v, w\in K$,

for all $u\in N\cap K$.

Assume that the method is consistent with the given problem on D (i.e. (4) holds) and that it is stable in the following sense:

(20) $\exists \gamma\in\mathbf{R}: \|\overline{C}(t, h, w)\| \leq 1+\gamma h, \quad \forall\in t[0, T], \quad \forall h\in [0, h_0], \quad \forall w\in K.$

Moreover, we suppose that there is a constant $r_0 < s$

with the property

$$(21) \qquad \left\| \prod_{\nu=0}^{n-1} \overline{C}(t_\nu, h, u_\nu)u_0 \right\| \le r_0$$

for all $u_0 \in U$, for all $n \in \mathbf{N}_0$, and for all $h \in [0, h_0]$ with $nh \in [0, T]$ (here, $\{u_\nu\}$ means the sequence arising from u_0 by the difference method, and (21) gives $\{u_\nu\} \subset K$), and that there is a constant $\varkappa_0$ with

$$(22) \qquad \varkappa\left(\prod_{\nu=0}^{n-1} C(t_\nu, h, u_\nu)u_0 \right) \le \varkappa_0 \qquad \text{for all} \quad u_0 \in U.$$

Then, under the assumptions of this chapter and using the sequences $\{n_j\}$ and $\{h_j\}$ introduced in chapter 2 and using the abbreviation Q_j which was also introduced in chapter 2, we have the following theorems:

THEOREM 1.

$$\| Q_j u_0 - Q_j v_0 \| \le e^{(\gamma + \varkappa_0)T} \| u_0 - v_0 \|, \qquad u_0, \ v_0 \in U.$$

THEOREM 2. *For* $u_0 \in D$, *the relation*

$$\| u_n - u(t_n) \| \le [e^{(\gamma + \varkappa_0)T} - 1] \frac{\varepsilon(h, u_0)}{h} \rightarrow 0$$

$$(for \quad h \rightarrow 0)$$

holds.

Theorem 1 shows that the operators Q_j fulfill condition (a) of Rinow's theorem, namely that these operators are equi- (lipschitz-) continuous on U.

Theorem 2 gives the pointwise convergence of the method (and the rate of this convergence) for all genuine solutions $E_0(t)u_0$ with $u_0 \in D$ and hence the property required by condition (b) of Rinow's theorem.

Thus, under the assumptions of this chapter, there
are generalized solutions for all $u_0 \in U$ and also these
generalized solutions can by numerically approximated by
the difference scheme.

The proofs of Theorem 1 and Theorem 2 are straightforward; the difficulties in connection with quasilinear
problems did not arise from these proofs but from a
suitable formulation of assumptions leading to theorems
of this type, particularly to a result of the type of
Theorem 1.

Finally, we must show that the class of problems
fulfilling the assumptions mentioned above does not only
include the linear problems already considered by L a x
and R i c h t m y e r.

In order to do this, we consider a quasilinear
hyperbolic example arising from fluid dynamics:

EXAMPLE (see [21]):

$$x = (x_1, \ldots, x_d) \in \mathbf{R}^d, \quad \| x \|_d = \max_{i=1,\ldots,d} |x_i|.$$

$$M = C_{2\pi}^0(\mathbf{R}^d) \times C_{2\pi}^0(\mathbf{R}^d) \times \ldots \times C_{2\pi}^0(\mathbf{R}^d) = [C_{2\pi}^0(\mathbf{R}^d)]^m,$$

$$\| f \|_M = \max_{\substack{j=1,\ldots,m \\ i=1,\ldots,m}} \max_{0 \le x_i \le 2\pi} |f_j(x_1, \ldots, x_d)|.$$

Hence: M is a Banach space.

In this space M we consider the problem

$$u_t(x, t) + (g(x, t, u(x, t)) \cdot \nabla) u(x, t) = 0,$$

$$(23) \qquad\qquad\qquad\qquad\qquad\qquad 0 \le t \le T$$

$$u(x, 0) = u_0(x)$$

with the ordinary inner product $(\cdot)$ on $\mathbf{R}^d$. The assumptions with respect to g are:

$$g = \begin{bmatrix} g_1 \\ \vdots \\ g_d \end{bmatrix}, \quad g_j(\cdot, t, w(\cdot)) \in C^0_{2\pi}(\mathbf{R}^d) \quad (j=1,\ldots,d),$$

$$\forall w \in M; \ \exists\, b, \ c:$$

$$\|g(x, t, z_1) - g(y, t, z_2)\|_d \leq b\|x-y\|_d + c\|z_1 - z_2\|_m$$

for all $x, y \in \mathbf{R}^d$, for all $z_1, z_2 \in \mathbf{R}^m$ with $\|z_i\|_m \leq s$ $(i=1,2)$ (using a suitable $s > 0$); g has continuous derivatives of first order with respect to all variables. We then have the following theorem: (R a u t m a n n 1974, [22]).

THEOREM. *There are genuine global solutions for all* $u_0 \in [C^1_{2\pi}(\mathbf{R}^d)]^m$.

The structure of the initial elements for which we want to show the existence and numerical computability of generalized solutions is described by

$$N := M \cap \mathrm{Lip}\ 1$$

(i.e. we will consider generalized solutions arising from certain initial elements which are not necessarily continuously differentiable but only lipschitz-continuous).

The preservation of this structure $(C(t, h):N \to N)$ is guaranteed p.e. by the method of C o u r a n t , I s a a c s o n and R e e s [23] (which does not preserve differentiability!) which will be stable if we realize the relation $\Delta x = g(h)$ of chapter 2 by

$$\frac{h}{\Delta x} =: \lambda = \mathrm{const}, \quad 0 < \lambda < \frac{1}{d\hat{g}}$$

with

$$\hat{g} := \max_{\substack{t \in [0, \ T] \\ 0 \le x_i \le 2\pi \\ i = 1, \ldots, d \\ z \in \mathbf{R}^m, \ \|z\|_m \le s}} \|g(x, \ t, \ z)\|_d .$$

We define

$$U := \{u_0 \in N, \ \|u_0\| \le r_0, \ \ y(t) \le \varkappa_0\}$$

with $\varkappa_0 > 0$ arbitrary and $r_0 > 0$ sufficiently small; $y(t)$ is the solution of the ordinary initial value problem

$$y' = by + cy^2, \qquad 0 \le t \le T$$

$$y(0) = L(u_0)$$

where $L(u_0)$ stands for the Lipschitz-constant of $u_0 \in N$.

REMARK. $y(t)$ is a majorant of the global error of the method.

If we finally define

$$D := U \cap [C_{2\pi}^2 (\mathbf{R}^d)]^m \subset [C_{2\pi}^1 (\mathbf{R}^d)]^m,$$

then the method is consistent with the given problem on D, D is dense in U (following from W e i e r s t r a ß' approximation theorem) and also the other previously mentioned assumptions are fulfilled. Thus, the existence and numerical computability of generalized solutions arising from $u_0 \in U$ are proven.

REMARK. There are also quasilinear parabolic examples; hence, also the type-independence of the classical
linear Lax-Richtmyer-theory is kept.

REMARK. Following from the results of this paper,
difference methods are obviously not very sensitive with
respect to irregularities of the type considered here;
this may help to replace the heuristical confidence in
difference methods (which can often be observed in real
applications) by mathematical arguments.

REFERENCES

[1]   P.D. Lax - R.D. Richtmyer, Survey of the stability
      of linear finite difference equations; *Comm. Pure
      Appl. Math.* 9(1956), 267-293.

[2]   L.W. Kantorowič, Functional Analysis and Applied
      Mathematics; *Uspeki Mat. Nauk.* 3(1948), 89-185,
      (English translation: National Bureau of Standards
      Report 1509, 1952).

[3]   B. Gustafson, Stability theory of difference ap-
      proximations for mixed initial boundary value prob-
      lems, II.; *Math. Comput.* 26(1972), 649-686.

[4]   D. Eisen, The equivalence of stability and con-
      vergence for finite difference schemes with singular
      coefficients; *Numer. Math.* 10(1967), 20-29.

[5]   H.M. Schultz, A generalization of the Lax equivalence
      theorem; *Proc. Amer. Math. Soc.* 17(1966), 1034-1035.

[6]   F. Stummel, Discrete convergence of mappings. In:
      *Proceedings of the Conference on Numerical Analysis
      Dublin 1972.* (ed. J. Miller) Academic Press, New
      York - London, 1973.

[7] H.J. Stetter, Anwendung des Äquivalenzsatzes von
P. Lax auf inhomogene Probleme; *ZAMM* 39(1959),
396-397.

[8] R. Ansorge, Der Äquivalenzsatz von Lax für halb-
lineare Probleme; *ZAMM* 46(1966), T35-T37.

[9] R. Ansorge, Konvergenz von Mehrschrittverfahren zur
Lösung halblinearer Anfangswertaufgaben; *Numer.
Math.* 10(1967), 209-219.

[10] M.N. Spijker, *Stability and Convergence of Finite
Difference Methods*, Thesis, Leiden University (1968).

[11] H. Kreth, Ein Äquivalenzsatz bei der numerischen
Lösung quasilinearer Anfangswertaufgaben; In: *Lecture
Notes in Mathematics*, vol. 395 (ed. R. Ansorge -
- W. Törnig) Springer, Berlin-Heidelberg-New York,
1974.

[12] H.v. Dein, Konvergenzbedingungen bei der numerischen
Behandlung nichtlinearer Anfangswertaufgaben mittels
Differenzenverfahren; In: *ISNM*, vol. 31 (ed. J.
Albrecht - L. Collatz) Birkhäuser, 1976.

[13] J. Peetre - V. Thomée, On the rate of convergence
for discrete initial-value problems; *Math. Scand.*
21(1967), 159-176.

[14] R. Ansorge - C. Geiger - R. Hass, Existenz und
numerische Erfaßbarkeit verallgemeinerter Lösungen
halblinearer Anfangswertaufgaben; *ZAMM* 52(1972),
597-605.

[15] L.W. Kantorowitsch - G.P. Akilov, *Funktionalanalysis
in Normierten Räumen*; Akademie Verlag, Berlin, 1964.

[16] R.J. Thompson, Difference approximations for inhomogeneous and quasilinear equations; *J. Soc.
Indust. Appl. Math.* 12(1964), 189-199.

[17] P. Reichelt, Eine Darstellung der Lösung von halb-
     linearen Anfangswetaufgaben in Integralform; *ZAMM*
     55(1975), 613.

[18] T. Kato, Abstract evolution equations of parabolic
     type in Banach and Hilbert spaces, *Nagoya Math.J.*
     19(1961), 93-125.

[19] W. Rinow, *Die innere Geometrie der metrischen Räume*,
     Springer, Berlin-Göttingen-Heidelberg, 1961.

[20] R. Ansorge, Zur Existenz verallgemeinerter Lösungen
     nichtlinearer Anfangswertaufgaben; In: *ISNM*, vol.
     12, 13-22, (ed. L. Collatz - H. Unger) Birkhäuser,
     Basel-Stuttgart, 1969.

[21] H. Kreth, Der Nachweis der Existenz verallgemeiner-
     ter Lösungen quasilinearer Anfangswertaufgaben
     mittels Differenzenverfahren; *Computing* 15(1975),
     251-261.

[22] R. Rautmann, Zur iterativen Lösung spezieller
     Systeme partieller Differentialgleichungen; *ZAMM*
     54(1974), T197-T199.

[23] R. Courant - E. Isaacson - M. Rees, On the solution
     of nonlinear hyperbolic differential equations by
     finite differences; *Comm. Pure App. Math.* 5(1952),
     243-255.

R. Ansorge
Universität Hamburg
Institut für Angewandte Mathematik
Bundesstraße 55
D-2000 Hamburg 13

THE USE OF WACHSPRESS PARAMETERS FOR THE NUMERICAL SOLUTION OF THREE-DIMENSIONAL ELLIPTIC PROBLEMS

G. AVDELAS - A. HADJIDIMOS

1. INTRODUCTION

For the last two decades many papers have been devoted to the numerical solution of the first boundary value problem for Laplace or Poisson equation in a parallelepiped by using Alternating Direction Implicit (A.D.I.) or Extrapolated (E.) A.D.I. methods. Among these papers we simply quote those by D o u g l a s and R a c h f o r d [3], D o u g l a s [2], S a m a r s k i i and An d r e e v [11] and [12], G u i t t e t [5], F a i r w e a t h e r , G o u r l a y and M i t c h e l l [4], H a d j i d i m o s [6], [7] and [9] and A v d e l a s [1]. The introduction of the extrapolation parameter together with the set of D o u g l a s or S a m a r s k i i and A n d r e e v parameters applied in the way described in H a d j i - d i m o s [8] and [9] has provided us with A.D.I. methods which were much more rapidly convergent compared with the previous unextrapolated ones. However, the problem

of finding the optimum three-dimensional A.D.I. scheme
in a way analogous to the one that W a c h s p r e s s
found for the two-dimensional scheme (see e.g. W a c h s-
p r e s s [13] and [14] and Y o u n g [15])is far
from having been solved.

In this paper we present a technique which has ena-
bled us to use the Wachspress parameters of the two-
dimensional problem in the three-dimensional one and thus
obtain A.D.I. schemes which, both theory and practice
show, converge much faster than the corresponding E.A.D.
I. ones.

2. STATEMENT OF THE PROBLEM

Consider the first boundary value problem for
Poisson equation

$$\frac{\partial^2 u}{\partial x_1^2} + \frac{\partial^2 u}{\partial x_2^2} + \frac{\partial^2 u}{\partial x_3^2} = f(x), \quad x \equiv (x_1, x_2, x_3) \in R$$

(2.1)
$$u(x) = g(x), \quad x \in \partial R,$$

where $R \equiv \{x \mid 0 < x_i < \ell_i \mid i = 1,2,3\}$ (i.e. the interior
of a parallelepiped with edges of lengths $\ell_i \mid i = 1,2,3$),
∂R the boundary of R and f and g known functions.
From now on and for the sake of simplicity we shall consider that the region involved is the unit cube i.e.
$\ell_1 = \ell_2 = \ell_3 = 1$. To solve numerically problem (2.1) we
impose a uniform grid of mesh size $h = 1/N$ $(N \geq 3)$
with N arbitrary integer and approximate it at each
internal node by a 7-point difference formula. The tota-
lity of difference equations thus obtained yealds a
linear system of the form

(2.2) $\quad (A_1 + A_2 + A_3) u = \varphi,$

where $A_i | i = 1,2,3$, are three known symmetric matrices that commute of order $(N-1)^3$, u is an unkown $(N-1)^3$-dimensional vector whose components $u_{i_1 i_2 i_3}$ are the approximate solutions at the nodes $(i_1 h, i_2 h, i_3 h)$ taken in their natural ordering and φ a known $(N-1)^3$-dimensional vector. The matrices $A_i | i = 1,2,3$ are of the following product forms

$$(2.3) \qquad A_1 = J \otimes J \otimes U, \quad A_2 = J \otimes U \otimes J, \quad A_3 = U \otimes J \otimes J,$$

with J being the unit matrix of order $N-1$, U a matrix of order $N-1$ given by

$$(2.4) \qquad U = \begin{bmatrix} 2 & -1 & & & \\ -1 & 2 & -1 & & \\ & & \ddots & & \\ & & -1 & 2 & -1 \\ & & & -1 & 2 \end{bmatrix}$$

and the symbol $\otimes$ denoting tensor product as is defined in H a l m o s [10]. The eigenvalues of $A_i | i = 1,2,3$ are obviously those of the corresponding U with each eigenvalue repeated $(N-1)^2$ times. Thus the different eigenvalues of A_i's are given by the expressions

$$(2.5) \qquad a_i = 4\sin^2 \frac{k_i \pi}{2N} | i = 1,2,3 \quad \text{and} \quad k_i = 1(1)N-1$$

3. AN A.D.I. - LIKE SCHEME

If we put

$$H = A_1, \quad V = A_2 + A_3$$

then system (2.2) can be written in the form

$$(3.1) \qquad (H+V)u = \varphi,$$

where the matrices H and V are obviously symmetric with different eigenvalues

$$a_1 = 4\sin^2 \frac{k_1 \pi}{2N} \Big| k_1 = 1(1)N-1,$$

$$b_{2,3} = 4\sin^2 \frac{k_2 \pi}{2N} + 4\sin^2 \frac{k_3 \pi}{2N} \Big| k_2, \ k_3 = 1(1)N-1$$

and commute. Therefore for the numerical solution of (3.1) the following A.D.I.-like scheme can be used

$$(3.2) \qquad \begin{aligned} (r_{m+1}^{(1)}I+H)u^{(m+1/2)} &= (r_{m+1}^{(1)}I-V)u^{(m)}+\varphi \\ (r_{m+1}^{(2)}I+V)u^{(m+1)} &= (r_{m+1}^{(2)}I-H)u^{(m+1/2)}+\varphi \end{aligned} \quad \Big| m=0,1,2,\ldots,$$

where I is the unit matrix of order $(N-1)^3$, $r_{m+1}^{(1)}$, $r_{m+1}^{(2)}$ are two sequences of positive acceleration parameters, $u^{(m)}$ is the mth iteration approximation to the solution u of (2.2) ($u^{(0)}$ arbitrary) and $u^{(m+1/2)}$ is an intermediate approximation to $u^{(m+1)}$. Scheme (3.2) is of exactly the same form as the one considered by W a c h s p r e s s (see W a c h s - p r e s s [13] and [14] and Y o u n g [15]). Therefore for a fixed total number of iterations m_0 optimum values for the two sequences of the acceleration parameters are given in terms of elliptic functions in the way described in the same references above.

During each iteration of scheme (3.2) we have to solve two linear systems. Because of the form of H, the matrix of the coefficients of the unknowns of the first system is the direct sum of tridiagonal matrices and therefore this system is easily solved. In the second system, however, the matrix of the coefficients of the unknowns is of a rather complicated form due to the form of V and therefore the corresponding system cannot be easily solved. This difficulty can be overcome if we work as follows.

The matrix of coefficients of the second equation of scheme (3.2) can be transformed in the way below

$$r_{m+1}^{(2)}I+V = r_{m+1}^{(2)}I+A_2+A_3 =$$

$$= r_{m+1}^{(2)}J \otimes J \otimes J + J \otimes U \otimes J + U \otimes J \otimes J =$$

$$= (H_{m+1}+V_{m+1}) \otimes J,$$

where we have set

$$H_{m+1} = J \otimes (\tfrac{1}{2}r_{m+1}^{(2)}J+U),$$

(3.3)

$$V_{m+1} = (\tfrac{1}{2}r_{m+1}^{(2)}J+U) \otimes J.$$

Thus the second equation of scheme (3.2) has the form

$$(3.4) \quad [(H_{m+1}+V_{m+1}) \otimes J]u^{(m+1)} = b^{(m+1)}$$

with

$$b^{(m+1)} = (r_{m+1}^{(2)}I-H)u^{(m+1/2)}+b.$$

System (3.4) can be split into $N-1$ systems of the form

$$(3.5) \quad (H_{m+1}+V_{m+1})v_i^{(m+1)} = d_i^{(m+1)} \mid i = 1(1)N-1,$$

where

$$v_i^{(m+1)} = (u_i^{(m+1)}, \ u_{i+(N-1)}^{(m+1)}, \dots, u_{i+(N-1)^2}^{(m+1)}, \dots,$$

$$u_{i+(N-2)(N-1)N}^{(m+1)})^T,$$

$$d_i^{(m+1)} = (b_i^{(m+1)}, \ b_{i+(N-1)}^{(m+1)}, \dots, b_{i+(N-1)^2}^{(m+1)}, \dots,$$

$$b_{i+(N-2)(N-1)N}^{(m+1)})^T$$

for $i=1(1)N-1$. It is apparent that systems (3.5) have the same matrix of coefficients and that the matrices H_{m+1}, V_{m+1} which constitute it are symmetric, positive definite and commute. Moreover H_{m+1} is a direct sum of tridiagonal matrices while V_{m+1} can be transformed, by permutation transformations, into the direct sum of tridiagonal matrices. Thus each of the systems (3.5) can now be solved numerically by using a new A.D.I. scheme of the form

$$(r_{m+1,\,n+1}^{(1)} I_1 + H_{m+1}) v_i^{(m+1,\,n+1/2)} =$$

$$= (r_{m+1,\,n+1}^{(1)} I_1 - V_{m+1}) v_i^{(m+1,\,n)} + d_i^{(m+1)}$$

$$(3.6) \qquad (r_{m+1,\,n+1}^{(2)} I_1 + V_{m+1}) v_i^{(m+1,\,n+1)} =$$

$$= (r_{m+1,\,n+1}^{(2)} I_1 - H_{m+1}) v_i^{(m+1,\,n+1/2)} + d_i^{(n+1)}$$

$$i=1(1)N-1, \ n=0,1,2,\dots,$$

where I_1 is the unit matrix of order $(N-1)^2$, $r_{m+1,\,n+1}^{(1)}$ and $r_{m+1,\,n+1}^{(2)}$ are two sequences of positive acceleration parameters, $v_i^{(m+1,\,n)}$ is the nth iteration approximation to the solution $v_i^{(m+1)}$ of (3.5)

$(v_i^{(m+1, \, 0)}$ arbitrary) and $v_i^{(m+1, \, n+1/2)}$ an intermediate approximation to $v_i^{(m+1, \, n+1)}$. For a fixed total number of iterations n_0 and for each value of $n = 0(1)n_0-1$ optimum values for the sets $r_{m+1, \, n+1}^{(1)}$ and $r_{m+1, \, n+1}^{(2)}$ can again be provided by the same analysis of Wachspress.

REFERENCES

[1] G. Avdelas, Contribution to the first three boundary value problems for the equations of Laplace, Poisson and Helmholtz (in Greek), *Ph. D. Thesis*, The University of Ioannina, Greece, 1975.

[2] J. Douglas, Alternating direction methods for three space variables, *Numer. Math.*, 4(1962), 41-63.

[3] J. Douglas - H.H. Rachford, On the numerical solution of heat conduction problems in two and three space variables, *Trans. Amer. Math. Soc.*, 82(1956), 421-439.

[4] G. Fairweather - A.R. Gourlay - A.R. Mitchell, Some high accuracy difference schemes with a splitting operator for equations of parabolic and elliptic type, *Numer. Math.*, 10(1967), 56-66.

[5] J. Guittet, Une nouvelle méthode de directions alternées à q variables, *J. Math. Anal. and Appl.*, 17(1967), 199-213.

[6] A. Hadjidimos, On a generalised alternating direction implicit method for solving Laplace's equation, *The Computer Journal*, 11(1968), 324-328.

[7] A. Hadjidimos, On some high accuracy difference schemes for solving elliptic equations, *Numer. Math.*, 13(1969), 396-403.

[8] A. Hadjidimos, Extrapolated alternating direction implicit methods, *BIT*, 10(1970), 465-475.

[9] A. Hadjidimos, Optimum extrapolated A.D.I. iterative difference schemes for the solution of Laplace's equation in three space variables, *The Computer Journal*, 14(1971), 179-183.

[10] P.R. Halmos, *Finite Dimensional Vector Spaces*, Van Nostrand, Princeton, 1958.

[11] A.A. Samarskii and V.B. Andreev, A difference scheme of higher accuracy for an equation of elliptic type in several space variables, *Ž. Vyčisl. Mat. i. Mat. Fiz.*, 3(1963), 1006-1013.

[12] A.A. Samarskii and V.B. Andreev, Alternating direction iterational schemes for the numerical solution of the Dirichlet problem, *Ž. Vyčisl. Mat. i. Mat. Fiz.*, 4(1964), 1025-1036.

[13] E.L. Wachspress, Extended application of alternating direction implicit iteration model problem theory, *J. Soc. Indust. Appl. Math.*, 11(1963), 994-1016.

[14] E.L. Wachspress, *Iterative Solution of Elliptic Systems*, Prencite Hall, New Jersey, 1966.

[15] D.M. Young, *Iterative Solution of Large Linear Systems*, Academic Press, New York, 1971.

G. Avdelas - A. Hadjidimos
Department of Mathematics,
University of Ioannina,
Ioannina,
Greece

COLLOQUIA MATHEMATICA SOCIETATIS JÁNOS BOLYAI
22. NUMERICAL METHODS, KESZTHELY (HUNGARY), 1977.

ON THE REPLACEMENT OF THE CONDITION OF BOUNDEDNESS FOR
CERTAIN SYSTEMS OF LINEAR ODE-S WITH REGULAR SINGULARITY

KATALIN BALLA

The problem of finding the bounded solutions of a
system of n linear ordinary differential equations of
first order with regular singularity was considered in
[1]. In this paper we give a simplified method for carrying the subspace of the bounded solutions out of the
point of singularity when handling systems of first or
second order. The method needs only a weak condition for
the smoothness of the coefficients, this is in contrast
to [1], but here the main matrix of the system has a
special form.

On the basis of these results one can solve boundary
value problems for linear systems with regular singularity. When the condition of boundedness is imposed on
the solution at the left end of the interval where the
system is supposed to have a regular singularity (this
point will be in the origin $t=0$) and a usual boundary
value is given at the right end of the interval $t = T$,
then the interval $(0,T]$ will be replaced by $[t_0, T]$,
where t_o is sufficiently small and a linear relation
will be obtained in $t = t_o$ instead of de-

manding the boundedness of the solution in $t = 0$.

<h3>§.1. AUXILIARY STATEMENTS</h3>

Here we will deal with systems of the form

$$(1) \qquad tY' = BY + h(t,Y) + g(t), \qquad 0 < t \leq t_0,$$

where B is a constant matrix, the eigenvalues of which have negative real part; $h(t,Y)$ is defined and continuous for small $|Y|$ and $0 \leq t \leq t_0$; $g(t)$ is defined and continuous for $0 \leq t \leq t_0$ and $g(t) \to 0$ when $t \to 0$.

The following lemmas were formulated in [2]:

LEMMA 1. *Let $h(t,Y)$ be a linear function of Y, i.e. $h(t,Y) = C(t)Y$ where $|C(t)| = o(1)$ if $t \to 0$. Let $\varphi(t)$ be a solution of (1) such that $\varphi(t) \neq 0$ if $t \to 0$. Then there exist positive constants M, $\tilde{t}$ and a such that*

$$\|\varphi(t)\| \geq Mt^{-a} \quad \text{for any} \quad t, \quad 0 < t \leq \tilde{t}.$$

COROLLARY. *If $g(t) \equiv 0$, then the statement holds for each solution of (1), except for $\varphi(t) \equiv 0$.*

LEMMA 2. *Let $h(t,0) \equiv 0$ and for any $\varepsilon > 0$, let $\delta_\varepsilon > 0$ and $t_\varepsilon > 0$ exist such that*

$$|h(t,Y_1) - h(t,Y_2)| \leq \varepsilon |Y_1 - Y_2|$$

for any $|Y_1| \leq \delta_\varepsilon$, $|Y_2| \leq \delta_\varepsilon$ and $t \leq t_\varepsilon$. Then there exists a solution $\Theta(t)$ of the system (1) such that

122

$\Theta(t) \to 0$ if $t \to 0$ and this solution is unique.

Moreover, if $h(t,Y)$ has continuous partial derivatives up to the N-th order with respect to t and each component of Y in the domain $\|Y\| < R_0$, $0 \le t \le t_0$, and $g(t)$ has N continuous derivatives on $0 \le t \le t_0$, then

$$\Theta(t) = \sum_{k=1}^{N} \Theta_k t^k + o(t^N), \qquad t \to 0,$$

where Θ_k can be defined formally from (1).

LEMMA 3. Let $h(t,Y)$ be the sum of linear and quadratic forms, i.e.

$$(2) \qquad h(t,Y) = V(t)Y + \sum_{k,\ell=1}^{n} Y_k Y_\ell G^{k,\ell}(t),$$

where Y_k, $k=1,\ldots,n$ are the components of Y. The matrix $V(t)$ and the vectors $G^{k,\ell}(t)$ and $g(t)$ can be expanded into power series for $t \le \tilde{t}$:

$$V(t) = \sum_{i=1}^{\infty} V_i t^i, \qquad G^{k,\ell}(t) = \sum_{i=0}^{\infty} G_i^{k,\ell} t^i,$$

$$g(t) = \sum_{i=1}^{\infty} g_i t^i.$$

a) Then there exists exactly one solution $\Theta(t)$ of the system (1) such that $\Theta(t) \to 0$ if $t \to 0$ and

$$(3) \qquad \Theta(t) = \sum_{i=1}^{\infty} \Theta_i t^i, \qquad t \to 0,$$

b) the series (3) can be majorated by the series

$$z(t) = \sum_{i=1}^{\infty} z_i t^i$$

which satisfies the equation

$$z(t) = c[\varphi(t)z(t) + \sum_{k,\ell=1}^{n} \psi^{k,\ell}(t)z^2(t) + \xi(t)] + |(E-B)^{-1}g_1|t, \qquad z(0) = 0,$$

where the scalar functions $\varphi(t)$, $\psi^{k,\ell}(t)$, $\xi(t)$ *are defined as*

$$\varphi(t) = \sum_{r=1}^{\infty} |V_r|t^r, \qquad \psi^{k,\ell}(t) = \sum_{r=0}^{\infty} |G_r^{k,\ell}|t^r,$$

$$\xi(t) = \sum_{r=2}^{\infty} |g_r|t^r$$

and

$$c = \max_{2 \leq i < \infty} |(iE-B)^{-1}|.$$

Besides the statements of [2] concerning the solutions of the system (1) we will need the following

LEMMA 4. *Let* $h(t,Y)$ *be of the form (2) in (1), where* $\sum_{k,\ell=1}^{n} \max_{0 \leq t \leq a} |G^{k,\ell}(t)| < \infty$ *and let the integrals*

$$\int_0^a \frac{|V(t)|}{t}dt \quad and \quad \int_0^a \frac{|g(t)|}{t}dt$$

be bounded. Then the unique solution $\theta(t)$ *of (1), that tends to zero if* $t \to 0$, *satisfies the condition*

$$\int_0^a \frac{|\theta(t)|}{t}dt < \infty.$$

§.2. LINEAR SYSTEMS OF FIRST ORDER

Consider the system of linear ordinary differential equations of first order

$$(4) \qquad tX' = A(t)X + f(t), \qquad 0 < t \le t_1,$$

where $A(t)$ is a continuous $n \times n$ matrix, $f(t)$ is a continuous vector with n entries for small t such that $\lim_{t \to 0} A(t) = A_0$ and $\lim_{t \to 0} f(t) = f_0$. Let A_0 be quasidiagonal:

$$(5) \qquad A_0 = \begin{bmatrix} A_- & 0 \\ 0 & A_+ \end{bmatrix},$$

where A_- and A_+ are quadratic, $r \times r$ and $(n-r) \times (n-r)$ matrices, respectively. The eigenvalues of matrix A_- lie on the left side of the imaginary axis $(\mathrm{Re}\lambda_{A_-} < 0)$ and those of matrix A_+ have nonnegative part. For an eigenvalue, lying on the imaginary axis, let the number of independent eigenvectors be equal to the multiplicity of that eigenvalue.

With the partitions

$$A(t) - A_0 = \begin{bmatrix} V_1(t) & V_2(t) \\ V_3(t) & V_4(t) \end{bmatrix}, \quad f(t) = \begin{bmatrix} f_-(t) \\ f_+(t) \end{bmatrix},$$

$$X(t) = \begin{bmatrix} X_-(t) \\ X_+(t) \end{bmatrix},$$

compatible to that of A_0, the following theorem holds:

THEOREM 1. *Let* A_0 *and* f_0 *be in the system* (4) *such that the equation* $A_+ q = f_{0+}$ *has solution and the integrals*

$$\int_0^a \frac{|V_i(t)|}{t}\, dt, \quad i=3,4 \quad and \quad \int_0^a \frac{|f_+(t)-f_{0+}|}{t}\, dt$$

are bounded. Then the condition

$$(6) \qquad |X(t)| = O(1) \qquad if \quad t \to 0$$

for the solutions of system (4) is equivalent to the condition

$$(7) \qquad X_-(t) = \alpha(t)X_+(t)+\beta(t)$$

for small t, where the rectangular matrix $\alpha(t)$ of order $r\times(n-r)$ is the unique solution of the Cauchy problem

$$(8) \qquad t\alpha' = A_-\alpha - \alpha A_+ - \alpha V_3 \alpha - \alpha V_4 + V_1 \alpha + V_2, \quad \lim_{t\to 0} \alpha(t) = 0;$$

and the vector $\beta(t)$ of r entries is the unique solution of the Cauchy problem

$$(9) \qquad t\beta' = A_-\beta + (V_1 - \alpha V_3)\beta - \alpha f_+ + f_- , \quad \lim_{t\to 0} \beta(t) = -A_-^{-1} f_{0-}.$$

Moreover, if $A(t)$ and $f(t)$ have a power series expansion with respect to t, asymptotic or converging, then the same holds for $\alpha(t)$ and $\beta(t)$. The coefficients in the asymptotic power series can be obtained by formal substitution of all series into the systems (8) and (9).

In order to prove the existence and uniqueness of the solution of the system (8), consider matrix α as an element of an $r\times(n-r)$ dimensional vector space and rewrite system (8) correspondingly. Then applying Lemma 2 yields the statement.

Let $\beta_0 = -A_-^{-1} f_{0-}$ and $\widetilde{\beta}(t) = \beta(t) - \beta_0$. Now one can apply Lemma 2 again to the system (9) which is rewritten for $\widetilde{\beta}(t)$.

The statement on the expansions into power series follows from the second part of Lemma 2 or from Lemma 3, respectively.

The equivalence of the conditions (6) and (7) can be obtained by considering vector $Z(t)$ defined as

$$(10) \qquad Z(t) = X_-(t) - \alpha(t) X_+(t) - \beta(t),$$

where $X(t)$ is a solution of (4), $\alpha(t)$ and $\beta(t)$ are the above mentioned solutions of (8) and (9). From (4), (8) and (9) one has

$$(11) \qquad t Z' = A_- Z + (V_1 - \alpha V_3) Z.$$

If $X(t)$ is bounded, then $Z(t)$ is bounded by definition (10) and applying the Corollary of Lemma 1 to the system (11), one gets $Z(t) \equiv 0$.

Now let $Z(t) \equiv 0$. Introducing the vector

$$Y = \begin{bmatrix} X_+ + q \\ c \end{bmatrix},$$

where c is a scalar and q is any solution of the equation $A_+ q = f_{0+}$, consider the system

$$t Y' = \begin{bmatrix} A_+ & 0 \\ 0 & 0 \end{bmatrix} Y + \begin{bmatrix} V_3 \alpha + V_4 & f_+ - f_{0+} - (V_3 \alpha + V_4) q + V_3 \beta \\ 0 \end{bmatrix} Y.$$

It follows from Theorem 1 of Chapter 2 in [3] that every solution of this system is bounded if $t \to 0$ and X_+ and $X_- = \alpha X_+ + \beta$ are also bounded.

§.3. LINEAR SYSTEMS OF SECOND ORDER

Consider the system of n equations

$$(12) \qquad t^2 X'' = A(t)X + tB(t)X' + f(t), \qquad 0 < t \leq t_0,$$

where $A(t)$, $B(t)$ and $f(t)$ are defined and continuous
for small t. Further on, we assume the followings:

(i) $\lim_{t \to 0} B(t) = B_0 = \mu E$, where μ is real;

(ii) The eigenvalues of the matrix $\lim_{t \to 0} A(t) = A_0$
for $\mu \neq -1$ do not lie on the left side of the
parabola

$$\Pi(\mu) = \{z : (1+\mu)^2 \operatorname{Re} z + (\operatorname{Im} z)^2 = 0\}.$$

For $\mu > -1$ the eigenvalues fulfil the condition

$$(1+\mu)^2 \operatorname{Re}\lambda_{A_0} + (\operatorname{Im}\lambda_{A_0})^2 > 0.$$

For $\mu < -1$: $\lambda_{A_0} \neq 0$

$$(1+\mu)^2 \operatorname{Re}\lambda_{A_0} + (\operatorname{Im}\lambda_{A_0})^2 \geq 0$$

hold. The number of eigenvectors belonging to an eigenvalue of A_0 that lies on the parabola $\Pi(\mu)$, $\mu < -1$,
is supposed to be equal to the multiplicity of the eigenvalue.

For $\mu = -1$, the eigenvalues of A_0 do not lie on
the nonpositive real half-axis $(-\infty, 0]$ i.e. $\operatorname{Im}\lambda_{A_0} \neq 0$
or $\operatorname{Re}\lambda_{A_0} > 0$.

With these assumptions and introducing the notations
$V(t) = A(t) - A_0$, $W(t) = B(t) - B_0$, $\lim_{t \to 0} f(t) = f_0$, $g(t) =$

$= f(t)-f_0$ one has the following

    THEOREM 2. *Let the integrals*

$$\int_0^a \frac{|V(t)|}{t}\,dt,\quad \int_0^a \frac{|W(t)|}{t}\,dt \quad and \quad \int_0^a \frac{|g(t)|}{t}\,dt$$

be bounded. Then the condition

$$|X(t)| = O(1) \qquad if \qquad t \to 0$$

for the solutions of system (12) *for small* t *is equivalent to the condition*

$$(13) \qquad tX'(t) = \alpha(t)X(t)+\beta(t),$$

where the quadratic matrix $\alpha(t)$ *is the unique solution of the Cauchy problem*

$$t\alpha'+\alpha^2-(E+B(t))\alpha-A(t) = 0$$

(14)
$$\lim_{t\to 0}\alpha(t) = \alpha_0 = \frac{1+\mu}{2}E+\sqrt{A_0+\frac{(1+\mu)^2}{4}E}\ ,$$

and the vector $\beta(t)$ *is the unique solution of the Cauchy problem*

$$t\beta'-(E+B(t)-\alpha(t))-f(t) = 0,$$

(15)
$$\lim_{t\to 0}\beta(t) = \beta_0 = \left(-\frac{1+\mu}{2}E+\sqrt{A_0+\frac{(1+\mu)^2}{4}E}\right)^{-1}f_0.$$

    Moreover, if the coefficients of system (12) *have expansion into power series with respect to* t, *asymptotic or converging, then the same holds for* $\alpha(t)$ *and* $\beta(t)$. *The coefficients in the asymptotic power series*

*can be obtained by formal substitution of all series
into the systems* (14) *and* (15).

Applying Lemma 4 to the equations (14) and (15), the
scheme of the proof remains the same as it was in Theorem 3 in [2].

§.4. APPROXIMATIVE SOLUTIONS

One usually solves the systems (8) and (9) or (14)
and (15) approximatively. In order to choose the point
where the condition of boundedness is transferred to,
i.e. where one of the boundary values (7) or (13) is
specified, an error estimate for the replacement of the
exact boundary value by the approximative one is needed.
The considerations of §.4 in [2] remain valid in these
more general cases, too.

The results of the paper were applied to the solution of practical problems, see [4].

REFERENCES

[1] A.A. Abramov, On transfer of boundedness condition...
     Ž. Vyčisl. Mat. i Mat. Fiz., 1961, I, N°4, 733-737.

[2] K. Balla, On error estimation in substitution of
     the condition of boundedness..., Ž. Vyčisl. Mat.
     i Mat. Fiz., 1978, I8, N°2, 370-378.

[3] R. Bellman, Stability Theory of Differential Equa-
     tions, McGraw-Hill, New York, 1953.

[4] K. Balla, On the evaluation of nuclear models by
     the method of hyperspherical functions, MTA Számi-
     tástechnikai és Automatizálási Kutató Intézete,
     Közlemények, 17/1976, 27-39.

Katalin Balla
Computer and Automation Institute
of the Hungarian Academy of Sciences
Budapest, Kende u. 13-17.
H-1111

NUMERICAL TREATMENT OF SOME SINGULAR BOUNDARY VALUE PROBLEMS

L.COLLATZ

SUMMARY

Singular boundary value problems whose differential equations and domains are not too complicated, often can be treated successfully by using approximation methods and monotonicity properties, admitting the possibility to get exact lower and upper bounds for the wanted solution. This paper presents some examples as text-examples, among them a free boundary value problem.

ZUSAMMENFASSUNG

Singuläre Randwertaufgaben mit nicht zu komplizierten Differentialgleichungen und Bereichen können oft numerisch mit Erfolg mit Approximationsmethoden und mit Hilfe von Monotonieprinzipien behandelt werden und man hat dann die Möglichkeit, die gesuchte Lösung in exakte untere und obere (punktweise gültige) Schranken einzuschließen. Die Methode wird an verschiedenen konkreten Testbeispielen vorgeführt, und bei einer freien Randwertaufgabe wird der freie Rand in einen garantierbaren Streifen eingeschlossen.

INTRODUCTION

An integral of a function $f(x)$ over a domain B of the x-space is called singular, if the domain B is unbounded or if $f(x)$ has a singularity at a finite point x_o of B, or if both occur. In the same way we define singularities of boundary value problems

1) if the domain B for the considered problem is unbounded, or

2) if a coefficient in the differential equation has a singularity at a point x_o of B or if both 1) and 2) occur.

But for boundary value problems also other types of singularities occur, for instance

3) the location of the singularity may not be known a priori,

4) free boundaries, etc.

One can apply different numerical procedures for calculating solutions, for instance discretization methods (like difference methods, finite elements, etc), or variational methods (Ritz-Galerkin-methods), or approximation methods, expansion in infinite series, assymptotic development, etc. But it is very important for each of these methods to take into account the type of singularity; otherwise the numerical results will be very unsatisfactory.

Of course the error analysis for singular problems is much more difficult than for regular problems and often the mathematical methods are not enough developed to guarantee bounds for the error. But there are cases for which error bounds easily can be guaranted. The method usually consists in using approximation theory (see e.g.

M e i n a r d u s [12]) in connection with monotonicity, in special cases with maximum principles. The progress in this field during the last few years will be illustrated by some simple examples, which may be considered as test examples. (For many other examples see e.g. *C o l l a t z* [4] [5] [6].)

1. DISCONTINUITY OF THE FUNCTION ALONG THE BOUNDARY ∂B, IN NONLINEAR PROBLEMS

Sometimes it is possible to subtract a function v, which has the same singularity as the prescribed boundary values for the wanted solution u. The difference $u-v$ is then continuous at the boundary ∂B. This is a wellknown and often used procedure e.g. for the Laplace equation, but is also helpful for nonlinear problems. (About general foundations compare *K a n t o r o w i t s c h* [11] [10] a.o.)

NUMERICAL EXAMPLE. A function $u(x,y)$ may satisfy the nonlinear differential equation

$$(1.1) \quad Tu = 0, \quad \text{where} \quad Tu = -\Delta u - \Phi(u) \quad \text{with}$$

$$\Delta u = \frac{\partial^2 u}{\partial x^2} + \frac{\partial^2 u}{\partial y^2} \quad \text{and} \quad \Phi(u) = 1 - u + u^2$$

in $B = \{(x,y), \ |x| < 1, \ |y| < 1\}$ and the boundary conditions

$$(1.2) \quad \begin{array}{ll} u=1 & \text{for} \quad |x|=1 \\ u=0 & \text{for} \quad |y|=1 \end{array} \quad \text{on} \quad \partial B$$

(See Fig. 1.) u may be interpreted as temperature distribution in a plate with prescribed values of the temperature along the boundary ∂B and with a heat source generated by nonlinear chemical reactions.

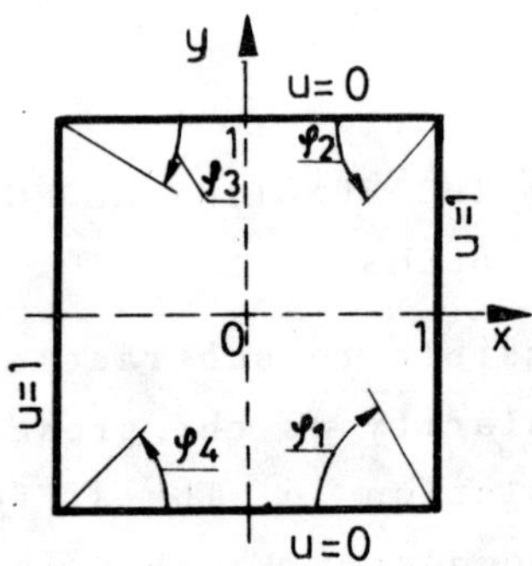

FIGURE 1.

We try to get a lower bound $\bar{w}$ and an upper bound $\hat{w}$ for u in the form

$$w = w(x,y,a_\nu) = w^* + \sum_{\nu=1}^{N} a_\nu \nu_\nu(x,y)$$

(1.3)

$$\text{with} \quad w^* = \frac{2}{\pi} \sum_{j=1}^{4} \varphi_j$$

φ_j are the angles at the four corners, Fig. 1, and w^* has at these corners the same singularities as the prescribed boundary values of u. We take

$$(1.4) \quad \bar{w} = w(x,y,a_\nu) \quad \text{and} \quad \hat{w} = w(x,y,\hat{a}_\nu) \, ,$$

furthermore ν_ν as polynomials which satisfy the symmetries of the problem:

$$\nu_1 = 1, \quad \nu_2 = x^2, \quad \nu_3 = y^2, \quad \nu_4 = x^4, \quad \nu_5 = x^2 y^2, \quad \nu_6 = y^4, \quad \nu_7 = y^6, \quad \nu_8 = x^6, \quad \nu_9 = x^4 y^2 + x^2 y^4.$$

We choose the parameters $\bar{a}_\nu$ such that $T\bar{w} \leq 0$ in B and $\bar{w} \leq u$ on ∂B, and correspondingly for $\hat{w}$: $T\hat{w} \geq 0$ in B and $\hat{w} \geq u$ on ∂B.

The parameters $\bar{a}_\nu$, $\hat{a}_\nu$ and a value γ are to be determined from the optimization problem

$$\gamma = \text{Min}$$

$$\left. \begin{array}{l} \gamma \geq w - \bar{w} \\[2mm] \gamma \geq -\sigma T\hat{w} \\[2mm] \gamma \geq \sigma T\bar{w} \end{array} \right\} \text{ in } B; \qquad \left. \begin{array}{l} \gamma \geq -\sigma(\hat{w} - u) \\[4mm] \gamma \geq \sigma(\bar{w} - u) \end{array} \right\} \text{ on } \partial B.$$

σ is a weightfactor; it was sufficient to choose $\sigma = 10^4$; greater values of σ have had no influence. With the values

ν	$\bar{a}_\nu$	$\hat{a}_\nu$
1	-1.2687	-1.2551
2	-0.2728	-0.2800
3	-0.1288	-0.1305
4	-0.0469	-0.0409
5	0.1915	0.2070
6	-0.00253	-0.00077
7	-0.0200	-0.0233
8	-0.0019	-0.0139
9	0.0201	0.0187

we get the error bounds: $0 \leq \hat{w} - \bar{w} \leq 0.01362$ in B, and for the arithmetical means $\left| \dfrac{\bar{w} + \hat{w}}{2} - u \right| \leq 0.00681$ in B.

2. DISCONTINUITY IN THE DERIVATIVES ALONG THE BOUNDARY ∂B FOR A MIXED PROBLEM

Another type of singularity occurs in the following problem. A function $u(x,y)$ satisfies the Laplace-equation

$$(2.1) \qquad \Delta u = 0 \quad \text{in } B, \quad B = \{(x,y), \ x^2 + 4y^2 < 4\}$$

and the boundary conditions (see Fig. 2.)

$$(2.2) \qquad u = -y \quad \text{on } \Gamma_1, \quad \Gamma_1 = \partial B \cap (y<0)$$

$$(2.3) \qquad \frac{\partial u}{\partial n} = 0 \quad \text{on } \Gamma_2, \quad \Gamma_2 = \partial B \cap (y<0).$$

n is the outer normal with the components:

$$n = \left(\frac{x}{\sqrt{16-3x^2}}, \ \sqrt{\frac{16-4x^2}{16-3x^2}} \right)$$

u may be interpreted as a temperature distribution in the cross-section of a swimming body; the outer temperature in the water-half-plane $y<0$ is increasing linearly.

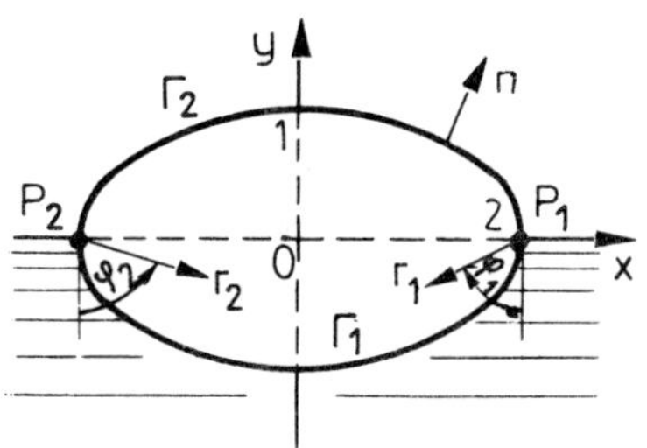

FIGURE 2.

There are two singular points: $P_1=(2,0)$, $P_2=(-2,0)$; we introduce polar coordinates r_j, φ_j at the point P_j ($j=1,2$) (Fig.2), and the functions v_1^*, v_2^* which have the required singular behaviour at P_j:

$$(2.4) \qquad v_k^*=r_1^{\alpha_k}\sin(\alpha_k\varphi_1)+r_2^{\alpha_k}\sin(\alpha_k\varphi_2),\; \alpha_1=\tfrac{1}{2},\alpha_2=\tfrac{3}{2}\;.$$

We take the approximate solutions $\bar{w}$ and $\hat{w}$ (analogously to (1.4)) in the form

$$w=w(x,y,a_k)=\sum_{k=1}^{2}a_k v_k^* + \sum_{k=3}^{p+2}a_k V_k(x,y)\qquad \text{with}$$

$$V_k(x,y)=\operatorname{Re}(y-ix)^{k-3}\qquad (k=3,4,\ldots)\;.$$

Here V_k are harmonic polynomials of degree $k-3$ with

139

$$V_k(-x,y) = V_k(x,y); \quad \text{e.g.}$$

$$V_3 = 1, \qquad V_4 = y, \qquad V_5 = y^2 - x^2, \quad \ldots$$

Now the classical maximum principle for harmonic functions is not useful for error bounds because the values of the function u are not known on the boundary Γ_2, but the monotonicity principle is applicable. We say that the operator R defined by

$$R\zeta = \begin{cases} -\Delta\zeta & \text{in} \quad B \\ \zeta & \text{on} \quad \Gamma_1 \\ \dfrac{\partial\zeta}{\partial n} & \text{on} \quad \Gamma_2 \end{cases}$$

is of monotonic type if

$$(2.5) \qquad R\zeta \leq R\Psi \quad \text{implies} \quad \zeta \leq \Psi \quad \text{in} \quad B \cup \partial B.$$

Here $R\zeta \leq R\Psi$ means that the same inequality holds point wise in B, Γ_1, Γ_2 for the three components of $R\zeta$ and $R\Psi$, respectively.

We determine parameters $\bar{a}_v$ with $R\bar{w} \leq Ru$ and values $\hat{a}_v$ with $R\hat{w} \geq Ru$; then we have the inclusion $\bar{w} \leq u \leq \hat{w}$ in all points of $B \cup \partial B$.

The calculation for this simultaneous approximation problem (compare *B r e d e n d i e k - C o l l a t z* [2]) uses, as in No. 1, optimization techniques for the following minimum problem:

$$0 \leq \hat{w} - \bar{w} \leq \sigma, \qquad \sigma = \text{Min}.$$

The following table gives the values of σ with or without using v_k^* and with polynomials V_k of different degrees:

Values of σ

using polynomials of degree	without ν_k^*	with ν_1^*	with ν_1^*, ν_2^*
$-$	$-$	1.618	1.071
0 $(p=1)$	1	1	1
1 $(p=2)$	1	0.898	0.132
2 $(p=3)$	1	0.166	0.101
5 $(p=6)$	0.747	0.0548	0.0329
8 $(p=9)$	0.591	0.0253	0.0184

We get therefore with ν_1^*, ν_2^* and polynomials of degree 8 the bound

$$0 \le \hat{w} - w \le 0.0184$$

and the error bound

$$|\tfrac{1}{2}(w+\hat{w})-u| \le 0.0092 \quad \text{in} \quad B \cup \partial B.$$

We observe, that the result without ν_1^*, ν_2^* is very bad ($\sigma=0.591$ instead of $\sigma=0.0184$); ν_1^* gives a good improvement, but the improvement by ν_2^* seems not to be very important for greater values of p.

3. DISCONTINUITIES IN THE DIFFERENTIAL EQUATIONS AND FINITE ELEMENTS

A function $u(x_1,\ldots,x_m)$ may be the solution of an elliptic boundary value problem with the differential equation of second order

$$(3.1) \qquad Tu = 0 \quad \text{in} \quad B,$$

for example

$$(3.2) \qquad Tu = -\Delta u + q(x_1, \ldots, x_m) = 0$$

We suppose B to be a bounded open connected domain in the m-dimensional space R^m; the boundary ∂B may consist of a part Γ_1 with prescribed values of u, of a part Γ_2 where $\frac{\partial u}{\partial n}$ is given and a part Γ_3 with prescribed values of a linear combination of u and $\frac{\partial u}{\partial n}$; n is the outer normal.

We subdivide (compare e.g. M i t c h e l l - W a i t [14], W h i t e m a n [18], [19], W e t t e r l i n g [17]) B into subdomains B_j ($j=1,\ldots k$) with inner normals v_j and with the interfaces Γ_{ij} between B_i and B_j (Fig.3).

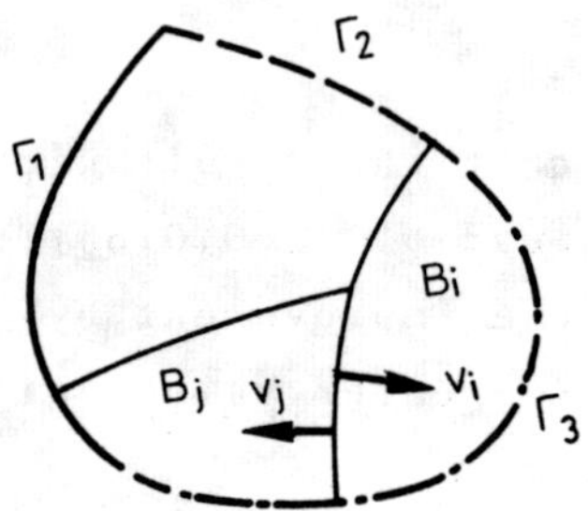

FIGURE 3.

Let u_j be the restriction of u to the domain B_j. We introduce the jump operator (e.g. C o l l a t z [3])

142

$$(3.3) \qquad S_{ij}u \;=\; \frac{\partial u_i}{\partial v_j} - \frac{\partial u_j}{\partial v_j} \quad .$$

Then the operator R defined by

$$(3.4) \qquad Ru \;=\; \begin{cases} Tu & \text{in} \quad B, \\[2mm] u & \text{on} \quad \Gamma_1 \\[2mm] \dfrac{\partial u}{\partial n} & \text{on} \quad \Gamma_2 \\[2mm] S_{ij}u & \text{on} \quad \Gamma_{ij} \end{cases}$$

is (with some additional weak assumptions) of monotonic type, that means (2.5) holds.

Originally this monotonicity principle was proved under strong conditions concerning the smoothness of u, but the principle was generalized to problems with discontinuities, for instance to discontinuous functions $q(x_1,\ldots,x_m)$ in (3.2) (H. W e r n e r [16], M e y n - - W e r n e r [13], N a t t e r e r [15]).

EXAMPLE: MODEL IN REACTOR PHYSICS

I thank Mr. K. H. M e y n for the following numerical example of a twodimensional model in reactor physics:

The concentration $u(x,y)$ in the domain $B(|x|<1,|y|<1)$ must satisfy the differential equation

$$(3.5) \qquad \Delta u(x,y) \;=\; \begin{cases} \varphi^2 u & \text{in} \quad B^* = (|x|<0,5, |y|<0.5) \\[2mm] 0 & \text{elsewhere, in} \quad B-B^* \end{cases}$$

(see fig. 4.) and the mixed boundary condition

$$u = 1 - \frac{1}{10.5} \frac{\partial u}{\partial n} \quad \text{on} \quad \partial B;$$

n is the outer normal. (The reaction takes place in a porous medium, in the interior of which there is a catalysator. Φ is the so called Thiele-modul; here $\Phi = 1.2$ is taken.

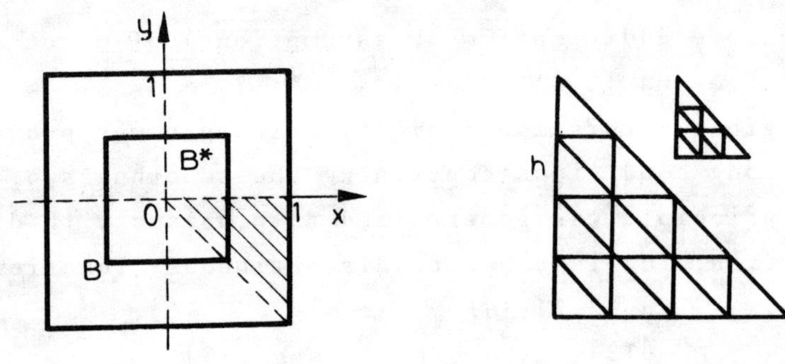

FIGURE 4.   FIGURE 5.

Only the shaded triangle on, fig. 4 was considered for reasons of symmetry. The finite element method was used with T-10-elements, fig. 5. The following table gives the numerical results for the meshsizes $h = \frac{1}{2}, \frac{1}{4}, \frac{1}{6}, \frac{1}{8}$ with the numbers of variables and inequalities, the percentage of the numbers in the starting tableau which are different from zero, lower and upper bounds for the value $u(0,0)$, and the difference δ between these two bounds.

h	Number of variables	inequal.	sparseness-percentage	lower bound for $u(0,0)$	upper	δ
$\frac{1}{2}$	22	36	35%	0.7379	0.8089	0.0710
$\frac{1}{4}$	61	120	14%	0.7611	0.7887	0.0276
$\frac{1}{6}$	120	254	8%	0.76685	0.78125	0.01440
$\frac{1}{8}$	199	438	5%	0.76875	0.77794	0.00919

4. FREE BOUNDARY VALUE PROBLEMS

Free boundary value problems occur in many different areas of applications, and many papers appeared about numerical calculation of free boundaries (B a i o c c h i [1], H o f f m a n n [9] a.o.). The most used methods are discretizations, difference and finite element methods. Using these methods one can get error bounds usually only with difficulties, but monotonicity principles allow, in not too complicated cases, to obtain bounds for the free boundaries. This was illustrated for the onedimensional Stefan Problem (C o l l a t z [7]), for the Obstacle- -Problem (C o l l a t z [8]) and will be shown here for a model of a biological phenomenon.

BIOLOGICAL MODEL. The concentration $c(x,y)$ must satisfy the Laplace equation

$$(4.1) \qquad -\Delta c = - \frac{\partial^2 c}{\partial x^2} - \frac{\partial^2 c}{\partial y^2} = -q(x,y) \quad \text{in } B$$

and the boundary conditions (fig. 6.):

$$(4.2) \qquad \begin{cases} c(0,y) = g(y) \quad \text{on} \quad \partial B_1 = (x=0, y_0 \leq y \leq y_1) \\ c = \frac{\partial c}{\partial x} = 0 \quad \text{on} \quad \partial B_2. \end{cases}$$

∂B_2 is the unknown "free boundary" which has to be calculated. (The physical boundary conditions on ∂B_2 are
$c = \frac{\partial c}{\partial n} = 0$, with n as outer normal, which is equivalent
to $c = \frac{\partial c}{\partial x} = 0$.) We suppose that ∂B_2 can be described
in the form

$$x = h(y) \quad \text{for} \quad y_0 \le y \le y_1$$

with $h(y_0) = h(y_1) = 0$, $h(y)$ positive and differentiable
for $y_0 \le y \le y_1$.

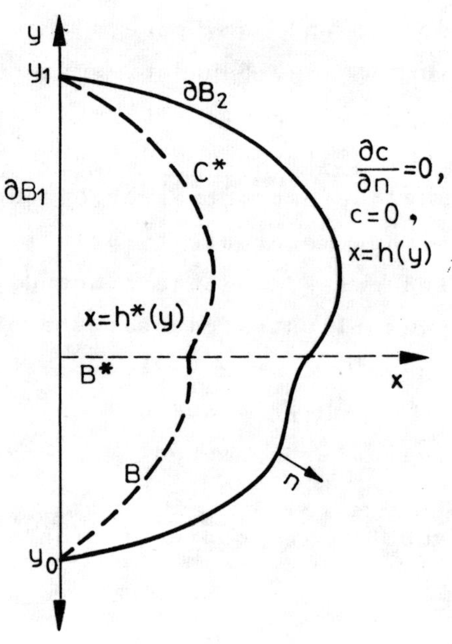

FIGURE 6.

Then the following monotonicity property holds:
let $c^*(x,y)$ be a "comparison" function for c, i.e.
there exists a curve $C^*: x = h^*(y)$ such that

146

i) h^* satisfies the same conditions as h,

ii) $c^*(x,y)=0$ on C^*,

iii) $\dfrac{\partial c^*}{\partial x}=0$ on C^*,

iv) $c^*>0$ in B^* (see Fig. 6);

then

$$c^*(x,y)\leqq g(y) \quad \text{on} \quad \partial B_1$$

and

$$-\Delta c^*\leqq -q \quad \text{in} \quad B^*$$

implies

$$h^*(y)\leqq h(y) \qquad (y_0\leqq y\leqq y_1)$$

In other words, the domain B^*, which is bounded by ∂B_1 and C^*, is contained in B.

Numerical example: ∂B_1 is the interval

$$-1\leqq y\leqq 1, \quad q(x,y)=20, \quad g(y)=\frac{1}{2}(1-y^2)^2.$$

We try to get an approximate solution in the form

$$(4.3) \quad c(x,y)\underset{\sim}{\sim}w(x,y,\alpha,\beta,\gamma)=\beta(1-y^2-\alpha x)^2(1-\gamma x).$$

We first take $\gamma=0, \beta=\frac{1}{2}$; the best value of α, which satisfies (4.2) is

$$\alpha=\sqrt{22} \text{ with } h^*(0)=\left|\frac{1}{22}\right.\approx 0,2132\leqq h(0).$$

This value can be slightly improved with

$$\beta = \frac{1}{2}, \quad \alpha = 4.5461, \quad \gamma = 0.14665,$$

we have $h(0) \geqq 0.2200$.

Analogously one obtains with $\gamma=0$ the upper bound

$$(4.4) \qquad \beta = \frac{1}{2}, \qquad \alpha = 4, \qquad h(0) \leq \frac{1}{4} = 0.25 .$$

Even slightly better results can be obtained with the form

$$(4.5) \qquad w(x,y,\alpha,\delta) = \frac{1}{2}(1-y^2-\alpha x-\delta x^2)^2$$

instead of (4.3).

We get for $\alpha=8/\sqrt{3}$, $\delta=-1/3$ a lower and for $\alpha=4$, $\delta=1$ an upper bound and the inclusion

$$(4.6) \qquad 4\sqrt{3} - 3\sqrt{5} \approx 0.2200 \leqq h(0) \leqq \sqrt{5} - 2 \approx 0.2361$$

instead of (4.4).

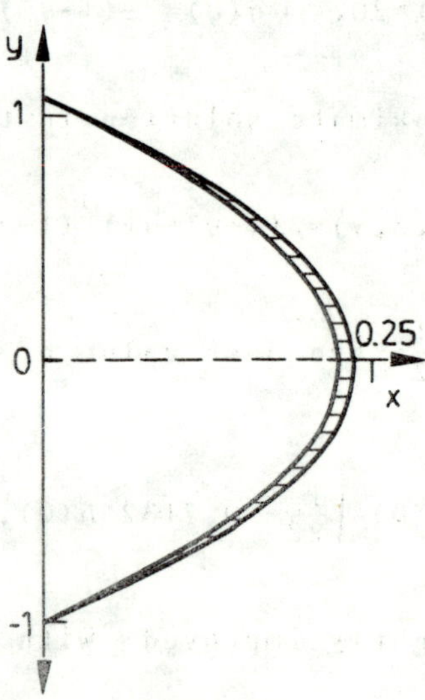

FIGURE 7.

Fig. 7 shows the lower and the upper bounds for $h(y)$ and the wanted free boundary is included in the hatched stripe.

I thank Mr. *U. G r o t h k o p f* for the numerical calculations on a computer.

REFERENCES

[1]   Baiocchi, Free boundary problems in the theory of fluid flow through porous media, *Proc. Intern. Congress of Math.* 1974, Vancouver, Vol. 2.

[2]   E. Bredendiek - L. Collatz, Simultan Approximation bei Randwertaufgaben, *Intern. Ser. Num. Math.* <u>30</u> (1976), 147-174.

[3]   L. Collatz, Monotonicity with discontinuities in partial differential equation, Proc. Conf. Ord. Part. Diff. Equ. Dundee 1974, *Lect. Notes Math.* Vol. 415 Springer (1974), 85-102.

[4]   L. Collatz, Nichtlineare Approximationen bei Randwertaufgaben, V. *IKM (Internationaler Kongreß für Anwendungen der Mathematik)* Weimar, herausgegeben von H. Matzke, 1969, S. 169-182.

[5]   L. Collatz, Some applications of approximation theory to differential equations, *Budapest Symp. Fourier Anal. and Approx. Th.* 15-21. Aug. 1976.

[6]   L. Collatz, The numerical treatment of some singular boundary value problems, Conference on Numer. Analysis, Dundee 1977, *Lect. Notes in Math.,* Vol. 630, Springer (1978), 41-50.

[7]   L. Collatz, Application of approximation to some singular boundary value problems, Proc. Conference

Num. Anal. Dundee, *Lect. Not. Math.* Vol. 63o,
Springer, (1978), 41-50.

[8] L. Collatz, Approximation methods for boundary value
problems with unbounded domains or free boundaries,
Proc. Symp. Num. Math. Plzen (C.S.S.R) 4-8. Sept.
1978, to appear.

[9] K.H. Hoffmann, Monotonie bei nichtlinearen Stefan-
-Problemen, *Internat. Ser. Num. Math.* 39 (1978),
162-190.

[10] L.W. Kantorowitsch - G.P. Akilow, *Funktionalanalysis
in normierten Räumen,* Akad. Verl. Berlin, 1964, 622 p.

[11] L.W. Kantorowitsch - W.I. Krylow, *Näherungsmethoden
der höheren Analysis,* VEB Berlin, 1956, 611 S.

[12] G. Meinardus, *Approximation of Functions, Theory and
Numerical Methods,* Springer Verlag, 1967, 198 S.

[13] K.-H. Meyn - B. Werner, Randmaximum- und Monotonieprinzipien für elliptische Randwertaufgaben mit
Gebietszerlegungen, to appear.

[14] A.R. Mitchell - R. Wait, *The Finite Element Method
in Partial Differential Equation,* 1977, 192 p.

[15] F. Natterer, Berechnung oberer Schranken für die Norm
der Ritz-Projektion auf finite Elemente, *Internat.
Ser. Num. Math.* 39 (1978), 236-245.

[16] B. Werner, Monotonie und finite Elemente bei elliptischen Differentialgleichungen, *Intern. Ser. Num.
Math.* 27 (1975), 393-401.

[17] W. Wetterling, Lecture on error bounds for solutions
of singular elliptic boundary value problems at
Mathem. Forschungsinstitut Oberwolfach, to appear in
Internat. Ser. Num. Math.

[18] J.R. Whiteman, *The Mathematics of Finite Elements and Applications*, Acad. Press 1973, 520 S.

[19] J.R. Whiteman - B. Schiff, Finite element approximation of singular functions, *Internat. Series Numer. Math.* 30 (1976), 319-333.

L. Collatz
Universität Hamburg
Institut für Angewandte Mathematik
Bundesstraße 55
D-2000 Hamburg 13.

FREE BOUNDARY PROBLEMS IN THE THEORY OF FLUID FLOW THROUGH POROUS MEDIA

T. DESPERAT

1. INTRODUCTION

Free boundary problems are boundary-value problems
in which part of the boundary, the free boundary, is
unknown and must be determined as a part of the problem.
Such problems often arise in the theory of fluid flow
through porous media. In this paper we study elliptic
free boundary problems from the numerical point of view.

2. FORMULATION OF THE PROBLEM

We shall study free boundary problems of the following class.

(i) It is required to find a twice continuously differentiable function u and a domain D such that

$$(1) \qquad Lu = (Au_x)_x + (Bu_y)_y = 0, \qquad (x, y) \in D,$$

where A and B are positive continuously differentiable
functions of x and y, while D is a bounded domain in

the xy-plane with a boundary C that has a continuously turning tangent except a finite number of corners. (ii) The solution u must satisfy the boundary condition:

$$(2) \qquad \ell u = \beta u_{\overline{\nu}} + \alpha u - \gamma = 0, \qquad (x, y) \in C,$$

where $\overline{\nu}$ is the unit conormal, α, γ are piecewise continuously differentiable functions of x, y and $\overline{\nu}$, while β is either 0 or 1. We set $C_1 = \{(x, y) \in C :$ $: \beta = 0\}$, $C_2 = \{(x, y) \in C : \beta = 1\}$ and assume that $\alpha \neq 0$ on C_1.
(iii) The final condition that must be satisfied is the one which characterizes the problem as a free boundary problem. Part of the boundary C, C_f, is unknown and it must be determined as part of the problem. On the free boundary C_f, the solution u must satisfy the following additional condition:

$$(3) \qquad \ell_f u = u - \gamma_f = 0,$$

where γ_f is of the same form as γ is.

Furthermore the free boundary C_f is required to satisfy certain "geometric restraints".

3. SOME REMARKS ON NUMERICAL ALGORITHMS

The question of whether a free boundary problem has a solution, and whether this solution is unique, is an extremely difficult mathematical problem. As far as we know the results of B a i o c c h i ([1], [2], [4]) concerning the problem of two-dimensional stationary flow of an incompressible fluid through an homogeneous porous medium are the only results in this field. The main idea in Baiocchi's method consists in reducing the problem to a variational inequality.

A general heuristic method for the solution of free boundary value problems related to stationary flow through porous media is the following.

We fix an initial approximation c_f^0 to the free boundary c_f. This curve together with the known part of the boundary defines a region D^0. We apply only one of the two conditions on c_f^0 which are given on the free boundary (generally condition (2)). On the remaining part of ∂D^0, we apply the given boundary conditions. We have then a boundary value problem for the elliptic equation on D^0. Let $u^{(0)}$ be the solution. By interpolating or extrapolating the values of $u^{(0)}$, a curve c_f^1 is determined on which the other free boundary condition and geometric restraints are satisfied, at least approximately. c_f^1 determines a domain D^1 to which the procedure is applied again. If the sequences $\{c_f^i\}$, $\{D^i\}$ and $\{u^{(i)}\}$ converge to $\tilde{C}_f$, $\tilde{D}$ and $\tilde{u}$ respectively, then the triplet $\{\tilde{u}, \tilde{D}, \tilde{C}_f\}$ is the solution of the problem. Some of the methods that have been used to solve free boundary problems are modifications of the presented algorithm. Methods along this line have been developed and applied by various authors (see for instance [3], [5]). As far as we know there is no mathematical proof of the convergence of the procedure. It must be remarked that the procedure is applied in the discrete. The curves c_f^0, c_f^1,... are computed only approximately (they are polygonals or elements of some class of curves). The functions $u^{(0)}$, $u^{(1)}$,... are computed numerically in the nodes of a point mesh. The procedure is repeated until the "distance" between two consecutive curves c_f^i, c_f^{i+1} is sufficiently small.

REFERENCES

[1] C. Baiocchi, Su un problema di frontiera libera connesso a questioni di idraulica, *Ann. di Mat. pura e appl.* 92(1972), 107-127.

[2] C.Baiocchi - V. Comincioli - E. Magenes - G. Pozzi, Free boundary problems in the theory of fluid flow through porous media: existence and uniqueness theorems, *Ann. di Mat. pura e appl.* 97(1973), 1-82.

[3] C. Baiocchi - V. Comincioli - L. Guerri - G. Volpi, Free boundary problems in the theory of fluid flow through porous media: numerical approach, *Calcolo*, 10(1973), 1-86.

[4] C. Baiocchi, Studio di una problema quasi-variazionale connesso a problemi di frontiera libera, *Boll. UMI* (4) 11(1975), 589-613.

[5] C.W. Cryer, On the approximate solution of free boundary problems using finite differences, *Journal ACM*, vol. 17, No. 3(1970), 397-411.

T. Desperat
00-132 Warszawa
Gzybowska 9m 702,
Poland

SOME NOTES ON ITERATIVE METHODS FOR EIGENVALUES AND EIGENVECTORS OF MATRICES

L. ELSNER

1. COMPUTING EIGENVALUES WITH THE POWER METHOD

Let A be a real or complex $n \times n$-matrix, diagonalizable with eigenvalues λ_i and eigenvectors x_i

$$
\begin{aligned}
A x_i &= \lambda_i x_i \qquad\qquad i = 1, \ldots, n \\
&|\lambda_1| > |\lambda_2| \geq \ldots \geq |\lambda_n|.
\end{aligned}
\tag{1}
$$

We assume

$$
u_0 = \sum_{i=1}^{n} d_i x_i, \qquad d_1 \neq 0.
\tag{2}
$$

This situation which we shall call "normal case" is considered to be the harmless case in studying the power method

$$
\begin{aligned}
v_s &:= A u_{s-1} \\
k_s &:= \ell(v_s) \qquad s = 1, 2, \ldots \\
u_s &:= k_s^{-1} v_s .
\end{aligned}
\tag{3}
$$

The following considerations show that this is not quite
true. Different authors (e.g. [2]) suppose that for

$$\ell(v) = \max(v), \qquad v \in C^n$$

where $\max(v)$ is the first component of v with maximal
modulus, (3) is convergent, i.e. that

$$(4) \qquad \lim k_s = \lambda_1, \quad \lim u_s = \frac{x_1}{\ell(x_1)}$$

hold. The example

$$A = \frac{1}{4}\begin{bmatrix} 1 & -3 \\ -3 & 1 \end{bmatrix} \qquad u_0 = \begin{bmatrix} 2 \\ 1 \end{bmatrix}, \quad \lambda_1 = 1, \quad x_1 = \begin{bmatrix} 1 \\ -1 \end{bmatrix}$$

$$\lambda_2 = -\frac{1}{2}, \quad x_2 = \begin{bmatrix} 1 \\ 1 \end{bmatrix} \quad \text{and} \quad A^s u_0 = \begin{bmatrix} 1+(-1)^{s+1}3 \cdot 2^{-(s+1)} \\ -1+(-1)^{s+1}3 \cdot 2^{-(s+1)} \end{bmatrix}$$

$$\lim k_s = -1$$

shows that this is incorrect.

We find by looking closely at the convergence proof
the following sufficient conditions on ℓ:

THEOREM 1. *Let the function* ℓ *satisfy*

(5)

    (a) $\ell : C^n \to C$

    (b) ℓ *continuous at* x_1

    (c) $\ell(\alpha x) = \alpha \ell(x), \quad \alpha \in C, \quad x \in C^n$

    (d) $\ell(x_1) \neq 0$.

*Then (3) converges in the normal case and (4) holds, if
the iteration does not break down. This is the case for*
u_0 *sufficiently near to* x_1.

We refrain from giving the elementary proof and remark only that in the example given above (5b) is not satisfied.

Theorem 1 is unsatisfactory as it requires some knowledge of the unknown eigenvector. One would have global convergence for any x_1 if ℓ satisfies

(6)

    (a) $\ell : C^n \to C$

    (b) ℓ continuous in C^n

    (c) $\ell(\alpha x) = \alpha \ell(x)$, $\alpha \in C$, $x \in C^n$

    (d) $\ell(x) = 0 \iff x = 0$.

Unfortunately these properties are contradictory.

THEOREM 2. *For* $n > 1$ *there exists no function* ℓ *satisfying* (6a)-(6d). *The same holds if one replaces* C *and* C^n *by* R *and* R^n *in* (6a)-(6d).

PROOF. We start with the real case. Let ℓ satisfy the "real" (6a)-(6d) and define $h(t) = \ell(1,\ldots,1,\ t)$. As

$$\frac{1}{t} h(t) = \ell(t^{-1},\ldots,t^{-1},\ 1) \to \ell(0,\ldots0,1) \neq 0$$

for $t \to \pm\infty$, $h(t)$ changes its sign, as t moves from $-\infty$ to $+\infty$. Hence the continuous function h has a zero, which contradicts (6d).

In the complex case we define $h(z)$ as above and consider the continuous function

$$s(z) = h(z)/|h(z)|$$

mapping C into the unit circle $C_1 = \{z \in C,\ |z| = 1\}$.

Assuming $\ell(0,\ldots,0,1) = 1$ we can prove as above

$$\lim_{r\to\infty} \frac{1}{r}\, e^{-i\varphi} h(re^{i\varphi}) = 1$$

uniformly in φ, and hence

$$\lim_{r\to\infty} s(re^{i\varphi}) = e^{i\varphi}.$$

From this and the continuity of s we can show that the image of the circle $C_r = \{z\in C,\ |z| = r\}$ under s is the whole of C_1 for large r. Deforming C_r continuously into one point yields a contradiction.

We have seen that there is no global function ℓ suitable for all cases. A remedy can be the use of varying functions ℓ_s instead of ℓ. An example is provided by

$$\ell_{s+1}(v) := \frac{u_s^H}{\|u_s\|}\cdot v$$

where $\|\cdot\|$ denotes the Euclidean norm and $u^H v = \sum_i \bar{u}_i v_i$.

In view of Theorem 1 one expects that in the normal case the modified algorithm

$$v_s = Au_{s-1}$$

$$k_s = \ell_s(v_s) = u_{s-1}^H Au_{s-1} / \|u_{s-1}\|$$

$$u_s = k_s^{-1} v_s$$

yields

$$\lim k_s = \lambda_1, \quad \lim u_s = x_1/\|x_1\|.$$

This can indeed be proved for any u_0 satisfying (2).

Let us finally remark that similar results hold for the inverse iteration

$$(A-\mu_s)v_{s+1} = u_s$$

$$k_{s+1} = \ell(v_{s+1})$$

(7)

$$\mu_{s+1} = \mu_s + k_{s+1}^{-1}$$

$$u_{s+1} = k_{s+1}^{-1} v_{s+1}.$$

2. NONNEGATIVE MATRICES. THE USE OF HOPF'S INEQUALITY FOR CONVERGENCE PROOFS

The problem described above can be avoided in case of the Perron root ρ and Perron vector $p > 0$ of a nonnegative irreducible matrix A

$$Ap = \rho p.$$

One can choose some positive linear functional ℓ which does not vanish at p.

Here will show that convergence for a general class of iterative methods can be proved completely in terms of the underlying order structure. The main tool is H o p f's inequality:

For $x, y \in R^n$, $y > 0$ *(i.e.* $y_i > 0$, $i=1,\ldots,n$) *and a matrix* $B > 0$, $B = (b_{ij})$ *(i.e.* $b_{ij} > 0$, $i, j=1,\ldots,n$)

$$(8) \qquad osc\left(\frac{Bx}{By}\right) \leq N(B)\ osc\left(\frac{x}{y}\right)$$

where

$$\operatorname{osc}\left(\frac{x}{y}\right) := \max\left(\frac{x}{y}\right) - \min\left(\frac{x}{y}\right) = \max_i \frac{x_i}{y_i} - \min_i \frac{x_i}{y_i}$$

and

$$N(B) = (\sqrt{k}-1)(\sqrt{k}+1)^{-1} < 1$$

$$k = k(B) = \max_{i,j,k,\ell} \frac{b_{ik}b_{j\ell}}{b_{jk}b_{i\ell}} .$$

Let $\|x\| = \max\left(\frac{x}{p}\right)$. We consider the general iteration procedure

$$v_s = B_{s-1}u_{s-1} \qquad s=1,2,\dots$$

$$(9) \qquad u_s = v_s / \|v_s\|$$

$$\underline{\lambda}_s = \min\left(\frac{Au_s}{u_s}\right) \qquad \overline{\lambda}_s := \max\left(\frac{Au_s}{u_s}\right).$$

Here the starting vector u_0 is positive and $\{B_s\}$ a sequence of positive matrices commuting with A.

Let us stress the fact that $\{\underline{\lambda}_s\}$, $\{\overline{\lambda}_s\}$ does not depend on the normalization of the $\{u_s\}$. It is well-known (see [1] for further references) that

$$\underline{\lambda}_s \le \underline{\lambda}_{s+1} \le \rho \le \overline{\lambda}_{s+1} \le \overline{\lambda}_s \qquad s=0,1,\dots$$

For the proof of convergence of the bounds $\{\underline{\lambda}_s\}$, $\{\overline{\lambda}_s\}$ to ρ the inequality

$$\overline{\lambda}_s - \underline{\lambda}_s = \operatorname{osc}\left(\frac{Au_s}{u_s}\right) = \operatorname{osc}\left(\frac{AB_{s-1}u_{s-1}}{B_{s-1}u_{s-1}}\right)$$

$$(10)$$

$$= \operatorname{osc}\left(\frac{B_{s-1}Au_{s-1}}{B_{s-1}u_{s-1}}\right) \le N(B_{s-1})\cdot\operatorname{osc}\left(\frac{Au_{s-1}}{u_{s-1}}\right) =$$

$$= N(B_{s-1})\{\overline{\lambda}_{s-1} - \underline{\lambda}_{s-1}\}$$

is crucial.

For the proof of convergence of $\{u_s\}$ to p we use the inequality

$$(11) \qquad \mathrm{osc}\left(\frac{u_{s+1}}{p}\right) \leq K_s \ \mathrm{osc}\left(\frac{u_s}{p}\right)$$

where

$$(12) \qquad K_s = \frac{N_s}{N_s t_s + (1-t_s)} < 1 \qquad N_s = N(B_s), \quad t_s = \mathrm{osc}\left(\frac{u_s}{p}\right),$$

$$0 \leq t_s < 1.$$

Let us prove (11)(12): From $AB_s = B_s A$ we get $B_s p = \gamma_s p$ for some γ_s. Hence

$$\mathrm{osc}\left(\frac{u_{s+1}}{p}\right) = \mathrm{osc}\left(\frac{B_s u_s}{\max\left(\frac{B_s u_s}{p}\right)p}\right) = \mathrm{osc}\left(\frac{B_s u_s}{B_s p}\right)\frac{1}{\max\left(\frac{B_s u_s}{B_s p}\right)} \ .$$

Writing the last denominator as $\mathrm{osc}+\mathrm{min}$ and using (8) and the monotonicity of $t \to \frac{t}{t+\alpha}$, $\alpha > 0$, we get

$$\leq \frac{N_s t_s}{N_s t_s + \min\left(\frac{B_s u_s}{B_s p}\right)} \leq K_s t_s,$$

where in the last inequality

$$\min\left(\frac{B_s u_s}{B_s p}\right) \geq \min\left(\frac{u_s}{p}\right) = \max\left(\frac{u_s}{p}\right) - \mathrm{osc}\left(\frac{u_s}{p}\right) = 1 - t_s$$

is used.

This shows that $\{t_s\}$ is decreasing. If $N_s \leq \gamma < 1$ for all s, (11)(12) yield $\lim t_s = 0$, i.e. the convergence of $\{u_s\}$ to p, and (10) yields $\lim \underline{\lambda}_s =$

$$= \lim \overline{\lambda}_s = \rho.$$

We consider three special cases:

Case 1: If $A > 0$ we can choose $B_s \equiv A$ and get a proof for the convergence of the power method.

Case 2: If $A \geq 0$ is primitive, there is an integer m with $A^m > 0$. Taking $B_s \equiv A$ and regarding m steps as one step with A^m, the analysis above gives the convergence of the power method also in this case.

Case 3: If $A \geq 0$, irreducible, we may choose

$$B_s = (\overline{\lambda}_s I - A)^{-1} > 0.$$

This is inverse iteration. Here we can show even quadratic convergence: Observe that

$$N(B_s) = N(\mathrm{adj}\,(\overline{\lambda}_s I - A))$$

converges for $\overline{\lambda}_s \to \rho$ to

$$N(\mathrm{adj}(\rho I - A)) = N(pq^t) = 0$$

where $q > 0$ is the suitably normalized Perron vector of A^T. Hence there is a number M such that

$$N(B_s) \leq M(\overline{\lambda}_s - \rho)$$

which together with (10)(11)(12) gives the announced result.

REMARK. The results of this second part and complete proofs may be found in [1]. This paper however does not contain the direct proof of the convergence of the $\{u_s\}$

via inequalities (11)(12).

REFERENCES

[1] L. Elsner, Inverse iteration for calculating the
     spectral radius of a non-negative irreducible mat-
     rix, *Lin. Alg. and its Appl.* 15(1976), 235-242.

[2] J.H. Wilkinson, *The Algebraic Eigenvalue Problem*,
     Clarendon Press, Oxford, 1965.

L. Elsner
Fakultät für Mathematik
Universität Bielefeld
Postfach 8640
4800 Bielefeld 1
Bundesrepublik Deutschland

ON UNIFORM CONVERGENCE AND STABILITY OF A FINITE- -DIFFERENCE SCHEME FOR WEAKLY NONLINEAR PARABOLIC EQUATIONS WITH CYLINDRICAL SYMMETRY

R.H. FARZAN - G. MOLNÁRKA

1. Modelling processes in chemical reactors one needs very often the solution of the following weakly nonlinear parabolic differential equation:

$$(1) \qquad \operatorname{div}(D \ \operatorname{grad} u) - \operatorname{div}(vu) - \frac{\partial u}{\partial t} + f(r,t,u) = 0,$$

where u is the density of a chemical component, D is the diffusion coefficient, v is the velocity vector of the stream and f is the source function. In case of cylindrical symmetry equation (1) is the following:

$$(2) \qquad \frac{1}{r} \frac{\partial}{\partial r}\left(rD\frac{\partial u}{\partial r}\right) - \frac{1}{r} \frac{\partial}{\partial r}(rwu) + \frac{\partial}{\partial z}\left(D\frac{\partial u}{\partial z}\right) - \frac{\partial}{\partial z}(vu) - \frac{\partial u}{\partial t} + f = 0$$

$$(r,z) \in \Omega = (0,R) \times (0,Z), \qquad t \in (0,T].$$

Suppose that the following assumptions

(i) $0 < D(r,z,t) \in C^{110}$;

(ii) $v_r = w(r,z,t) \in C^{100}$;

(iii) $v_z = v(r,z,t) \in C^{010}$;

(iv) $f \in C$, $|f(r,z,t,u_1) - f(r,z,t,u_2)| \leq L_f |u_1 - u_2|$;
$(r,z,t) \in \bar{Q} = \Omega \times [0,T]$

are satisfied by the functions in (2). The initial and boundary conditions are

$$u(r,z,0) = u_o(r,z), \qquad (r,z) \in \bar{\Omega};$$

$$(3) \qquad
\begin{aligned}
u(r,0,t) &= u_1(r,t), \\
&\qquad\qquad\qquad\qquad r \in [0,R], \quad t \in (0,T], \\
u(r,Z,t) &= u_2(r,t),
\end{aligned}$$

$$\frac{\partial u}{\partial r} - \sigma(z)u = v(z), \qquad \sigma(z) \geq 0, \quad r=R, z \in (0,Z), t \in (0,T].$$

Note that in practical cases the boundary conditions are often boundary conditions of the third type for $z=Z$. This fact leads to complications when setting up an approximate equation, but it does not influence the stability and convergence of the scheme.

We have applied the finite-difference method. The grid points are defined as follows:

$$z_i = ih_z, \qquad i=0,1,\ldots,M, \qquad h_z = \frac{Z}{M};$$

$$r_j = (j+\tfrac{1}{2})h_r, \qquad j=0,1,\ldots,N, \qquad h_r = \frac{R}{N+\frac{1}{2}};$$

$$t_k = k\tau, \qquad k=0,1,\ldots,K, \qquad \tau = \frac{T}{K}; \qquad \bar{\Omega}_h = \{i,j\}, \quad \bar{Q}_h = \{i,j,k\}$$

and the grid functions are denoted by $B_{ij}^k = B(r_j, z_i, t_k)$. In writing finite-difference equations, integer and half-

-integer indexes will be used. For example:

$$D^k_{i+\frac{1}{2},j} = D(r_j, z_{i+\frac{1}{2}}, t_k), \quad \text{where} \quad z_{i+\frac{1}{2}} = (i+\frac{1}{2})h_z.$$

Take $y_k = \{y^k_{ij}\}_{ij \in \Omega_h}$ and $f_k(y_k) = \{f^k_{ij}(y^k_{ij})\}_{ij \in \overline{\Omega}_h}$ as vectors. The differential operators of (2) are approximated by the finite-difference operators Λ^k_r and Λ^k_z, where

(4)
$$(\Lambda^k_z y_k)_{ij} = \frac{1}{h_z}(D^k_{i+\frac{1}{2}j} \frac{y^k_{i+1,j} - y^k_{ij}}{h_z} - D^k_{i-\frac{1}{2}j} \frac{y^k_{ij} - y^k_{i-1,j}}{h_z})$$

$$- \frac{1}{2}(v^k_{i+\frac{1}{2},j} \frac{y^k_{i+1,j} - y^k_{ij}}{h_z} - v^k_{i-\frac{1}{2},j} \frac{y^k_{ij} - y^k_{i-1,j}}{h_z}), \quad i,j \in \Omega_h;$$

(5)
$$(\Lambda^k_r y_k)_{ij} = \frac{1}{h_r}(\frac{r_{j+\frac{1}{2}}}{r_j} D^k_{i,j+\frac{1}{2}} \frac{y^k_{i,j+1} - y^k_{ij}}{h_r}$$

$$- \frac{r_{j-\frac{1}{2}}}{r_j} D^k_{i,j-\frac{1}{2}} \frac{y^k_{ij} - y^k_{i,j-1}}{h_r}) - \frac{1}{2}(\frac{r_{j+\frac{1}{2}}}{r_j} w^k_{i,j+\frac{1}{2}} \frac{y^k_{i,j+1} - y^k_{ij}}{h_r}$$

$$+ \frac{r_{j-\frac{1}{2}}}{r_j} w^k_{i,j-\frac{1}{2}} \frac{y^k_{ij} - y^k_{i,j-1}}{h_r}), \quad i,j \in \Omega_h.$$

This scheme slightly generalizes the scheme suggested by F r i a z i n o v [1].

The finite difference operators, Λ^k_r and Λ^k_z, approximate the corresponding differential operators with the accuracy of $O(\frac{h_r^2}{r})$ and $O(h_z^2)$. After dividing by τ the difference equation

$$(E - \tau \Lambda_z^k - \tau \Lambda_r^k + \tau^2 \Lambda_z^k \Lambda_r^k) y_k = \Delta y_{k-1} + \tau \tilde{f}(y_k), \quad k = 1, \ldots, K$$

(6)

$$\tilde{f}_{i,j}^k = f_{i,j}^k, \quad i,j \in \Omega_h,$$

approximates equation (2) with the accuracy of $O(\tau + \frac{h_r^2}{r} + h_z^2)$. The initial condition for the grid function is

$$y_0 = u_0 = \{u_0(r_j, z_i)\}_{i,j \in \bar{\Omega}_h}.$$

Introduce the operator Δ for approximating the boundary conditions in (6) as

$$(\Delta y)_{i,j} = \begin{cases} y_{i,j}; & i \neq 0 \wedge i \neq M \\ \\ 0; & i = 0 \vee i = M. \end{cases}$$

The above introduced operators Λ_r^k and Λ_z^k were defined on the interior grid points. Assume that for points on the boundary

$$(\Lambda_z^k y_k)_{o,j} = (\Lambda_r^k y_k)_{o,j} = 0, \quad \tilde{f}_{o,j}^k = \frac{1}{\tau} u_1(r_j, t_k),$$

$$j = 0, 1, \ldots, N$$

(7)

$$(\Lambda_z^k y_k)_{M,j} = (\Lambda_r^k y_k)_{M,j} = 0, \quad \tilde{f}_{M,j}^k = \frac{1}{\tau} u_2(r_j, t_k),$$

$$(\Lambda_r^k y_k)_{i,o} = \frac{z}{h_r} D_{i,\frac{1}{2}}^k \frac{y_{i,1}^k - y_{i,o}^k}{h_r} - w_{i,\frac{1}{2}}^k \frac{y_{i,1}^k + y_{i,o}^k}{h_r}, \quad \tilde{f}_{i,o}^k = f_{i,o}^k;$$

$$(\Lambda_r^k y_k)_{i,N} = -\frac{1}{h_r}\left(2\frac{r_{N-\frac{1}{2}}}{r_N} D^k_{i,N-\frac{1}{2}} \frac{y^k_{i,N}-y^k_{i,N-\frac{1}{2}}}{h_r} + D^k_{i,N}\sigma_i y^k_{i,N}\right)$$

$$(7) \qquad -w^k_{i,N}\frac{y^k_{i,N}-y^k_{i,N-1}}{h_r} - \frac{1}{h_r}\frac{r_N w^k_{i,N}-r_{N-\frac{1}{2}} w_{i,N-\frac{1}{2}}^k}{r_N} y^k_{i,N} \, ,$$

$$\tilde{f}^k_{i,N} = f^k_{i,N} + \frac{2}{h_r} D^k_{i,N}\nu_i$$

When $j=0$ and $j=N$, the operator Λ_z^k is expressed by formula (4).

Taking into account (7) the system of vector equations approximates the initial boundary value problem with the accuracy of $O(\tau+\frac{h_r^2}{r}+h_z^2)$ on every grid points of $\bar{Q}_h$

One can solve the nonlinear system (6) by an iterative method using the alternating direction method in each step of the following iteration:

$$(8) \qquad (E-\tau\Lambda_z^k)(E-\tau\Lambda_r^k)y_k^{(s)}=\Delta y_{k-1}+\tau\tilde{f}_k(y_k^{(s-1)}), \quad s=1,2,\ldots$$

2. In this paper we are going to find sufficient limitations for τ, h_z and h_r, which guarantee the consistency of the difference scheme and the convergence of the iterative processes. The consistency conditions are determined in the C norm.

The conditions of the stability of the alternating direction method will be examined. Suppose that the right side of (8) is known and denote it by F_k. So

$$(E-\tau\Lambda_z^k)(E-\tau\Lambda_r^k)y_k = F_k.$$

This equation decomposes into the following vector equations:

$$(9) \qquad (E-\tau\Lambda_r^k)y_k = x_k ,$$

$$(10) \qquad (E-\tau\Lambda_z^k)x_k = F_k$$

and both of these equations decompose into systems of linear equations. In particular, equation (10) decomposes into $N+1$ systems for every different value of j. Actually the equations for the componensts x_k can be written in the following form (for the simplicity the index k is omitted):

$$
\begin{aligned}
(11) \quad &- \frac{\tau}{h_z^2}\left(D_{i-\frac{1}{2},j} + \frac{h_z}{2} v_{i-\frac{1}{2},j}\right) x_{i-1,j} + \left[1 + \frac{\tau}{h_z^2}\left(D_{i-\frac{1}{2},j}\right.\right. \\
&\left.\left. - \frac{h_z}{2} v_{i-\frac{1}{2},j} + D_{i+\frac{1}{2},j} + \frac{h_z}{2} v_{i+\frac{1}{2},j}\right)\right] x_{i,j} - \frac{\tau}{h_z^2}\left(D_{i+\frac{1}{2},j}\right. \\
&\left. - \frac{h_z}{2} v_{i+\frac{1}{2},j}\right) x_{i+1,j} = F_{i,j} ,
\end{aligned}
$$

where $0\le j\le N$, $0\le i\le M$ and

$$x_{o,j} = u_1(r_j,t_k), \qquad x_{M,j} = u_2(r_j,t_k) .$$

Assume that:

$$D^k_{i-\frac{1}{2},j} + \frac{h_z}{2} v^k_{i-\frac{1}{2},j} > 0,$$

$$D^k_{i+\frac{1}{2},j} - \frac{h_z}{2} v^k_{i+\frac{1}{2},j} > 0$$

hold for every i, j, k. These conditions are clearly fulfilled when

$$(12) \qquad h_z < \min_Q \frac{2D}{|v|} .$$

Then the coefficients of $x_{i,j}$ in equation (11) are greater than unity. The system of equations (11) can be solved by the factorization method.

For the convergence of the factorization method of the equation

$$a_i x_{i-1} - b_i x_i + c_i x_{i+1} = d_i$$

when $a_i > 0$, $b_i > 1$ and $c_i > 0$, the sufficient conditions (see [2]) are

$$b_i - a_i - c_i > 0, \qquad i = 1, 2, \ldots, N-1 .$$

For equation (11) this means that

$$1 + \frac{\tau}{h_z^2} \left(h_z v_{i+\frac{1}{2},j} - h_z v_{i-\frac{1}{2},j} \right) > 0,$$

that obviously holds if

$$(13) \quad \tau < \frac{1}{L_v} \,, \qquad L_v = \max_{\bar{Q}} |v_z'|$$

and the inequalities are true for every k.

Similarly, if

$$(14) \quad h_r < \min_{\bar{Q}} \frac{2D}{|w|}$$

and

$$(15) \quad \tau < \frac{1}{L_w} \,, \qquad L_w = \max_{\bar{Q}} \max_{|\theta|<1} \frac{\left| (rw)_r' \right|}{r + \dfrac{h_r}{2}\theta}$$

hold then the factorization method is convergent for the equation (9).

Note that the presence of h_r does not involve that τ depends on h_r, because when using (14) one can choose L_w in the form of

$$L_w = \max_{\bar{Q}} \max_{\theta^*} \frac{\left| (rw)_r' \right|}{r + \theta^*} \,, \quad \text{where } |\theta^*| \leq \min \frac{D}{|w|}.$$

Thus the solutions of (9) and (10) with the factorization method are absolutely stable, when (12), (13) and (15) hold and these conditions can be easily met.

3. In this part of the paper an estimation for the norm of the inverse operators to (9) and (10) will be given. For estimating the norm of the operator $(E-\tau\Lambda_z^k)^{-1}$ let i_o, j_o be so that

$$|x_{i_o,j_o}| = \max_{\overline{\Omega}_h} |x_{i,j}|.$$

If there are several i_o, j_o then take the smallest one at first in i_o and then in j_o. There is no loss of generality by assuming that

$$\text{(i)} \quad 0 < i_o < M$$

(16)

$$\text{(ii)} \quad x_{i_o,j_o} > 0.$$

If putting x_{i_o,j_o} instead of $x_{i_o \pm 1,j}$ in (11) then due to (12) one gets

$$(17) \qquad \left(1 + \tau \, \frac{v_{i_o+\frac{1}{2},j_o} - v_{i_o-\frac{1}{2},j_o}}{h_z}\right) x_{i_o,j_o} \leq F_{i_o,j_o},$$

where, in consequence of (13), the value of the expression in the parenthesis is positive. The norm

$$\|x_k\|_k = \max_{\overline{\Omega}_h} |x_{i,j}^k|$$

will be called the "layer norm". Obviously one can rewrite equation (12) in the following form:

$$x_k = (E - \tau \Lambda_z^k)^{-1} F_k.$$

As $F_{i_o,j_o} \leq \|F_k\|_k$ and taking into account (17)

$$(18) \qquad \| (E-\tau \Lambda_z^k)^{-1} \|_k \;\leq\; \frac{1}{1-\tau L_v}$$

where the matrix norm is consistent with the "layer norm".

Observe that conditions (16) can be easily omitted. Indeed, if $x_{i_o,j_o} < 0$ then the inequality (17) will turn to the opposit, but in this case $\| x_k \| = -x_{i_o,j_o}$, so (18) is fulfilled as well. The case of $u_1 = u_2 = 0$ is not considered hence the case when $x_{i_o,j_o} = 0$ can be omitted. Moreover, if $i_o = 0$ or $i_o = M$ then one gets the estimation

$$\| (E-\tau \Lambda_z^k)^{-1} \| \leq 1$$

when the inequality (18) is obviously fulfilled. Similarly one can also omit the assumptions (16).

For the inverse operator of (9), assuming (14) and (15) one gets

$$(19) \qquad \| (E-\tau \Lambda_r^k)^{-1} \| \leq \frac{1}{1-\tau L_w}$$

in the same way. The case when (i_o, j_o) is on the boundary needs a separate investigation. If $j_o = N$ then

$$\| (E-\tau \Lambda_r^k)^{-1} \| \leq \left[1-\tau L_w + \frac{\tau}{h_z} \min_{\overline{\Omega}} (D\sigma) \right]^{-1}.$$

It follows from this estimation that (19) will hold as D and σ are greater than zero. The estimations (18) and (19) do not depend on k and this is a very important feature.

4. In this section the conditions of stability and convergence of the iterative process (8) will be exanined. The iterative process

$$z^{(s)} = g(z^{(s-1)}), \quad s=1,2,\ldots$$

is stable and convergent for every initial value $z^{(o)}$, if there exists $q<1$, so that (see [3, p 407])

$$(20) \qquad \rho(g(z*),g(z**)) \leq q\,\rho(z*,z**).$$

For getting the function $g(y_k^{(s-1)})$, one can rewrite equation (8) into the explicit form

$$(21) \qquad y_k^{(s)} = (E-\tau\Lambda_r^k)^{-1}(E-\tau\Lambda_z^k)^{-1}[\Delta y_{k-1} + \tau\tilde{f}_k(y_k^{(s-1)})].$$

Let the distance be measured by the "layer norm" as follows:

$$\rho(z*,z**) = \|z*-z**\|$$

Applying this metrics to equation (21) and making use of the linearity of operators one has

$$\|g(z*)-g(z**)\| = \|(E-\tau\Lambda_r^k)^{-1}(E-\tau\Lambda_z^k)^{-1}\tau(\tilde{f}_k(z*)-\tilde{f}_k(z**)\| \leq$$

$$\leq \|(E-\tau\Lambda_r^k)^{-1}\|\,\|(E-\tau\Lambda_z^k)^{-1}\|\,\tau L_f\,\|z*-z**\|.$$

On comparing with (20), clearly

$$q = \tau\|(E-\tau\Lambda_r^k)^{-1}\|\,\|(E-\tau\Lambda_z^k)^{-1}\|L_f\ .$$

From (18) and (19) it follows that

$$q < \frac{\tau L_f}{(1-\tau L_v)(1-\tau L_w)} \, ,$$

hence the inequality $q<1$ holds when

$$(22) \qquad \tau \leq \frac{1}{L_v + L_w + L_f}$$

and in this case, the iterative process (8) is stable and convergent. There are no h_r or h_z, in the estimation for τ, therefore the process converges absolutely.

5. Here the consistency of the difference scheme (6) will be discussed. It follows from the existence of the inverse operators of (9) and (10) and from the convergence of the iterative process that a unique solution of (6) exists. We are going to show the stability of the solution of (6) which corresponds to perturbations in the initial data and in the right hand side of the equation. Let y_k be the solution of the following finite-difference equation:

$$(E-\tau \Lambda_z^k)(E-\tau \Lambda_r^k)\hat{y}_k = \Delta \hat{y}_{k-1} + \tau \tilde{\hat{f}}_k(\hat{y}_k)$$

$$\hat{y}_0 = \hat{u}_0 \, .$$

Suppose that $\tilde{\hat{f}}_k$ satisfies the Lipschitz condition with the constant $\hat{L}_f$. Denote $y_k - \hat{y}_k$ by $\hat{z}_k$. Thus for $\hat{z}_k$ one has the equation:

(23)
$$(E-\tau \Lambda_z^k)(E-\tau \Lambda_r^k)z_k = \Delta z_{k-1} + \tau[\tilde{f}_k(y_k)-\hat{\tilde{f}}_k(\hat{y}_k)] \;,$$

$$z_0 = u_0 - \hat{u}_0$$

or in explicit form

$$\hat{z}_k = (E-\tau\Lambda_r^k)^{-1}(E-\tau\Lambda_z^k)^{-1}[\Delta\hat{z}_{k-1}+\tau(\tilde{f}_k(y_k)-\hat{\tilde{f}}_k(y_k)+\hat{\tilde{f}}_k(y_k)-\hat{f}_k(\hat{y}_k))]\;.$$

Knowing that $\|\Delta\|=1$, one can get an upper bound for the norm of z_k as

$$\|\hat{z}_k\| \leq \|(E-\tau\Lambda_r^k)^{-1}\| \; \|(E-\tau\Lambda_z^k)^{-1}\| \; [\|\hat{z}_{k-1}\|+\tau\|\tilde{f}_k(y_k)-$$

$$\hat{\tilde{f}}_k(\hat{y}_k)\|+\tau\|\hat{\tilde{f}}_k(y_k)-\hat{f}_k(\hat{y}_k)\|]\;.$$

Assume that

$$\tau < (L_v+L_w+\hat{L}_f)^{-1}$$

hence from (18), (19) and from the Lipschitz condition for $\hat{\tilde{f}}_k$ one gets

(24)
$$\|\hat{z}_k\| \leq \hat{R}(\tau)[\|\hat{z}_{k-1}\|+\tau d]\;,$$

where

$$\hat{R}(\tau) = [1-\tau L_v+L_w+\hat{L}_f]^{-1} \quad \text{and} \quad d = \max_k \|\tilde{f}_k(y_k)-\hat{\tilde{f}}_k(y_k)\|.$$

Substituting $\|\hat{z}_{k-1}\|$ by $\|\hat{z}_{k-2}\|$ and continuing this process up to $\|z_0\|$, one has

179

$$\| \hat{z}_k \| \leq \hat{R}^k(\tau) \| \hat{z}_o \| + \tau\, d \sum_{n=o}^{k-1} \hat{R}^n(\tau) .$$

Assume that

$$(25) \qquad \tau \leq \frac{1}{2} (L_v + L_w + \hat{L}_f)^{-1} .$$

Then one gets estimation

$$(26) \qquad \| \hat{z}_k \| \leq e^{2T(L_v + L_w + \hat{L}_f)} \| u_o - \hat{u}_o \| + T e^{2T(L_v + L_w + \hat{L}_f)} d .$$

Therefore the finite-difference equation is consistent, if (25) is valid.

6. Using the estimations (18) and (19), one can easily give conditions for τ to guarantee the convergence of the solution of the difference scheme (6) to the solution of the differential equation (2) and (3).

Let $y_k - u_k$ be denoted by z_k. Then one gets the following equation for z_k:

$$(27) \qquad (E - \tau \Lambda_z^k)(E - \tau \Lambda_r^k) z_k = \Delta z_{k-1} + \tau [\tilde{f}_k(y_k) - \tilde{f}(u_k)] + \tau\, \varphi_k ,$$

where

$$\| \varphi_k \| = O(\tau + \frac{h_r^2}{r} + h_z^2) .$$

For z_k of (27) one has the following estimation:

$$\| z_k \| \leq \| (E - \tau \Lambda_r^k)^{-1} \| \; \| (E - \tau \Lambda_z^k)^{-1} \| [\| z_{k-1} \| + \tau L_f \| z_k \| + \tau \| \varphi_k \|]$$

and by (18), (19) and (22)

$$(28) \qquad \| z_k \| \leq R(\tau) \left[\| z_{k-1} \| + \tau \| \varphi_k \| \right],$$

where

$$R(\tau) = \left[1 - \tau (L_v + L_w + L_f) \right]^{-1}.$$

In the inequality (28) one can substitute $\| z_{k-2} \|$ for $\| z_{k-1} \|$ and so on. Finally

$$\| z_k \| \leq R^k(\tau) \| z_o \| + \tau \max_k \| \varphi_k \| \sum_{n=o}^{k-1} R^n(\tau).$$

Assuming that

$$(29) \qquad \tau \leq \frac{1}{2} (L_v + L_w + L_f)^{-1}$$

and knowing that $\| z_o \| = 0$, the following estimation is valid:

$$\| z_k \| \leq T \exp \left[2T (L_v + L_w + L_f) \right] \max_k \| \varphi_k \| \leq O \left(\tau + \frac{h_r^2}{r} + h_z^2 \right).$$

Thus the proof of convergence is complete.

7. In the investigation of the stability and the convergence of the finite-difference scheme for the equation (2) the conditions (12) and (14) for h_z and h_r were made use of. For τ, there were given several conditions and the strongest one was given by (29). It is a very important result that τ, h_z and h_r do not depend on each other, so the convergence is absolute in

all case. In the investigation the C norm was used. The
results of this paper are rather sharp and for the
concrete equation (2) the generalization is not very easy.
As we have shown above, the systems of equation (6),
after dividing by τ, approximate the equation (2) with
the accuracy of $O(\tau+\frac{h_r^2}{r}+h_z)$. Using the general theory
of [4], it is easy to find a finite-difference scheme
with the accuracy of $O(\tau^2+\frac{h_r^2}{r}+h_z^2)$. For example instead
of the right hand side of (6) one can apply the next
formula:

$$(E+\tau\Lambda_z^{k-1})(E+\tau\Lambda_r^{k-1})y_{k-1}+\tau\tilde{f}_{k-\frac{1}{2}}(\tfrac{1}{2}(y_k+y_{k-1})).$$

The new difference scheme is absolutely by convergent in
the energy norm. But in the C norm, one has a restriction,
for the convergence, namely, $\tau\le O(h^2)$. It means that this
last scheme is not absolutely convergent in the C norm.

REFERENCES

[1]   I. V. Friazinov, On difference schemes for the
       Poisson's equation in polar, cylindrical and
       spherical coordinate systems, *Ž. Vycisl. Mat. i
       Mat. Fiz.*, vol 11, fasc. 5 (1971) 1219–1228
       (in Russian)

[2]   S. K. Godunov – V. S. Riabenkii, *Difference Schemes*
       (in Russian), Nauka, Moscow 1973.

[3]   N. S. Bakhvalov, *Numerical Methods* (in Russian),
       Nauka, Moscow 1973.

[4]   A. A. Samarskii, *Theory of the Difference Schemes*
       (in Russian), Nauka, Moscow 1977.

R. H. Farzan
G. Molnárka

Dept. of Numerical Methods and
Computer Sciences of the
Eötvös Loránd University
Budapest, Muzeum krt. 6-8.
H-1088

SOME COMBINATORIAL ASPECTS IN MATRIX THEORY AND NUMERICAL ALGEBRA

M. FIEDLER

We shall be interested here in those properties of matrices or numerical procedures which do not depend (or do not depend much) on the magnitude of the non--zero entries but only (or mostly) on the structure of the zero- and non-zero entries. An example of such situation is the classical König's theorem:

Let A be an $n \times n$ matrix. If A contains, maybe after independent permutations of rows and columns, a $p \times q$ block of zeros with $p+q > n$ then $\det A = 0$.

We shall deal first the spectral properties of matrices. Therefore, only simultaneous permutations of rows and columns will be allowed.

The next to the situation in König's theorem is that A contains a $p \times q$ block of zeros with $p+q = n$, maybe after some simultaneous permutations of rows and columns. If this zero block contains no diagonal entry, such a matrix, as is well known, is called *reducible*. If A is not reducible, A is *irreducible*.

The notion of irreducibility plays, of course, an important role in the theory of nonnegative matrices.

According to the classical Perron-Frobenius theorem, an
irreducible nonnegative matrix has always a positive
eigenvector corresponding to the spectral radius which
is an eigenvalue of that matrix.

Also the so-called imprimitivity or primitivity of a
nonnegative matrix is of a purely combinatorial character and has a striking influence on the geometrical configuration of the eigenvalues.

However, we shall turn to the following question:
What can be said about the signs of coordinates of the
remaining eigenvectors of a nonnegative irreducible matrix? Since these eigenvectors should be real, it is natural to restrict the investigations to the symmetric
case.

The example

$$A = ee^T + \varepsilon vv^T$$

where

$$e = (1,1,\ldots,1)^T \, ,$$

$$v = (v_i) ,$$

$$v_i = \frac{1}{|M_1|} \quad \text{if} \quad i \in M_1 ,$$

$$v_i = - \frac{1}{|M_2|} \quad \text{if} \quad i \in M_2 ,$$

$$v_i = 0 \quad \text{otherwise},$$

(ε is a small positive number, M_1, M_2 are disjoint
non-void subsets of $N = \{1,2,\ldots,n\}$ and T stands
for transposition) shows that one cannot expect any
conditions on the sign-pattern of the coordinates of the

eigenvector corresponding to the second largest eigenvalue for a positive symmetric matrix except that it is
neither nonnegative, nor nonpositive.

However, if the matrix contains some zero entries,
the following theorem (cf. [4]) yields conditions re-
lating the sign-pattern of this eigenvector to the non-
-zero-entries pattern of the matrix:

THEOREM 1. *Let* A *be an* $n \times n$ *nonnegative symmet-*
ric irreducible matrix with eigenvalues $\lambda_1 \geq \lambda_2 \geq \ldots$
$\ldots \geq \lambda_n$ *(in fact,* $\lambda_1 > \lambda_2$*). Let* $v = (v_i)$ *be an*
eigenvector of A *corresponding to* λ_2*. If* $M =$
$= \{i \in N; v_i \geq 0\}$ *then the principal submatrix* $A(M, M)$
(with row- and column-indices from M*) is irreducible.*

To prove this, let $M = \{1, 2, \ldots, m\}$ and suppose
that $A(M, M)$ is reducible:

$$
A(M, M) = \begin{bmatrix}
A_{11} & 0 & \cdots & 0 \\
0 & A_{22} & \cdots & 0 \\
\multicolumn{4}{c}{\cdots\cdots\cdots\cdots\cdots\cdots\cdots} \\
0 & 0 & \cdots & A_{rr}
\end{bmatrix},
$$

where $r \geq 2$ and the matrices $A_{11}, \ldots, A_{rr}$ are already
irreducible.

Thus,

$$
A = \begin{bmatrix}
A_{11} & 0 & \cdots & 0 & A_{1,r+1} \\
0 & A_{22} & \cdots & 0 & A_{2,r+1} \\
\multicolumn{5}{c}{\cdots\cdots\cdots\cdots\cdots\cdots\cdots\cdots\cdots} \\
0 & 0 & \cdots & A_{rr} & A_{r,r+1} \\
A_{1,r+1}^T & A_{2,r+1}^T & \cdots & A_{r,r+1}^T & A_{r+1,r+1}
\end{bmatrix}.
$$

Let $\begin{bmatrix} v^{(1)} \\ \vdots \\ v^{(r)} \\ v^{(r+1)} \end{bmatrix}$ be the corresponding partitioning of

the eigenvector v so that

$$v^{(1)} \geq 0, \ v^{(2)} \geq 0, \ldots, v^{(r)} \geq 0, \ v^{(r+1)} < 0.$$

We have then

$$(1) \qquad (A_{kk} - \lambda_2 I_k) v^{(k)} = -A_{k,r+1} v^{(r+1)}, \qquad k = 1, \ldots, r.$$

The matrix

$$\widetilde{A} = \lambda_2 I(M) - A(M, M)$$

is a principal submatrix of the matrix $\lambda_2 I - A$ (I, $I(M)$ etc. being, throughout the paper, identity matrices). Since $\lambda_2 I - A$ has exactly one negative eigenvalue, $\widetilde{A}$ has at most one negative eigenvalue. Therefore, r being greater than one, some of the diagonal blocks in $\widetilde{A}$, say $\lambda_2 I_k - A_{kk}$, has only nonnegative eigenvalues. Thus $\lambda_2 I_k - A_{kk}$ is a (singular or nonsingular) M-matrix (cf. [7]). If $\lambda_2 I_k - A_{kk}$ were singular, there would exist a positive vector, say $z^{(k)}$, for which

$$(\lambda_2 I_k - A_{kk}) z^{(k)} = 0.$$

By (1) for $i = k$,

$$(z^{(k)})^T A_{k,r+1} v^{(r+1)} = 0$$

which would imply $A_{k,r+1} = 0$, a contradiction with the

188

irreducibility of A.

Thus $\lambda_2 I_k - A_{kk}$ is nonsingular and its inverse positive (cf. [7]). From (1) for $i = k$,

$$v^{(k)} = (\lambda_2 I_k - A_{kk})^{-1} A_{k,r+1} v^{(r+1)}.$$

The right-hand-side is nonpositive, the left-hand-side nonnegative. Consequently,

$$(\lambda_2 I_k - A_{kk})^{-1} A_{k,r+1} v^{(r+1)} = 0$$

so that $A_{k,r+1} = 0$ again, a contradiction. The proof is complete.

Much stronger results about the sign properties of all the eigenvectors can be obtained for a rather special class of symmetric matrices, the so called acyclic matrices. A symmetric irreducible $n \times n$ matrix $A = (a_{ik})$, $n \geq 2$, is called *acyclic* if

$$a_{i_1 i_2} a_{i_2 i_3} a_{i_3 i_4} \cdots a_{i_r i_1} = 0$$

whenever $r \geq 3$ and $i_1, \ldots, i_r$ are distinct indices. This means, in the graph-theoretical formulation, that the graph of A is a tree.

Examples of acyclic matrices are symmetric irreducible tridiagonal matrices and matrices obtained by bordering a diagonal matrix in a symmetric manner by one row and one column with non-zero entries.

We shall prove now the following theorem (cf. [5]) on eigenvectors of acyclic nonnegative matrices (in fact, the assumption of nonnegativity is not much restrictive since any real acyclic matrix can be brought to a non-negative acyclic matrix by a diagonal similarity whose matrix has diagonal entries $+1$ and -1).

THEOREM 2. *Let* $A = (a_{ik})$ *be an* $n \times n$ *nonnegative acyclic matrix with eigenvalues*

$$\lambda_1 \geq \lambda_2 \geq \ldots \geq \lambda_n .$$

If $u = (u_i)$ *is a (real) eigenvector corresponding to* λ_r *and* $u_i \neq 0$ *for all* i *then the number of all pairs* (i, k), $i \neq k$, *for which* $a_{ik} \neq 0$ *and* $u_i u_k < 0$ *is exactly* $r-1$.

Moreover, the multiplicity of λ_r *is one.*

PROOF. Denote $A - \lambda_r I = B = (b_{ik})$. Since $Bu = 0$, we have identically

$$(2) \qquad \sum_{i,k=1}^{n} b_{ik} x_i x_k = \sum_{i<k} (-b_{ik} u_i^{-1} u_k^{-1})(x_i u_k - x_k u_i)^2 .$$

The sum on the right-hand-side has exactly $n-1$ non--zero terms (a tree with n vertices has $n-1$ edges) and the linear forms $x_i u_k - x_k u_i$ for (i, k), $i < k$ corresponding to these terms can easily be proved to be linearly independent.

Let us denote by p the number of positive, by q the number of negative and by s the number of zero eigenvalues of B so that

$$p \leq r-1, \quad q \leq n-r, \quad p+q+s = n.$$

By the Sylvester's law of inertia, p is also equal to the number of positive coefficients $(-b_{ik} u_i^{-1} u_k^{-1})$, q to the number of negative coefficients in (2). Therefore,

$$p+q = n-1$$

so that

$$p = r-1, \quad s = 1,$$

which is equivalent to the assertions.

In the sequel, we shall be interested in qualitative properties in the inversion of matrices and in Gauss elimination. We shall be using Boolean matrices, the entries of which are zeros and ones and the operations with which will also be Boolean; we shall also use inequalities among Boolean matrices in the usual sense.

The set of all Boolean $n \times n$ matrices will be denoted by $\mathcal{B}_n$, the set of all real $n \times n$ matrices by M_n. As usual, a matrix P is called a permutation matrix iff in each row and in each column of P exactly one entry is different from zero and all its non-zero entries are ones.

If $A = (a_{ik}) \in M_n$, we shall denote by A_B the Boolean matrix $A_B = (b_{ik})$ defined by $b_{ik} = 1$ if $a_{ik} \neq 0$, $b_{ik} = 0$ if $a_{ik} = 0$.

If $C \in \mathcal{B}_n$, we denote

$$N(C) = \{A \in M_n; \; A_B = C, \; \det A \neq 0\}.$$

The following theorem is immediate.

THEOREM 3. *Let* $C \in \mathcal{B}_n$. *Then* $N(C)$ *is nonvoid iff there exists a permutation matrix* $P \leq C$.

We shall call a Boolean matrix C *combinatorially nonsingular* (shortly *c-nonsingular*) iff $N(C) \neq 0$.

If $C \in \mathcal{B}_n$ is nonsingular, we define the *combinatorial inverse* (*c-inverse*) of C as the matrix (*Boolean sum*)

$$C^{-1} = \sum_{A \in N(C)} (A^{-1})_B.$$

Before we shall state now a theorem chracterizing
the c-inverse to C we define, for $C \in B_n$ c-nonsingular
and P a permutation matrix, $P \le C$, a matrix
$K(P, C) \in M_n$ by

$$K(P, C) = P - (n+1)^{-1} (C-P).$$

Clearly, $P^T K(P, C)$ is a nonsingular M-matrix so that
(cf. [7])

$$(K(P, C))^{-1} \ge 0,$$

$$(3) \qquad (K(P, C))^{-1} = \left(\sum_{k=0}^{\infty} ((n+1)^{-1} (P^T C - I))^k \right) P^T.$$

THEOREM 4. *Let $C \in B_n$ be c-nonsingular. Let $D = (d_{ik}) \in B_n$. Then the following are equivalent:*

$1°$ $D = C^{-1}$;

$2°$ $d_{ik} = 1$ *iff* $C(N \backslash \{k\}, N \backslash \{i\})$ *is c-nonsingular;*

$3°$ *for every permutation matrix $P \le C$ and for*
$\hat{C} = C-P,$ $D = P^T + P^T \hat{C} P^T + P^T \hat{C} P^T \hat{C} P^T + \ldots + \underbrace{P^T \hat{C} P^T \hat{C} \ldots \hat{C} P^T}_{2n-1 \text{ factors}}$;

$4°$ *there exists a permutation matrix $P_0 \le C$ such that*
for $C_0 = C - P_0$

$$D = P_0^T + P_0^T C_0 P_0^T + \ldots + \underbrace{P_0^T C_0 P_0^T C_0 \ldots C_0 P_0^T}_{2n-1 \text{ factors}};$$

$5°$ *for every permutation matrix $P \le C$,*
$D = ((K(P, C))^{-1})_B$;

$6°$ *there exists a permutation matrix $P_0 \le C$ such that*
$D = ((K(P_0, C))^{-1})_B$.

PROOF. Let us denote the matrix D defined in $1°$ by D_1, in $2°$ by D_2, in $3°$ by $D_3(P)$ (it depends on P), in $5°$ by $D_5(P)$. We shall show that

$$D_1 \leq D_2 \leq D_3(P) \leq D_5(P) \leq D_1$$

for any permutation matrix $P \leq C$. This will in fact prove the theorem since then all these matrices are equal (for all P's) and, moreover, $4°$ and $6°$ are also equivalent with the remaining since $P_0 \leq C$ exists by c-nonsingularity of C.

Thus, let us show that $D_1 \leq D_2$. Let (in the obvious notation) $(D_1)_{ik} = 1$. This means that there exists a matrix $A \in N(C)$ such that $(A^{-1})_{ik} \neq 0$ which is equivalent that the entry (i, k) of the adjoint matrix of A is different from zero. Hence $\det A(N \setminus \{k\}, N \setminus \{i\}) \neq 0$ so that, by Theorem 3, $C(N \setminus \{k\}, N \setminus \{i\})$ is c-nonsingular. Therefore, $(D_2)_{ik} = 1$.

To show that $D_2 \leq D_3(P)$ for any permutation matrix $P \leq C$, assume that $(D_2)_{ik} = 1$. This means that there exists an $(n-1) \times (n-1)$ permutation matrix P_1 such that $P_1 \leq C(N \setminus \{k\}, N \setminus \{i\})$. If completed by the entry 1 in (k, i) and zeros elsewhere, P_1 will become an $n \times n$ permutation matrix which we shall denote by $\hat{P}_1$. If the matrix E_{ki} with the only non-zero entry $e_{ki} = 1$ satisfies $E_{ki} \leq P$ then (in similar notation) $E_{ik} = E_{ki}^T \leq P^T$ and $(D_3(P))_{ik} = 1$. Assume thus that $E_{ki} \not\leq P$. The matrix $Q = P^T \hat{P}_1$ is a permutation matrix for which $P^T E_{ki} = E_{ti}$ and $t \neq i$ (since otherwise $P^T E_{ki} \leq I$ and $E_{ki} \leq P$, a contradiction). Therefore, the corresponding permutation to Q decomposed into cycles contains a cycle $(t\,i\,j_1\,j_2\ldots j_s)$. Consequently, $E_{ij_1} \leq Q$, $E_{j_1 j_2} \leq Q, \ldots, E_{j_s t} \leq Q$, and in these inequalities Q can even be replaced by $\hat{Q} = P^T(\hat{P}_1 - E_{ki})$ which satisfies $\hat{Q} = P^T\hat{C}$. Moreover, t, i, $j_1, \ldots, j_s$ are distinct indices so that $s \leq n-2$. Thus

$$E_{ik} = E_{ij_1} E_{j_1 j_2} \cdots E_{j_s t} P^T \leq (P^T\hat{C})^{s+1} P^T \leq D_3(P).$$

To prove the inequality $D_3(P) \leq D_5(P)$, it suffices to observe that in (3) all summands are nonnegative. Therefore,

$$D_3(P) \leq ((K(P, C))^{-1})_B = D_5(P).$$

The inequality $D_5(P) \leq D_1$ follows from the definition of C^{-1} since $D_5(P) = (A^{-1})_B$ for $A = K(P, C) \in N(C)$. The proof is complete.

Some trivial corollaries follow such as:

COROLLARY 1. *If* $C_1 \in \mathcal{B}_n$, $C_2 \in \mathcal{B}_n$, $C_1 \leq C_2$ *and* C_1 *is* *c-nonsingular then* C_2 *is* *c-nonsingular and* $C_1^{-1} \leq C_2^{-1}$.

COROLLARY 2. *If* $C \in \mathcal{B}_n$ *is* *c-nonsingular then* C^{-1} *is* *c-nonsingular and* $C \leq (C^{-1})^{-1}$.

We shall turn now to the notion of the Boolean Schur complement. As is well known (cf. [2], [1]), the Schur complement of the non-singular submatrix A_{11} in the matrix

$$\begin{bmatrix} A_{11} & A_{12} \\ A_{21} & A_{22} \end{bmatrix}$$

is the matrix $A_{22} - A_{21} A_{11}^{-1} A_{12}$. Its significance in numerical algebra is based on the fact that if A_{11} is non-

-singular then the system

$$A_{11}x_1 + A_{12}x_2 = b_1,$$

$$A_{21}x_1 + A_{22}x_2 = b_2$$

is equivalent (after "elimination" of x_1 using the first equation) to

$$A_{11}x_1 + A_{12}x_2 = b_1,$$

$$(A_{22} - A_{21}A_{11}^{-1}A_{12})x_2 = b_2 - A_{11}^{-1}A_{21}b_1.$$

Analogously to the usual Schur complement, we define the *Boolean Schur complement* $[A/A_{11}]$ of the c-nonsingular submatrix A_{11} in the Boolean partitioned matrix

$$A = \begin{bmatrix} A_{11} & A_{12} \\ A_{21} & A_{22} \end{bmatrix}$$

as

$$[A/A_{11}] = A_{22} + A_{21}A_{11}^{-1}A_{12}$$

(with Boolean operations).

THEOREM 5. *Let* $C \in \mathcal{B}_n$,

$$C = \begin{bmatrix} C_{11} & C_{12} \\ C_{21} & C_{22} \end{bmatrix}$$

be such that C_{11} *is* c-nonsingular. Let G *be a Boolean matrix of the same type as* C_{22}. *Then the following are equivalent:*

1° $G = [C/C_{11}]$;

2° *for every permutation matrix* $P \leq C_{11}$, $G = (A_{22} - A_{21}A_{11}^{-1}A_{12})_B$ *where*

$$\begin{bmatrix} A_{11} & A_{12} \\ A_{21} & A_{22} \end{bmatrix} = \begin{bmatrix} K(P, C_{11}) & C_{12} \\ C_{21} & -C_{22} \end{bmatrix};$$

3° *there exists a permutation matrix* $P_0 \leq C_{11}$ *such that* $G = (\hat{A}_{22} - \hat{A}_{21}\hat{A}_{11}^{-1}\hat{A}_{12})_B$ *where*

$$\begin{bmatrix} \hat{A}_{11} & \hat{A}_{12} \\ \hat{A}_{21} & \hat{A}_{22} \end{bmatrix} = \begin{bmatrix} K(P_0, C_{11}) & C_{12} \\ C_{21} & -C_{22} \end{bmatrix};$$

4° $G = \sum (A_{22} - A_{21}A_{11}^{-1}A_{12})_B$
where the Boolean summation is extended over all matrices $A \in M_n$ *such that, with the corresponding*

partitioning $A = \begin{bmatrix} A_{11} & A_{12} \\ A_{21} & A_{22} \end{bmatrix}$, $A_B = C$ *and*

$\det A_{11} \neq 0$.

PROOF. We denote by G_1 the matrix G from 1°, by $G_2(P)$ the matrix defined in 2° (depending on P) and by G_4 the matrix defined in 4°. We shall prove the theorem by showing that

$$(4) \qquad G_1 \leq G_2(P) \leq G_4 \leq G_1$$

from which the assertion follows since also $G_3(P_0)$ from 3° which clearly exists is equal to $G_2(P_0)$ and thus to all the remaining G's.

Let us prove first that $G_1 \leq G_2(P)$. This (even with equality) follows from the facts that on one side,

$((K(P, C_{11}))^{-1})_B = C_{11}^{-1}$ by Theorem 4 and that $A_{22} \leq 0$, $A_{21} \geq 0$, $A_{11}^{-1} \geq 0$, $A_{12} = 0$ so that no terms in $A_{22} - A_{21}A_{11}^{-1}A_{12}$ cancel.

The inequality $G_2(P) \leq G_4$ being immediate, it remains to prove $G_4 \leq G_1$. Let thus $(G_4)_{ik} = 1$. Thus there exists a matrix $A = \begin{bmatrix} A_{11} & A_{12} \\ A_{21} & A_{22} \end{bmatrix}$ such that $\det A_{11} \neq 0$, $A_B = C$ and $(A_{22} - A_{21}A_{11}^{-1}A_{12})_{ik} \neq 0$. If $(A_{22})_{ik} \neq 0$, we have $(G_1)_{ik} = 1$. Assume thus $(A_{22})_{ik} = 0$. Then the entry $(A_{21}A_{11}^{-1}A_{12})_{ik}$ is different from zero so that there exist indices j, t such that $(A_{21})_{ij} \neq 0$, $\alpha_{tj} \neq 0$ and $(A_{12})_{jk} \neq 0$ where α_{tj} is the entry of A_{11}^{-1}. By Theorem 4, $(C_{11}^{-1})_{tj} = 1$ so that really $(C_{21}C_{11}^{-1}C_{12})_{ik} = 1$ and $(G_1)_{ik} = 1$. This completes the proof of (4) and of the whole theorem.

The following two theorems (which hold in matricial form with equalities [2]) will be presented without proofs. These may be found in [6] (in the graph-theoretical form). Let us just mention that they are based on the following lemma.

LEMMA. *If $P \in B_n$ is a permutation matrix, $Q \in B_n$ a matrix which can be completed (by changes of some zeros into ones) to a permutation matrix then the submatrix of the c-inverse A^{-1} of the matrix $A = P+Q$ in the square block symmetric ("transpose") to the block formed by intersecting the zero rows and zero columns of Q is a permutation matrix. Moreover, if Q has its non-zero entries exactly in the block corresponding to A_{11} in the decomposition $A = \begin{bmatrix} A_{11} & A_{12} \\ A_{21} & A_{22} \end{bmatrix}$ then the Boolean Schur complement $[A/A_{11}]$ is also a permutation matrix.*

THEOREM 6. *Let* $C = \begin{bmatrix} C_{11} & C_{12} \\ C_{21} & C_{22} \end{bmatrix} \in \mathcal{B}_n$ *be c-nonsingular and let* $C^{-1} = \begin{bmatrix} \Gamma_{11} & \Gamma_{12} \\ \Gamma_{21} & \Gamma_{22} \end{bmatrix}$. *If* C_{11} *is c-nonsingular then the Schur complement* $[C/C_{11}]$ *is c-nonsingular and*

$$\Gamma_{22} \leq [C/C_{11}]^{-1}.$$

REMARK. The converse is not true: If $C = \begin{bmatrix} 1 & 1 & 1 \\ 1 & 0 & 0 \\ 1 & 0 & 0 \end{bmatrix}$ and C_{11} is 1×1 then $[C/C_{11}] = \begin{bmatrix} 1 & 1 \\ 1 & 1 \end{bmatrix}$ is c-nonsingular whereas C is c-singular.

THEOREM 7. *Let* $C \in \mathcal{B}_n$,

$$C = \begin{bmatrix} C_{11} & C_{12} & C_{13} \\ C_{21} & C_{22} & C_{23} \\ C_{31} & C_{32} & C_{33} \end{bmatrix},$$

let C_{11} *as well as* $\begin{bmatrix} C_{11} & C_{12} \\ C_{21} & C_{22} \end{bmatrix}$ *be ç-nonsingular. If we denote*

$$\hat{C} = [C/C_{11}] = \begin{bmatrix} \hat{C}_{22} & \hat{C}_{23} \\ \hat{C}_{32} & \hat{C}_{33} \end{bmatrix}$$

then $\hat{C}_{22}$ *is nonsingular and*

$$(5) \qquad [C / \begin{bmatrix} C_{11} & C_{12} \\ C_{21} & C_{22} \end{bmatrix}] \leq [\hat{C}/\hat{C}_{22}] .$$

REMARK. In (5), equality does not hold in general as the following example shows:

$$C = \begin{bmatrix} 1 & 1 & 1 & 1 \\ 1 & 0 & 1 & 0 \\ 1 & 0 & 1 & 0 \\ 0 & 0 & 0 & 1 \end{bmatrix},$$

C_{11} is 1×1, C_{22} also 1×1. Then

$$\hat{C} = [C/C_{11}] = \begin{bmatrix} 1 & 1 & 1 \\ 1 & 1 & 1 \\ 1 & 1 & 1 \end{bmatrix},$$

$$[\hat{C}/\hat{C}_{22}] = \begin{bmatrix} 1 & 1 \\ 1 & 1 \end{bmatrix}$$

but

$$\begin{bmatrix} C / \begin{bmatrix} C_{11} & C_{12} \\ C_{21} & C_{22} \end{bmatrix} \end{bmatrix} = \begin{bmatrix} 1 & 0 \\ 0 & 1 \end{bmatrix} .$$

It may thus happen that in the block elimination we obtain a zero coefficient while eliminating one unknown in each step, a theoretically vanishing, but practically maybe non-vanishing coefficient appears. A simple example of this situation is

$$c_{11}x_1 + c_{12}x_2 + c_{13}x_3 + c_{14}x_4 = d_1,$$
$$c_{21}x_1 \qquad\quad + c_{23}x_3 \qquad\qquad = d_2,$$
$$c_{31}x_1 \qquad\quad + c_{33}x_3 \qquad\qquad = d_3,$$
$$c_{41}x_1 + c_{42}x_2 + c_{43}x_3 + c_{44}x_4 = d_4$$

where in the resulting system after two eliminations with pivots $(1, 1)$ and $(2, 2)$ the entry c_{34} should vanish.

Let us conclude with the remark that this situation cannot happen (i.e., in (5) equality holds) if in C all diagonal entries are ones (cf. [3]).

REFERENCES

[1] R.W. Cottle, Manifestations of the Schur complement, *Lin. Alg. Appl.* 8(1974), 189-211.

[2] D.E. Crabtree - E.V. Haynsworth, An identity for the Schur complement of a matrix, *Proc. AMS* 22(1969), 364-366.

[3] M. Fiedler, Some applications of the theory of graphs in matrix theory and geometry. In: *Theory of Graphs and its Applications*, Proc. Symp. Smolenice 1963. Acad. Publ. House, Prague 1964, 37-41.

[4] M. Fiedler, A property of eigenvectors of nonnegative symmetric matrices and its application to graph theory, *Czech. Math. J.* (100) 25(1975), 619--633.

[5] M. Fiedler, Eigenvectors of acyclic matrices, *Czech. Math. J.* (100) 25(1975), 607-618.

[6] M. Fiedler, Inversion of bigraphs and connections with the Gauss elimination. In: *Graphs, Hypergraphs and Block Systems*, Proc. Symp. Zielona Góra 1976, 57-68.

[7] R.S. Varga, *Matrix Iterative Analysis*, Prentice
     Hall, 1962.

Miroslav Fiedler
Mathematical Institute of Academy
Žitná 25
115 67 Praha 1
Czechoslovakia

NEW STABILITY PROPERTY CONCERNING STIFF METHODS
A. GALÁNTAI

1. INTRODUCTION

In this paper the notions of A, $A(\alpha)$ and A_o-stability are extended to stiff differential systems of the form

$$(1) \qquad y'=Ay + g(t) \; ; \qquad y(0) = y_o,$$

where A is an $m \times m$ matrix and $g(t) \in C([0,+\infty])$.

The necessity of such extensions is known from many papers and books ([1], [5], [7], [9-13]). Their main feature is that the homogeneous stiff system

$$(2) \qquad y' = Ay; \qquad y(0) = y_o$$

is described only by the concept of A-stability and its weaker versions.

Introduce the following definition:

DEFINITION 1. Let $g: R^+ \to C$ be a fixed continuous
function. Then a numerical method is said to be $A[\alpha,g]$-
stable if $\|y_n - y(t_n)\| \to 0$ as $n \to +\infty$ $(t_n = nh)$ when it is
applied with arbitrary fixed positive h to any diffe-
rential equation of the form

$$(3) \qquad y' = \lambda y + g(t) \; ; \quad y(0) = 1,$$

where $\lambda \in W(\alpha) = \{z \in C \mid |\arg z - \pi| < \alpha\}$ $(0 \leq \alpha \leq \pi/2)$.

Clearly, in case of $g \equiv 0$ one gets the concept of
A, $A(\alpha)$ and A_o-stability. More precisely, the test
equation $y' = \lambda y$ $(\lambda \in C_-)$ is asymptotically (even globally)
stable in the sense of L y a p u n o v and the
requirement of A-stability demands the same condition
for the arising difference equation.

For nonvanishing $g(t)$ and $\lambda \in C_-$, the solution of
(3) remains asymptotically stable but its behaviour at
infinity may be rather different for each perturbation.

If the notions of L y a p u n o v stability
theory (see, e.g. [2]) are applied to the discretization
method we have conditions only for the sequence
$\{\|y_n - \hat{y}_n\|\}_{n=0}^\infty$, where $\{y_n\}_{n=k}^\infty$ is the solution of the
difference equation with respect to (w.r.t) the initial
value $\{y_j\}_{j=0}^{k-1}$ and $\{\hat{y}_n\}_{n=k}^\infty$ is the solution w.r.t. the
perturbed initial value $\{\hat{y}_j\}_{j=0}^{k-1}$. These conditions
may be, e.g. uniform boundedness of $\{\|y_n - \hat{y}_n\|\}$ in case
of stability or $\|y_n - \hat{y}_n\| \to 0$ $(n \to +\infty)$ in case of asympto-
tical stability, e.t.c.

This possibility was investigated by D a h l q u i s t
[4] and S t e t t e r [10], [11] and also appears in
the concept of B-stability due to B u t c h e r [3].

However one can see in this approach that there is
no immediate connection between the exact and the approximate solutions. The concept of $A[\alpha,g]$-stability has

such a connection, because it expresses the convergence of the discretization method over infinite intervals. It also provides economical computations on long intervals.

2. EXISTENCE RESULTS

We have considered two classes of methods. The first class consists of multistep - multiderivative methods of the form

$$(4) \qquad \sum_{i=0}^{k} \alpha_{i0} y_{n+i} + \sum_{i=0}^{k} \sum_{j=1}^{l_i} h^j \alpha_{ij} y^{(j)}_{n+i} = 0$$

with $\alpha_{ij} \in R$, $\alpha_{k0} \neq 0$, $\sum_{i=0}^{k} \alpha_{i0} = 0$, $l_i \geq 0$, $k>0$ and $l = \max_i l_i > 0$.

The second investigated set consists of implicit Runge-Kutta processes of the form

$$y_{n+1} - y_n = h \sum_{i=1}^{m} c_i k_i$$

$$(5)$$

$$k_i = f(x + a_i h, y + h \sum_{j=1}^{m} b_{ij} k_j) \qquad (i=1,\ldots,m)$$

where $\sum_{i=1}^{m} c_i = 1$.

Let $\mathfrak{M}$ be the class of all real continuous functions $g(t)$ $(\mathscr{D}(g) \supset [0,+\infty))$ which have finite limit at infinity $(+\infty)$ and let $\mathfrak{M}^{(j)} = \{g(t) \in \mathfrak{M} \mid g^{(i)} \in \mathfrak{M}, \quad i=1,\ldots,j\}$.

The following results are then established.

THEOREM 1.

(a) *A multistep - multiderivative method of the form* (4) *is* $A[\alpha,g]$-*stable for all* $g \in \mathfrak{M}^{(l-1)}$ *if and only if the process is A(α)-stable.*

(b) *An implicit Runge-Kutta process is A[α,g]-stable
for all g∈𝔐 if and only if the method is A(α) -stable
(0≤α≤π/2).*

PROOF. (a) By simple calculations the method (4) ge-
nerates the difference equation

$$\sum_{i=0}^{k} [\alpha_{i0} + \sum_{j=1}^{l_i} (\lambda h)^j \alpha_{ij}] y_{n+i} =$$

(6)

$$= - \sum_{i=0}^{k} \sum_{j=1}^{l_i} h^j \alpha_{ij} \sum_{s=1}^{j} \lambda^{j-s} g^{(s-1)}(t_{n+i}).$$

The condition $g(t) \in \mathfrak{M}^{(l-1)}$ implies $\lim_{t \to +\infty} g^{(j)}(t) = 0$

$(j=1,\ldots,l-1)$. Thus the right side of (6) tends to

$$- \sum_{i=0}^{k} \sum_{j=1}^{l_i} \alpha_{ij} \lambda^{j-1} h^j c \quad \text{as } n \to +\infty \quad (c = \lim_{t \to +\infty} g(t)).$$

Now a lemma is needed for the solution of the inhomogeneous difference equation

(7) $$\sum_{i=0}^{k} a_i y_{n+i} = c_{n+k-1} \qquad (a_k=1),$$

LEMMA. *If every zeros of* $q(z) = \sum_{i=0}^{k} a_i z^i$ *has modulus
less than 1 and* $c_n \to c$, *then for the solution of* (7) *the
relation*

(8) $$\lim_{n \to +\infty} y_n = c \Big/ (\sum_{i=0}^{k} a_i)$$

holds.

In order to prove (8) the well-known relation (see,

e.g. in [8], [11])

$$(9) \qquad y_n = \sum_{i=0}^{k-1} \varphi_i(n) y_i + \sum_{j=k-1}^{n-1} \varphi_{k-1}(n+k-j-2) c_j$$

can be used where $\varphi_j(n)$ denotes the solution of the homogeneous difference equation

$$(10) \qquad \sum_{i=0}^{k} a_i y_{n+i} = 0$$

w.r.t. the initial condition

$$(11) \qquad y_i = \begin{cases} 1 & (i=j) \\ 0 & (i \neq j, \quad 0 \leq i \leq k-1). \end{cases}$$

Let m be the number of different zeros of $q(z)$ and let m_i be the multiplicity of the root z_i. Then

$$(12) \qquad \varphi_j(n) = \sum_{l=1}^{m} q_l(n) z_l^n \qquad (n=0,1,\ldots) ,$$

where $q_l(x)$ is some polynomial of degree $q_l(x) \leq m_l - 1$.

The first sum in (9) tends to zero by the condition $\max_i |z_i| < 1$. The second sum can be decomposed as follows

$$(13) \qquad \sum_{j=k-1}^{n-1} c_j \varphi_{k-1}(n+k-j-2) = \sum_{j=k-1}^{n-1} [c+(c_j-c)] \varphi_{k-1}(n+k-j-2).$$

Since for zero sequences $\{c_j^*\}_{j=0}^{\infty}$ the sum

$$(14) \qquad S_n = \sum_{j=k-1}^{n-1} (n+k-j-2)^r z_i^{n+k-j-2} c_j^* \to 0$$

as $n \to +\infty$ ($r \geq 1$, $1 \leq i \leq m$ are fixed), the second part of (13) also tends to zero by (12). Using the generating function method one has

$$(15) \qquad \sum_{j=0}^{\infty} \varphi_{k-1}(j) x^j = x^k / q(x) \qquad \left(\min_i \frac{1}{|z_i|} > |x| \right) ,$$

that implies the requested relation (8).

The zeros of the characteristic equation of (6) are of modulus less than 1 by $A(\alpha)$-stability.

When applying the lemma to the difference equation (6), one gets $y_n \to -c/\lambda$. Using a theorem of [2] one can prove that $y(t)$ also tends to $-c/\lambda$ as $t \to +\infty$. This proves part (a) of the theorem.

(b) Applying the Runge-Kutta process to the test equation (3) one obtains

$$(16) \qquad y_{n+1} = \alpha(z) y_n + \beta_n(z) ,$$

where $\alpha(z) = 1 + c^T (I - zB)^{-1} ze$ $(z = \lambda h)$, $\beta_n(z) = hc^T (I - zB)^{-1} g_n$, $B = (b_{ij})_{i,j=1}^{m}$, $c^T = (c_1, \ldots, c_m)$, $e^T = (1, \ldots, 1)$ and $g_n^T = (g(t_n + a_1 h), \ldots, g(t_n + a_m h))$. For function $\alpha(z)$

$|\alpha(z)| < 1$ holds by $A(\alpha)$ — stability for all $z \in W(\alpha)$.

Hence, $\beta_n(z) \to hc^T (I - zB)^{-1} ed$ $(d = \lim_{t \to +\infty} g(t))$ and the lemma implies the statement.

Q.E.D.

Note that the condition $g(t) \in \mathfrak{M}^{(l-1)}$ of part (a) cannot be weakened as it can be shown by simple examples.

A much sharper result can be proven for implicit Runge-Kutta processes.

THEOREM 2. *If the implicit Runge-Kutta process* (5)
satisfies $a_i = \sum_{j=1}^{m} b_{ij}$ $(i=1,\ldots,m)$ *and is* $A(\alpha)$ - *stable*
$(0 \leq \alpha \leq \pi/2)$, *then it is* $A[\alpha,g]$-*stable for all perturbation*
$g(t)=at+b+s(t)$, *where* $a,b \in C$ *and* $s(t) \in \mathfrak{M}$.

The proof is published in [6].

3. NONEXISTENCE RESULTS

For bounded $g(t)$ and $g^{(j)}(t)$ $(j=1,\ldots,1-1)$, the $A(\alpha)$-stability $(0 \leq \alpha \leq \pi/2)$ of the methods (5)-(6) imply the boundedness of the global error sequence $\{\|y_n - y(t_n)\|\}_{n=0}^{\infty}$ for all $\lambda h \in W(\alpha)$. So it is reasonable to expect further existence results for "regular" perturbations $g(t) \notin \mathfrak{M}$. However the results of the section prove the contrary of this expectation.

Consider the linear k-step methods of the form

$$(17) \qquad \sum_{i=0}^{k} \alpha_i y_{n+i} - h \sum_{i=0}^{k} \beta_i y'_{n+i} = 0$$

with $\alpha_i, \beta_i \in R$, $\sum_{i=0}^{k} \alpha_i = 0$, $\alpha_k = 1$ and $\sum_{i=0}^{k} \beta_i \neq 0$.

THEOREM 3. *Any* $A(\alpha)$-*stable* $(0 \leq \alpha \leq \pi/2)$ *linear k-step method of the form* (17) *cannot be* $A[\alpha,\hat{g}]$-*stable, if* $\hat{g}$ *is equal to one of the functions*

$$(18) \qquad \begin{array}{l} g_1(t)=a \sin bt, \quad g_2(t)=a \cos bt, \quad g_3(t)=a \sin^2 bt, \\[2mm] g_4(t)=a \cos^2 bt \quad (a,b \neq 0). \end{array}$$

The statement can be proven by the lemma given in

Theorem 1 and by an easy calculation if one choses the steplength h to be equal to the period of the function $g_i(t)$ $(i=1,2,3,4)$.

For implicit Runge-Kutta methods a much sharper result can be obtained.

DEFINITION 2. A numerical method is said to be locally $A[\alpha,g]$-stable if there exist $h_o,r_o>0$ such that for all $h\in(0,h_o)$ and for all $\lambda\in\{z\in W(\alpha)\,|\,|z|<r_o\}$ the condition $\|y_n-y(t_n)\|\to 0$ (as $n\to+\infty$) is fulfilled concerning the test equation (3).

THEOREM 4 ([6]). *There exist no $A(\alpha)$-stable Runge- -Kutta process which is locally $A[\alpha,g_i]$-stable $(i=1,2)$.*

All the investigated methods are linear in the perturbation $g(t)$ if the process is $A(\alpha)$-stable. More precisely, if an $A(\alpha)$-stable method is $A[\alpha,g]$, and $A[\alpha,\hat{g}]$-stable then it is $A[\alpha,\mu g+\tau\hat{g}]$-stable $(\mu,\tau\in C)$. Similarly, if an $A(\alpha)$-stable method is $A[\alpha,g]$-stable but not $A[\alpha,g^*]$-stable, then it cannot be $A[\alpha,g+g^*]$- -stable. From this result and the above theorems it follows that requiring $A[\alpha,g]$-stability of an $A(\alpha)$-stable process is not a well posed problem w.r.t. the norm

$$\|g\|=\max_{t\in[0,+\infty]}|g(t)|.$$

Since any change in the perturbation $g(t)$ may cause rather different stability effects, the present theory surveyed in [5,13] is not sufficient to describe the real structure of the inhomogeneous linear case. This fact has important consequences concerning the nonlinear stiff systems, too (see [5]).

ACKNOWLEDGEMENT. The author is indebted to Mrs. Maria Strehó for many helps and discussions.

REFERENCES

[1]   N.S. Bakhvalov, *Numerical Methods I.* (in Russian)
        Nauka, Moscow, 1973.

[2]   E.A. Barbashin, *Introduction in the Theory of
        Stability*, (in Russian), Nauka, Moscow, 1967.

[3]   J.C. Butcher, A stability property of Runge-Kutta
        methods, *BIT* 15 (1975), 358-361.

[4]   G. Dahlquist, A special stability problem for li-
        near multistep methods, *BIT* 3 (1963), 27-43.

[5]   G. Dahlquist, Recent work on stiff equations,
        Report, TRITA-NA-7512, *Dept. of Inf. Proc. of the
        Royal Institute of Technology*, Stockholm, 1975.

[6]   A. Galántai, The linear stiff problem (in Hungarian)
        Alkalmazott Matematikai Lapok (to appear)

[7]   C.W. Gear, *Numerical Initial Value Problems in
        Ordinary Differential Equations*, Prentice-Hall Inc.,
        Englewood Cliffs, New Yersey, 1971.

[8]   P. Henrici, *Discrete Variable Methods in Ordinary
        Differential Equations*, John Wiley and Sons, New
        York, 1962.

[9]   A. Prothero - A. Robinson, On the stability and
        accuracy of one-step methods for solving stiff
        systems of ordinary differential equations, *Math.
        Comp.* 28 (1974), 145-162.

[10]   H.J. Stetter, *Stability of Discretizations on In-
        finite Intervals*, Lecture Notes in Math., No. 228,
        Springer, Berlin, 1971.

[11]   H.J. Stetter, *Analysis of Discretization Methods
        for Ordinary Differential Equations*, Springer,
        Berlin, 1973.

[12]   J.G. Verwer, *S*-stability properties for generalized
       Runge-Kutta methods, *Numer. Math.* 27 (1977),
       359-370.

[13]   R. Willoughby, (ed), *Stiff Differential Systems*,
       Plenum Press, New York, 1974.

A. Galántai

University of Agricultural Sciences,
Mathematical Institute, 2103 Gödöllő,
Hungary

ON A STABILIZATION METHOD OF MIELE (CONVERGENCE AND APPLICATIONS)

L. GERENCSÉR

INTRODUCTION

The aim of this paper is to show how some widely discussed stabilization problems of nonlinear optimization can be solved with the help of a simple, new result of stability theory. We give a unique treatment and an extension of previous results due to M i e l e [1]. The first part of the paper describes the concrete methods, while the theoretical background is given at the end of the paper.

When describing algorithms the differential equation approach will be used, which is a standard technique. The idea of this method is the following. Suppose that an algorithm of the form

$$(1) \qquad x^{n+1} = x^{n} + \alpha^{n} p(x^{n})$$

is given, where $p(x^{n})$ is a direction, α^{n} is a suitable stepsize. The convergence properties of algorithms of

this type are in strong connection with the stability
properties of the differential equation

$$(2) \qquad x = p(x) .$$

We say that x^* is a stable equilibrium point of the latter
differential equation if

$$(3) \qquad p(x^*) = 0 ,$$

and for all eigenvalues λ_i of the Jacobian $p_x(x^*)$ we have

$$(4) \qquad \mathrm{Re}\, \lambda_i < 0 .$$

These notions are well-known from the theory of differential
equations.

The importance of stability theory in optimization
lies in the fact that if x^* is a stable equilibrium point
of (2) then the sequence x^n defined by (2) converges to
x^*, whenever the stepsizes are suitably chosen, e.g.
$\alpha^n = \alpha$, where α is small.

There are many other possibilites for discretizing
(2) but the elaboration of effective discrete methods is
still a task for future research.

1. STABILIZATION OF THE REDUCED GRADIENT METHOD

As a first example, consider the constrained optimization problem

$$(1.1) \qquad \min f(x,y)$$

$$(1.2) \qquad g_i(x,y) = 0, \qquad\qquad i = 1,2,\ldots,m ,$$

where $x=(x_1,\ldots,x_n)'$ and $y=(y_1,\ldots,y_m)'$. A local solution is (x^*,y^*). The variables are separated in such a way that the reduced gradient method is applicable. That is, y has exactly m components and thus it can be expressed from (1.2) as a function of x. We denote this function by $Y(x)$. In a compact form it is defined by

$$(1.3) \quad g(x,Y(x)) = 0 ,$$

where $g=(g_1,\ldots,g_m)'$. Then the problem (1.1) (1.2) is reduced to the unconstrained problem

$$(1.4) \quad \min \varphi(x) ,$$

where

$$(1.5) \quad \varphi(x) = f(x,Y(x)) .$$

Both problems (1.3) and (1.4) can only be solved numerically. Possible methods in differential equation form are

$$(1.6) \quad \dot{y} = Gg(x,y)$$

and

$$(1.7) \quad \dot{x} = F\varphi_x .$$

The matrices G and F are chosen in such a way that stability is ensured. For example, they may be good approximations to $-g_y^{-1}$ and $-\varphi_{xx}^{-1}$. The pair of differential equations (1.6) and (1.7) is unsuitable for numerical computations, because φ_x is available only after integrating (1.6) for all x. This difficulty has been experienced in practice by J. M a y e r [3]. His extensive experimental work stimulated the development of an improved scheme.

The method, that we present, had been used previously by A. M i e l e [1] for the solution of control problems. Our convergence theory seems to be new. There is also a recent contribution by H. Y a m a s h i t a [3]. He developed a differential equation which is stable in (x^*,y^*). He also investigated the problems of numerical integration of the proposed differential equation.

Now we briefly describe our method. The presentation will contain heuristic elements, but an exact description and convergence proof is given at the end of this paper.

We are going to develop a single system of differential equations which has a stable equilibrium point (x^*,y^*). This has been done in an earlier paper of mine [2] in which the following scheme was proposed: Define the matrix

$$(1.8) \qquad A(x,y) = -g_x\, g_y^{\,-1}.$$

If $Y(x)$ is substituted for y then

$$(1.9) \qquad A(x,Y(x)) = \partial Y(x)/\partial x.$$

Thus φ_x can be expressed as

$$(1.10) \qquad \varphi_x = f_x + A f_y.$$

The right hand side can be evaluated for anarbitrary pair of points (x,y). The proposed system of differential equations is:

216

$$(1.11) \quad \dot{x} = F(f_x + Af_y) \, ,$$

$$(1.12) \quad \dot{y} = Gy + A^* \dot{x} \, .$$

The arguments (x,y) must be substituted on the right hand side.

It is obvious that (x^*, y^*) is an equilibrium point of (1.11) and (1.12). The correction term $A^* \dot{x}$ ensures that this equilibrium point is stable.

2. THE ESTIMATION OF A NONLINEAR PARAMETER OF A DYNAMIC SYSTEM

We shall describe a method for the estimation of a nonlinear parameter of a dynamic system. Our approach is a generalization of the finite dimensional procedure. The objective is to develop an algorithm with an emphasize on its elements and its structure, without a rigorous convergence proof.

Let us consider the initial value problem

$$(2.1) \quad \dot{x} = f(t,x,p) \, , \qquad x(o) = x_o \, ,$$

where p is an unknown parameter. There is a set of time instants at which $x(t)$ is measured. These instants will be denoted by $t_1, \ldots, t_k$, the measured values are $y_1, \ldots, y_k$. As measurements may include errors, the y_i-s do not determine p even if k is very large. Therefore a least square estimation will be used. That is, we introduce the error function:

$$(2.2) \quad \Phi(p) = \sum_{i=1}^{k} (x(t_i,p) - y_i)'(x(t_i,p) - y_i)$$

and determine p by minimizing $\Phi(p)$. Here we used the notation $x(t,p)$, which denotes the solution of (2.1) for a fixed parameter value p.

The dimensionality of x is n, that of p is m. Accordingly the components of x are $x_1,\ldots,x_n$, those of p are $p_1,\ldots,p_m$. The gradient of Φ is easily computed:

$$(2.3) \qquad -\delta p = \Phi'_p = \sum_{i=1}^{k} \frac{\partial x(t_i,p)}{\partial p}(x(t_i,p)-y_i).$$

Thus a replacement of p for $p+\delta p$ would decrease the error function. Unfortunately the evaluation of Φ_p is very difficult. We have to solve (2.1) in order to obtain $x(t_i,p)$. Then we have to solve the variational differential equation:

$$(2.4) \qquad \delta\dot{z} = f_x(t,x,p)\delta z + f_p(t,x,p) \qquad \delta z(o) = 0$$

in order to get $\delta z(t_i)=\partial x(t_i,p)/\partial p$. Note that (2.4) is a matrix differential equation.

All these steps can be performed simultaneously which is likely to reduce the time of computation. The j-th step of the iteration yields a vector p^j and a function $x^j(t)$ with $x(0)=x^o$. Therafter the following elementary steps should be performed. Solve the matrix differential equation (2.4) after having substituted $x=x^j$ and $p=p^j$. The solution will be denoted by $z^j=z^j(t)$. In the next step evaluate δp^j in analogy with (2.3)

$$(2.5) \qquad \delta p^j = -\sum_{i=1}^{k} z^j(t_i)(x^j(t_i,p)-y_i).$$

The next step is to ensure that the difference of $x^j(t)$ and $x(t,p^j)$ remains unchanged after a displacement δp^j. That is, we compute a neccessary correction of $x^j(t)$ corresponding to δp^j. This is simply

$$(2.6) \qquad \delta x^{j^1}(t) = \delta z^j(t)\delta p^j.$$

Finally we add a correction to $x^j(t)$ in order to reduce the error between $x^j(t)$ and the true solution $x(t,p^j)$. This may be easily done by quasilinearization, i.e. by solving:

$$(2.7) \qquad \delta \dot{x}^{j^2} = f_x(t,x^j,p^j)\,\delta x^{j^2} + f(t,x^j,p^j) - \dot{x}^j, \quad \delta x^{j^2}(o)=0.$$

It is worth noting that this linear differential equation has the same coefficient-matrix as (2.4). Finally define

$$(2.8) \qquad \delta x^j = \delta x^j(t) = \delta x^{j^1}(t) + \delta x^{j^2}(t).$$

The next approximations are

$$(2.9) \qquad p^{j+1} = p^j + \delta p^j, \qquad x^{j+1} = x^j + \delta x^j.$$

Under appropriate conditions, the convergence of analogous algorithms in finite dimension will be proved. The proof is based on Lyapunov's stability theory, thus it is likely that a generalization for infinite - dimensional problems is correct.

The above procedure can be used without difficulty for problems where two-point boundary conditions are given. Suppose e.g. that we are given

$$(2.10) \qquad
\begin{aligned}
\dot{x} &= f(t,x,y,p), & x(0) &= x^o, \\
\dot{y} &= g(t,x,y,p), & y(T) &= y^T.
\end{aligned}$$

Then the differential equation for the variations (2.4)

transforms into

$$(2.11) \quad \begin{aligned} \delta\dot{x} &= f_x\delta x + f_y\delta y + f_p, & \delta x(0) &= 0, \\ \delta\dot{y} &= g_y\delta x + g_y\delta y + g_p, & \delta y(T) &= 0. \end{aligned}$$

The course of the computation is unchanged except for the quasilinearization step (2.7) which is now

$$(2.12) \quad \begin{aligned} \delta\dot{x} &= f_x\delta x + f_y\delta y + f - \dot{x}, & \delta x(0) &= 0, \\ \delta\dot{y} &= g_x\delta x + g_y\delta y + g - \dot{y} & \delta y(T) &= 0. \end{aligned}$$

More complicated problems can also be handled.

3. A THEOREM OF STABILITY THEORY

A general result will be described and proved. Let

$$(3.1) \quad \dot{x} = p(x,Y(x)) = q(x)$$

be a differential equation, which is stable in x^*. We can express this fact as follows: For all solutions λ_i of

$$(3.2) \quad \det(q_x(x^*) - \lambda I) = 0$$

Re $\lambda_i < 0$ holds. Here $\det(A)$ denotes the determinant of the matrix A. We shall say in these cases that $q_x(x^*)$ is a negative matrix.

Let us suppose that $Y(x)$ is defined by

$$(3.3) \quad r(x,Y(x)) = 0$$

and that this equation is given in a stable form. This

means that the differential equation

(3.4) $\dot{y} = r(x,y)$

is stable in $y = Y(x)$.

Consequently, we have $\mathrm{Re}\mu_i < 0$ for all solutions μ_i of

(3.5) $\det(r_y(z^*) - \mu I) = 0$

with $z^* = (x^*, Y(x^*))$.

In order to develop a joint system of differential equations, one is tempted to put

(3.6) $\dot{x} = p(x,y)$

(3.7) $\dot{y} = r(x,y)$.

In order to analyze the Jacobian of the right hand side, let us rewrite condition (3.2). The matrix q_x can be expressed as

(3.8) $q_x = p_x + (\frac{\partial Y}{\partial x})p_y = p_x - r_x r_y^{-1} p_y$.

The Jacobian of the right hand side of the system (3.6), (3.7) equals

(3.9) $D = \begin{bmatrix} p_x & r_x \\ p_y & r_y \end{bmatrix}$

The information we have is that r_y and the determinant-like expression of q_x in (3.8) are negative matrices. For a symmetric D we could conclude that D is negative. This result is a generalization of the well-known result of S y l v e s t e r . Unfortunately, for nonsymmetric

221

matrices no such conclusion on D is possible. Counterexamples for 2x2 matrices can easily be given. (The
idea is to choose a very small off-diagonal element.)

This difficulty can be circumvented if we go back
from the matrix-theoretical problem to the original
formulation. The trouble with this formulation lies in the
fact that the correction of x may drive off the feasible
surface $(x,Y(x))$. In order to prevent this, we add a
correction term p_2 to the second equation (3.7). The
requirement is that the vector (p,p_2) should be tangential
to the surface $(x,Y(x))$ if (x,y) lies on this surface.
In other words,

$$(3.10) \qquad p_2 = A^* p,$$

where the matrix $A=A(x,y)$ is an extension of $\partial Y(x)/\partial x$,
i.e.

$$(3.11) \qquad A(x,Y(x)) = \partial Y(x)/\partial x = -r_x r_y^{-1}.$$

The suggested modification is

$$(3.12) \qquad \dot{x} = p(x,y),$$

$$(3.13) \qquad \dot{y} = r(x,y)+A^* p.$$

The stability of this equation was proved in an earlier
paper [2] by the Lyapunov-function method. We shall
give now a different proof based on the first method of
Lyapunov. The advantage of this proof is that the eigenvalues that characterize the convergence rate are explicitly found. Furthermore an interesting result of matrix
theory is obtained.

To prove that (3.12), (3.13) is strongly stable in

$z^*=(x^*,Y(x^*))$, let us compute the Jacobian of the right hand side. We obtain

$$(3.14) \qquad D = \begin{bmatrix} p_x & r_x + A^* p_x \\ p_y & r_y + A^* p_y \end{bmatrix}.$$

To obtain the characteristic polynomial of D let us compute $\det(D-\lambda I)$:

$$(3.15) \qquad \det(D-\lambda I) = \det \begin{bmatrix} p_x - \lambda I & r_x + A^* p_x \\ p_y & r_y + A^* p_y - \lambda I \end{bmatrix}.$$

The value of the determinant remains unchanged if we substract A^* times the first "column" from the second "column". Thus we get

$$(3.16) \qquad \det(D-\lambda I) = \det \begin{bmatrix} p_x - \lambda I & r_x + \lambda A^* \\ p_y & r_y - \lambda I \end{bmatrix}.$$

To evaluate this determinant we make use of the following identity: if

$$(3.17) \qquad M = \begin{bmatrix} R & S \\ Q & T \end{bmatrix}$$

where R, T are square matrices, then

$$(3.18) \qquad \det(M) = \det(T) \cdot \det(R - S T^{-1} Q)$$

if T^{-1} exists.

In the present case we get: if

$$(3.19) \qquad \det(r_y - \lambda I) = 0$$

then by hypothesis we get $\mathrm{Re}\lambda < 0$. On the other hand, if $T = r_y - \lambda I$ is invertible, then $\det(M)$ may become zero only if the second term in (3.18) vanishes. That is, we must have

$$(3.20) \qquad \det(p_x - \lambda I - (r_x + \lambda A^*)(r_y - \lambda I)^{-1} p_y) = 0.$$

Substituting $-r_x r_y^{-1}$ for A, the second term in the bracket reduces to $-r_x r_y^{-1} p_y$. Thus we get

$$(3.21) \qquad \det(p_x - r_x r_y^{-1} p_y - \lambda I) = 0.$$

Again, by hypothesis, for all solutions of this equation we have $\mathrm{Re}\lambda < 0$. Thus the stability of (3.13) (3.14) is proved.

REFERENCES

[1] A. Miele, Gradient methods in optimal control theory. In *Optimization and Design* (ed. M.Avriel-M.J.Rijckaert- - D.J.Wilde) Prentice-Hall, Englewood Cliffs, 1973.

[2] L. Gerencsér, Computational methods in game theory. *J. Computational Math. and Math. Phys*, <u>15</u> (1975) No. 3. (In Russian)

[3] *Survey of Mathematical Programming* (Vol. I-III.), Proceedings of the 9th Mathematical Programming Symposium, Budapest, August 23-27, 1976. (ed. Prékopa András - Székely Gábor)

L. Gerencsér
Computer and Automation Institute
of the Hungarian Academy of Sciences
Budapest, Kende u. 13-17.   H-1111

COLLOQUIA MATHEMATICA SOCIETATIS JÁNOS BOLYAI

22. NUMERICAL METHODS, KESZTHELY (HUNGARY), 1977.

MATRIX INVERSION AND SOLUTION OF LINEAR AND NONLINEAR SYSTEMS BY THE METHOD OF BORDERING

J. GERGELY

ABSTRACT

In this paper equivalence of the method of bordering and the Gauss-Jordan elimination is proven. Thereafter the method of bordering is used for solving linear and nonlinear systems of equations. Finally some consequences are investigated.

INTRODUTION

A great number of methods is known for matrix inversion. Among these the most frequently applied ones are the Gaussian elimination, the Gauss-Jordan elimination, th Crout's method and the method of bordering. According to several sources in the literature, the first three of the above mentioned methods are equivalent (cf. e.g. [3], [4]). We shall prove here that the method of bordering is also equivalent with the others, in particular, we are going to prove the identity of the Gauss-Jordan elimination and the method of bordering.

225

Let matrix $A = \{a_{ij}\}$, $i,j=1,\ldots,n$ be nonsingular
and let A_k denote the left upper sector of k-th order
of the matrix A, that is, the minor matrix consisting of
the elements a_{ij}, $i,j=1,2,\ldots,k$.

The method of bordering is the following: (cf. [1]).
Let us look for the inverse of the matrix

$$A_k = \begin{bmatrix} A_{k-1} & u_k \\ v_k^T & \alpha_k \end{bmatrix}$$

in the form

$$A_k^{-1} = \begin{bmatrix} P_{k-1} & r_k \\ q_k^T & \beta_k \end{bmatrix},$$

where $A_1 = a_{11}$, $(A_1^{-1} = 1/a_{11})$. The inverse can then be
computed with the following formulae

$$\beta_k = \frac{1}{\alpha_k - v_k^T A_{k-1}^{-1} u_k} \,, \quad r_k = -\beta_k A_{k-1}^{-1} u_k, \quad q_k^T = -\beta_k v_k^T A_{k-1}^{-1}$$

$$(1)$$

$$P_{k-1} = A_{k-1}^{-1} + \beta_k A_{k-1}^{-1} u_k v_k^T A_{k-1}^{-1}.$$

If we compute the formulae (1) for $k=2,\ldots,n$, then
$A_n^{-1} = A^{-1}$, that was sought.

The following formulae yield the i-th step of the
Gauss-Jordan elimination

$$a_{ii} := 1/a_{ii};$$

$$a_{ij} := -a_{ij} a_{ii}, \quad j=1,\ldots,n, \quad j \neq i;$$

$$(2) \quad a_{k\ell} := a_{k\ell} + a_{ki} a_{i\ell}, \quad k,\ell=1,\ldots,n, \quad k,\ell \neq i;$$

$$a_{ji} := a_{ji} a_{ii} \quad j=1,\ldots,n, \quad j \neq i.$$

The symbol $:=$ means that we should calculate the value of the expression standing on the right side of the symbol and assign it to the variable standing on the left side. If we compute relations (2) for $i=1,\ldots,n$, then we obtain the inverse A^{-1} in the place of the original matrix $A = \{a_{ij}\}$.

THEOREM. *In the course of the inversion of matrix A, the method of bordering calculated from (1) and the method of Gauss-Jordan elimination calculated from (2) carry out identical operations.*

PROOF. It is obvious that procedure (1) yields matrix A_k^{-1} after its k-th ($2 \leq k \leq n$) step, that is the inverse of the left upper sector of k-th order of the matrix A. We obtain exactly the same inverse after the k-th step of procedure (2) in the left upper sector of k-th order of matrix A. Namely, procedure (1) and (2) yield the same matrix after the k-th ($2 \leq k \leq n$) step in the left upper sector of matrix A. We wish to prove that when we carry out the $k+1$-st step from the k-th state, $k=m-1 \leq n-1$ by computing (1) and (2), we perform the same operations.

1. Let $m-1 < n$ and change from the $m-1$-st step to the m-th step. It is known (e.g. cf. [2]) that (2) results in

$$(3) \qquad a_{mm} := 1/a_{mm} = \frac{\det(A_{m-1})}{\det(A_m)}$$

after the m-th step. Let us calculate β_m according to (1). This is $(A_m^{-1})mm$, the element with indices m,m of matrix A_m^{-1}, therefore according to to the known relation of matrix theory

$$(4) \qquad \beta_m = (A_m^{-1})_{mm} = \frac{1}{\det(A_m)}(\text{adj } A_m)_{mm} = \frac{\det(A_{m-1})}{\det(A_m)} \; .$$

Thus by comparison of (3) and (4)

$$(5) \qquad \beta_m = \frac{\det(A_{m-1})}{\det(A_m)} = a_{mm} \; .$$

2. Solve the equation $Ax+b = 0$ with the Gauss-Jordan method. This can be achieved by adjoining vector b to the matrix A so that it is assumed to be its $n+1$-st column. Compute formulae (2) also for $j=n+1$. Let us stop computing (2) at $i=m-1$, then the first $m-1$ element of the $n+1$-st column contain the solution of the equation system

$$A_{m-1} x_{m-1} + b_{m-1} = 0,$$

that is

$$x_{m-1} = -A_{m-1}^{-1} b_{m-1},$$

where x_{m-1} and b_{m-1} are the vectors containing the first $m-1$ elements of vectors x and b resp. While performing (2), we carry out the same operations for all columns of matrix A with indices larger than m as we have done for the $n+1$-st column. By collecting the first $m-1$ elements of the m-th column into vector u_m, the vector $-A_{m-1}^{-1} u_m$ occupies the locus of u_m after the completion of the $i=m$-th step. In the m-th step according to the fourth row of (2), this is multiplied by a_{mm} which, on the other hand, is identical with β_m according to 1. Comparing with formulae (1), it can be seen that procedure (2) puts vector r_m in the m-th step in the locus of the first $m-1$ elements of the m-th column.

3. In computing the inverse, the columns and rows of the matrix play similar roles, therefore when computing (1) and (2), the same type of operations have to be performed for the rows and columns. Hence, valid statements for the identity of the columns can also be carried over for the rows. This could also be justified independently from the above considerations.

4. Procedure (2) yields new values for A_{m-1} that are given by the third row of (2) with subscripts $k, \ell = 1, \ldots, m-1$. Write the first term of the right side in matrix form as A_{m-1}^{-1}. The first factor of the second term - as we have seen in point 2 - is $-A_m^{-1} u_m$. The second factor of the second term according to point 3 is $-\beta_m v_m^T A_{m-1}^{-1}$. Thus procedure (2) yields also the matrix P_{m-1} in the place of the left upper sector of $m-1$-st order. Hereby we have completed the proof of the theorem.

SOLUTION OF LINEAR SYSTEMS OF EQUATIONS

A method for the solution of linear systems of equations by the method of bordering is discussed in book [3]. Now we formulate the method in another way from which it is easy to see the connection between the Gauss-Jordan elimination method and the method of bordering. Thereafter we give a generalization of the method for the nonlinear case.

Let the system of the linear equations be

$$(6) \qquad \sum_{j=1}^{n} a_{ij} x_j = b_i, \qquad i = 1, \ldots, n.$$

Let $x^0 = 0$, $x^k = (x_1^k, x_2^k, \ldots, x_k^k, 0, \ldots, 0)^T$, $1 \le k \le n$. First we consider the first equation of (6) as the function of only one unknown x_1. Let the change in x_1 be δx_1 and let $a_{11}\delta x_1 = b_1$, from which $\delta x_1^0 = b_1/a_{11}$ and let $x_1^1 = \delta x_1^0$. Then let us solve the second equation of (6) for δx_2 (the change of x_2), supposing that the linear change of the first equation of (6) is equal to 0, that is

$$a_{11}\delta x_1 + a_{12}\delta x_2 = 0,$$

from where

$$\delta x_1 = c_1^1 \delta x_2, \qquad c_1^1 = -\frac{a_{12}}{a_{11}},$$

and the second equation

$$a_{21}(x_1^1 + c_1^1 \delta x_2) + a_{22}\delta x_2 = b_2.$$

From the last equation (as c_1^1 is known)

$$\delta x_2^0 = \frac{b_2 - a_{21}x_1^1}{a_{22} + a_{21}c_1^1}.$$

Then let

$$x_1^2 = x_1^1 + c_1^1 \delta x_2^0, \qquad x_2^2 = \delta x_2^0.$$

In general, we have in the $k+1$-st step

$$\begin{aligned}
&\sum_{j=1}^{k} a_{ij} c_j^k \delta x_{k+1} + a_{i\,k+1}\delta x_{k+1} = 0 \qquad i=1,\ldots,k, \\
&\sum_{j=1}^{k} a_{k+1,j}(x_j^k + c_j^k \delta x_{k+1}) + a_{k+1\,k+1}\delta x_{k+1} = b_{k+1}.
\end{aligned}$$

(7)

The vector c^k can be computed from the first equations of (7) as

$$(8) \qquad \begin{bmatrix} c^k_1 \\ \vdots \\ c^k_k \end{bmatrix} = - \begin{bmatrix} a_{11} & \cdots & a_{1k} \\ \vdots & & \vdots \\ a_{k1} & \cdots & a_{kk} \end{bmatrix}^{-1} \begin{bmatrix} a_{1\,k+1} \\ \vdots \\ a_{k\,k+1} \end{bmatrix} .$$

Substituting c^k for the second equation of (8) we get a linear equation. By solving this we get

$$(9) \qquad \delta x^0_{k+1} = \frac{b_{k+1} - \sum\limits_{j=1}^{k} a_{k+1\,j} x^k_j}{a_{k+1\,k+1} + \sum\limits_{j=1}^{k} a_{k+1\,j} c^k_j} .$$

Then let

$$(10) \qquad \begin{aligned} x^{k+1}_i &= x^k_i + \delta x^0_{k+1} c^k_i , \qquad i \leq k . \\[2mm] x^{k+1}_{k+1} &= \delta x^0_{k+1} . \end{aligned}$$

When carrying out the composition for $k=0,\ldots,n-1$, we obtain that x^n is the solution sought.

According to the steps of the method described above, the computation requires the inversion of a coefficient matrix by the method of bordering. The number of the multiplicative operations is n^3. However, an algorithm can be given that requires less operations by taking advantage of the equivalence of the methods for inversion.

Let us perform the Gauss–Jordan elimination for matrix A:

$$(11) \qquad \begin{aligned} a_{ii} &:= 1/a_{ii} , \\[2mm] a_{ij} &:= a_{ij} \cdot a_{ii} , \qquad j > i , \\[2mm] a_{k\ell} &:= a_{k\ell} - a_{ki} a_{i\ell} , \qquad k \neq i , \quad \ell > i , \quad i=1,2,\ldots,n . \end{aligned}$$

The elements of the matrix are denoted during and after the elimination as they had been before the elimination. The diagonal element a_{kk} should remain unchanged in the k-th step. By using the notation of the introduction, (8) and (9) can be set

$$c^k = -A_k u_k,$$

(12)

$$\delta x^0_{k+1} = \frac{b_{k+1} - v_k^T x^k}{\alpha_k - v_k^T A_k^{-1} u_k} = (b_{k+1} - v_k^T x^k) a_{k+1,k+1}$$

vector $-c^k$ appears in the $k+1$-st column of the matrix (in the place of the first k elements) after the elimination.

THE ALGORITHM FOR THE SOLUTION

The following is the algorithm of the method of bordering for solving the linear system (6):
1. We carry out Gauss–Jordan elimination for matrix A by (11);
2. We carry out for $k=0,1,\ldots,n-1$ the following:
   a) we calculate δx^0_{k+1} from (12);
   b) we calculate the vector x^{k+1} from (10).
The vector x^n will be the solution of the system (6).

EXAMPLE 1. Let us solve the system of the equations

$$5x+2y+z = -12$$

$$-x+4y+2z = 20$$

$$2x-3y+10z = 3$$

that is

$$A = \begin{bmatrix} 5 & 2 & 1 \\ -1 & 4 & 2 \\ 2 & -3 & 10 \end{bmatrix}, \quad b = \begin{bmatrix} -12 \\ 20 \\ 3 \end{bmatrix},$$

Let us carry out the Gauss-Jordan elimination by (11) and let us write the initial matrix elements into the under-diagonal part of the matrix

$$\begin{bmatrix} .2 & .4 & 0 \\ -1 & 1/4.4 & .5 \\ 2 & -3 & 1/11.5 \end{bmatrix}.$$

Let us carry out the secon point of algorithm for $k= =0,1,2$

$$x_1^1 = \delta x_1^0 = a_{11} b_1 = -2.4, \quad x^1 = [-2.4,0,0]^T$$

$$x_2^2 = \delta x_2^0 = (b_2 - a_{21} x_1^1) a_{22} = 4, \quad x_1^2 = -4, \quad x^2 = [-4,0,0]^T$$

$$x_3^3 = \delta x_3^0 = (b_3 - a_{31} x_1^2 - a_{32} x_2^2) a_{33} = 2, \quad x_1^3 = -4, \quad x_2^3 = 3.$$

The solution is $x_3 = [-4,3,2]^T$.

SOLUTION OF NONLINEAR SYSTEMS OF EQUATIONS

Let the system of nonlinear equations be

$$(13) \qquad f_i(x) = 0, \quad i=1,\ldots,n, \quad x \in R^n.$$

First we wolve (13) by Newton's method. Let Jacobi's matrix of the system be

$$A = [a_{ij}], \quad a_{ij} = \left.\frac{\partial f_i}{\partial x_j}\right|_{x=x^0} \quad i,j=1,\ldots,n.$$

Starting from a first approximation x^0, we look for the new approximation x^1 in the form $x^1 = x^0 + u$, where u is the solution of the linear system

$$(14) \qquad Au = -f(x^0), \quad f = (f_1, \ldots, f_n).$$

We solve the system (14) by the method of bordering. Let $u^k = (u_1^k, u_2^k, \ldots, u_k^k, 0, \ldots, 0)$,

$$u_i^{k+1} = u_i^k + c_i^k \delta u_{k+1}, \quad i \leq k, \quad u_{k+1}^{k+1} = \delta u_{k+1},$$

then the solution of (14) implies the solution of the equations

$$(15) \qquad \sum_{j=1}^{k} a_{ij} c_j^k \delta u_{k+1} + a_{ik+1} \delta u_{k+1} = 0, \quad i \leq k,$$

$$(16) \qquad \sum_{j=1}^{k} a_{k+1\,j} (u_j^k + c_j^k \delta u_{k+1}) + a_{k+1\,k+1} \delta u_{k+1} = -f_{k+1}(x^0)$$

for $k=0, \ldots, n-1$. u^n is the solution of the system (14).

THE MODIFICATION OF NEWTON'S METHOD

We modify Newton's method so that we look for the new approximation x^1 now in form $x^1 = x^0 + v^n$, where v^n is the solution of the equations

$$(17) \qquad \sum_{j=1}^{k} a_{ij} c_j^k \delta v_{k+1} + a_{i\,k+1} \delta v_{k+1} = 0, \quad i \leq k$$

$$(18) \qquad f_{k+1}(x^0 + v^k + (c^k, 1) \delta v_{k+1}) = 0$$

$$\left. \right\} \; k=0, \ldots, n-1,$$

$$v^k = (v_1^k, \ldots, v_k^k, 0, \ldots, 0), \quad v_i^{k+1} = v_i^k + c_i^k \delta v_{k+1}, \quad v_{k+1}^{k+1} = \delta v_{k+1}.$$

The difference between the system (17)-(18) and the

system (15)-(16) is that while (16) is a linear equation
for δu_{k+1}, now (18) is a nonlinear equation for δv_{k+1}.
(Equations (15) and (17) are the same).

The system (17)-(18) yields better approximation
for nonlinear equations (13) than the system (15)-(16),
so it is to be expected that x^0+v^n will be a better
approximation for the solution of the system (13) than
x^0+u^n is.

In the paper [5] an iteration method was discussed.
It is easy to see that the iteration method corresponds
to the modified Newton's method written up above. We
have presented in paper [5] an example, showing that the
convergence of the method applied there has been faster
then the convergence of Newton's method.

Now we give another example for the use of the
modified Newton's method.

EXAMPLE 2. Let us solve the nonlinear system

$$f_1(x,y) \equiv 4x^2+y^2+2xy-y-2 = 0$$

$$f_2(x,y) \equiv 2x^2+y^2+3xy-3 = 0$$

$$x_0 = 0.4, \quad y_0 = 0.9.$$

Newton's method gives the following approximations:

$$x_1 = 0.514 \qquad y_1 = 1.000$$

$$x_2 = 0.50017 \qquad y_2 = 0.99986.$$

The use of the modified Newton's method means the
following:

a) We solve the equation

$$4(0.4+\delta v_1)^2+0.9^2+2(0.4+\delta v_1)0.9-0.9-2 = 0$$

for δv_1. The solution is $\delta v_1=0.13205$ and $x_1^1=x_0+\delta v_1=0.53205$.

b) From the equation

$$df_1 = (8x_0+2y_0)c_1^1\delta v_2+(2y_0+2x_0-1)\delta v_2 = 0$$

$c_1^1 = -0.32$. Then we solve the second equation:

$$2(0.53205-0.32\delta v_2)^2+(0.9+\delta v_2)^2+$$

$$+3(0.53205-0.32\delta v_2)(0.9+\delta v_2)-3 = 0.$$

The solution is $\delta v_2=0.09987$ and

$$y_1 = y_0+\delta v_2 = 0.99987, \quad x_1 = x_1^1+c_1^1\delta v_2 = 0.50009.$$

The exact solution of the system is

$$x = 0.5, \quad y = 1.$$

It appears from example 2 that the modified Newton's method gives a better solution in one step than Newton's method gives in two steps.

A good illustration can be given to the methods: the vector $(c^k,1)$ (defined by (15) in a linear case and by (17) in a nonlinear case) determines a direction in the $k+1$ dimension space. We have to solve the $k+1$-st equation (equation (16) in a linear case and equation (18) in a nonlinear case) in this direction.

CONSEQUNCES

1. As it is well-known, the Gauss-Jordan elimination cannot be continued if for some i $a_{ii}=0$ and it yields an inexact result of the a_{ii} element becomes small in absolute value. In accordance with the earlier proved theorem, the same problem arises also with the method of bordering if $\beta_k=0$ or the absolute value of β_k is small.

2. If the matrix to be inverted is not singular, then it is always applicable and numerically stable if the Gauss-Jordan elimination is carried out by pivoting (cf. [2]). It is also possible to choose a pivotal element in the method of bordering. In the m-th step this means that we compute the quantity β_m by chosing all rows and columns with greater subscripts than m and of these we select the row and column with which the denominator of β_m will be the greatest. This row and column must be interchanged with the m-th row and column and the calculation continues. This choice of a pivotal element is very laborious, much more laborious than in the Gauss-Jordan method. So it is only a possibility in principle, but in practice cannot be used.

3. If the matrix to be inverted is positive definite then the Gauss-Jordan elimination method is numerically stable (cf. [2]). Thus according to the preceding theorem inversion by the method of bordering is also numerically stable for a positive definite matrix.

4. The Gauss-Jordan elimination can be carried out by chosing pivotal elements in the algorithm when solving linear or nonlinear equations, too. In this case we note the order of the rows and columns of the matrix after the elimination. When computing (9) and (10), we take into consideration that the change of the rows involves the change of the equations and the change of

the columns involves the change of the variables.

REFERENCES

[1] D.K. Faddeev – V.N. Faddeeva, *Computational Methods of Linear Algebra*, Fizmatgiz., Moscow – Leningrad 1963. (In Russian)

[2] J.H. Wilkinson, *The Algebraic Eigenvalue Problem*, Clarendon Press, Oxford 1965.

[3] V.V. Voevodin, *Numerical Methods of Algebra*, Nauka Moscow, 1966. (In Russian)

[4] R.P. Tewarson, *Sparse Matrices*, Academic Press, New York – San Francisco – London, 1973.

[5] J. Gergely, Numerical solution of a nonlinear boundary value problem, *Colloquia Math. Soc. János Bolyai* 15. *Differential Equations*, 1975.

J. Gergely
Computer and Automation Institute
of the Hungarian Academy of Sciences
H-1111 Budapest, Kende u. 13-17.

ENERGY CONSERVING DISCRETIZATIONS OF DIFFUSION EQUATIONS
R. GORENFLO

SUMMARY.

Under conditions either of periodicity or of reflecting walls equations like

$$\gamma(x)u_t = (a(x)u)_{xx} + (b(x)u)_x \quad \text{in} \quad [0,1]\times[0,\infty)$$

for diffusion of energy with density $\gamma(x)u(x,\,t)$ conserve energy:

$$\int_0^1 \gamma(x)u(x,\,t)dx = \int_0^1 \gamma(x)u(x,\,0)dx \quad \text{for all} \quad t > 0.$$

Sometimes, particularly in long-time calculations, computing physicists and engineers are disturbed by their difference schemes not sharing a discrete analogue of this property. In this paper we present a method for constructing explicit and implicit schemes discretely conserving energy. In a natural way, interface problems and problems with additional source term and inhomogeneous boundary conditions can also be treated; in the

latter case conservation of energy extends to the energy
generated by the source term and flowing in through the
boundaries.

ZUSAMMENFASSUNG

Unter Bedingungen der Periodizität oder reflektie-
render Wände haben Gleichungen wie

$$\gamma(x)u_t = (a(x)u)_{xx} + (b(x)u)_x \quad \text{in} \quad [0,1] \times [0,\infty)$$

für die Diffusion von Energie mit Dichte $\gamma(x)u(x,\,t)$
die Eigenschaft

$$\int_0^1 \gamma(x)u(x,\,t)dx = \int_0^1 \gamma(x)u(x,0)dx \quad \text{für alle} \quad t>0$$

der Energie-Erhaltung. Rechnende Physiker und Ingenieure
klagen manchmal (vor allem bei Langzeitrechnungen)
darüber, daß die von Ihnen benutzten Differenzenschemata
nicht einem diskreten Analogon dieser Eigenschaft genügen. In der vorliegenden Arbeit wird eine Methode zur
Konstruktion expliziter und impliziter energie-erhalten-
der Differenzenschemata vorgestellt. In natürlicher Weise
können auch Probleme mit inneren Übergangsbedingungen
und Probleme mit zusätzlichem Quellterm und inhomogenen
Randbedingungen behandelt werden; dabei wird die durch
den Quellterm erzeugte Energie und die über die inhomogenen Randbedingungen einströmende Energie durch das
Differenzenschema voll erfasst.

1. LINEAR ENERGY-DIFFUSION EQUATION

Let $w(x,\,t) = \gamma(x)u(x,\,t)$ for $0 \leq x \leq 1$, $t \geq 0$, be
density of energy (interpretation: $u(x,\,t)$ is temperature, $\gamma(x) > 0$ is specific heat). We consider the

energy diffusion equation:

$$(1.1) \qquad w_t(x, t) = (Lu)(x, t) + f(x, t) \quad \text{in} \quad [0,1] \times [0,\infty)$$

with

$$(1.2) \qquad (Lu)(x, t) = (a(x)u(x, t))_{xx} + (b(x)u(x, t))_x.$$

We assume a, b, γ, f *sufficiently smooth*, at least piecewise, furthermore

$$(1.3) \qquad 0 < \check{a} \le a(x, t) \le a^* < \infty, \quad 0 < \check{\gamma} \le \gamma(x) \le \gamma^* < \infty,$$
$$|b(x)| \le b^* < \infty.$$

As *initial condition* we take

$$(1.4) \qquad u(x, 0) = g(x) \quad \text{for} \quad 0 \le x \le 1.$$

The energy stored at time t in the interval $[\alpha_1, \alpha_2] \subset [0,1]$ is

$$(1.5) \qquad S(t; \alpha_1, \alpha_2) = \int_{\alpha_1}^{\alpha_2} w(x, t)\,dx,$$

and with

$$(1.6) \qquad (Bu)(x, t) = (a(x)u(x, t))_x + b(x)u(x, t) \quad \text{and}$$

$$S' = dS/dt$$

we have

$$S'(t; \alpha_1, \alpha_2) = \int_{\alpha_1}^{\alpha_2} w_t\,dx = \int_{\alpha_1}^{\alpha_2} (Lu+f)\,dx,$$

hence

— 16 —

$$S'(t; \alpha_1, \alpha_2) = -(Bu)(\alpha_1, t)+(Bu)(\alpha_2, t)+$$

(1.7)
$$+ \int_{\alpha_1}^{\alpha_2} f(x, t)dx.$$

Assuming boundary conditions

(1.8) $-(Bu)(0, t) = \varphi(t), \quad (Bu)(1, t) = \Psi(t)$ for $t > 0$

we have

(1.9) $S'(t; 0,1) = \varphi(t)+\Psi(t)+\int_0^1 f(x, t)dx.$

    In each of the following three problems (P), (R), (I) *the total energy* $S(t) = S(t; 0,1)$ *is constant, i.e. does not depend on* t.

(P) *Periodicity:* (1.1), (1.4), $f(x, t) \equiv 0$,

        $u(0, t) = u(1, t), \quad (Bu)(0, t) = (Bu)(1, t)$ for

        $t > 0.$

(R) *Reflecting Walls:* (1.1), (1.4), $f(x; t) \equiv 0,$ and

        (1.8) with $\varphi(t) \equiv 0, \quad \Psi(t) \equiv 0.$

(I) In case of a *jump discontinuity of* $a, b,$ *or* γ *at an interior point* $\xi \in (0,1)$ *we take as interface condition*

        $u(\xi-0, t) = u(\xi+0, t), \quad (Bu)(\xi-0, t) = (Bu)(\xi+0, t)$

        for $t > 0$ in addition to (1.1), (1.4) and

        $f(x, t) \equiv 0.$

The interface condition is to be combined either with (P) or (R).

    In 2. we derive general algebraic properties of energy conserving difference schemes. In 3. we first treat these three problems, thereby developing our method

of construction, before describing how to handle the
general inhomogenous problem

$$(G) \ = \ \{(1,1), \ (1.4), \ (1.8)\}$$

which can be combined with an interface condition as in
(I).

To avoid the tedious job of having to determine the
minimum number of derivatives required in each discretization let us agree on the following *assumptions*.

In problem (P) let a, b, γ, g be extendible to
functions in $C^\infty(\mathbf{R})$ with period 1, so that $u \in C^\infty(\mathbf{R} \times [0,\infty))$.
In problem (R) let a, b, γ, g be in $C^\infty[0,1]$ and
$u \in C^\infty([0,1] \times [0,\infty))$.

In problem (I) let a, b, γ, g be of clas C^∞ in
each of the intervals $[0,\xi]$ and $[\xi,1]$ where at $x = \xi$
limits are to be taken from the left or from the right,
for consideration in $[0,\xi]$, or $[\xi,1]$, respectively. Let compatibility conditions be fulfilled such that u is of class
C^∞ in each of the half strips $[0,\xi] \times [0,\infty)$ and
$[\xi,1] \times [0,\infty)$, with the same convention about one-sided
limits at $x = \xi$. In problem (G) we require φ, Ψ, $f \in C^\infty$
and $u \in C^\infty$ with corresponding relaxed conditions if there is
an additional interface condition. For a, b, γ, g we
require the same conditions to be fulfilled as stated
for problems (R) and (I), respectively.

These smoothness assumptions can, of course, be
considerably weakened, but they are reasonable in applications where there usually occur functions very smooth
piecewise.

DESIDERATA. To imitate discretely *constancy* (i.e.
conservation) *of total energy* $S(t)$ in problems (P),
(R), (I). In problem (G) to take *exact account of energy*

flowing in through the boundaries with rates $\varphi(t)$ and $\Psi(t)$ *and of energy being generated in the interior* with rate $f(x, t)$. In addition, one may want to imitate another important property of the diffusion process, namely *conservation of non-negativity*: if $g(x)$, $f(x, t)$, $\varphi(t)$, $\Psi(t)$ are ≥ 0, then $w(x, t) \geq 0$; here always $0 \leq x \leq 1$, $0 \leq t < \infty$ (see Walter [6] for the Nagumo-
-Westphal lemma).

REMARK 1.1. Instead of taking the operators L and B as in (1.2) and (1.6) one may, with $\beta(x) = a'(x) + b(x)$, equivalently define

$$(1.2') \qquad (Lu)(x, t) := (a(x)u_x(x, t))_x + (\beta(x)u(x, t))_x,$$

$$(1.6') \qquad (Bu)(x, t) := a(x)u_x(x, t) + \beta(x)u(x, t).$$

REMARK 1.2. In the particular case $\gamma(x) \equiv 1$ (1.1) is the equation for *diffusion of mass* with density $w(x, t) = u(x, t)$. The basic theory of mass conserving difference schemes for this special case has been given by the author in [4]. The previous results are generalized in this paper to non-constant γ.

2. GENERAL PROPERTIES OF ENERGY CONSERVING SCHEMES

We discretize: $m \in \mathbf{N}$, $h = 1/m$, $\mu > 0$, $\tau = \mu h^2$. With $n+1 \in \mathbf{N}$ and

$$(2.1) \qquad j \in J := \{\tfrac{1}{2}, \tfrac{3}{2}, \tfrac{5}{2}, \ldots, m - \tfrac{1}{2}\}$$

we take grid points $x_j = jh$ on $[0,1]$ and $t_n = n\tau$ on $[0,\infty)$.

NOTE: The x_j are the midpoints of adjacent intervals into which $[0,1]$ is equidistantly partioned.

Let $x_j = jh$ also for values $j \notin J$.

We put $a_j = a(x_j)$, $b_j = b(x_j)$, $\gamma_j = \gamma(x_j)$ or alternatively $\gamma_j = \frac{1}{h} \int_{x_{j-1/2}}^{x_{j+1/2}} \gamma(x)dx$ (with $O(h^2)$-accuracy), and try to derive difference schemes for calculation of approximate values

$$\hat{u}_{j,n} \quad \text{for} \quad u(x_j, t_n), \quad \hat{w}_{j,n} = \gamma_j \hat{u}_{j,n} \quad \text{for}$$

$$w(x_j, t_n).$$

In problem (I) at $\xi = x_r$ take $\gamma_r = \frac{1}{2}(\gamma(\xi-0)+\gamma(\xi+0))$.

REMARK 2.1. In the light of what follows it is more appropriate to consider $\hat{w}_{j,n}$ as an approximation to

$$\frac{1}{h} \int_{x_{j-1/2}}^{x_{j+1/2}} w(x, t)dx \quad \text{which itself is}$$

$$w(x_j, t_n)+O(h^2).$$

We form column vectors

$$\tilde{u}_n = (\hat{u}_{1/2,n}, \hat{u}_{3/2,n}, \ldots, \hat{u}_{m-1/2,n})^T,$$

$$\tilde{w}_n = (\hat{w}_{1/2,n}, \hat{w}_{3/2,n}, \ldots, \hat{w}_{m-1/2,n})^T,$$

$$\tilde{f}_n = (\hat{f}_{1/2,n}, \hat{f}_{3/2,n}, \ldots, \hat{f}_{m-1/2,n})^T.$$

By $h\tau\hat{f}_{j,n}$ we denote the amount of energy being generated in the space-interval $[x_j - \frac{h}{2}, x_j + \frac{h}{2}]$ in the time interval $[t_n, t_{n+1}]$. For the boundary indices $j = 1/2$ and $j = m-1/2$ we have to take care of the inhomogenities

$\varphi(t)$ and $\Psi(t)$. With regards to (1.7) this leads us to

$$(2.2) \quad h\tau\hat{f}_{j,n} = \int_{t_n}^{t_{n+1}} \int_{x_{j-1/2}}^{x_{j+1/2}} f(x, t)\,dx\,dt + c_{j,n}$$

where $c_{j,n} = 0$ for $j \in \{3/2, 5/2, \ldots, m-3/2\}$,

$$c_{1/2,n} = \int_{t_n}^{t_{n+1}} \varphi(t)\,dt, \quad c_{m-1/2,n} = \int_{t_n}^{t_{n+1}} \psi(t)\,dt.$$

We imagine the amount $h\hat{w}_{j,n}$ of energy to be concentrated in grid point x_j at time t_n and simulate the diffusion of energy by redistribution of these discrete amounts and of the amounts $h\tau\hat{f}_{j,n}$ at each time step from t_n to t_{n+1}. We discretize the total energy $S(t_n) = S(t_n; 0,1)$, see (1.5), by the midpoint quadrature rule

$$(2.3) \quad \hat{S}_n = h \sum_{j \in J} \hat{w}_{j,n} = h \sum_{j \in J} \gamma_j \hat{u}_{j,n}.$$

One-step difference schemes (we are not interested in more complicated schemes) have the structure

$$(2.4) \quad h\hat{w}_{j,n+1} = \sum_{k \in J} p_{jk} h\hat{w}_{k,n} + \sum_{k \in J} q_{jk} h\tau\hat{f}_{k,n} \quad \text{for}$$

$$j \in J, \ n+1 \in \mathbf{N},$$

or, with $P = (p_{jk})$ and $Q = (q_{jk})$; $j \in J, k \in J$ as matrices of coefficients, in matrix-vector notation,

$$(2.4') \quad \widetilde{w}_{n+1} = P\widetilde{w}_n + \tau Q\widetilde{f}_n.$$

Conservation of non-negativity is now seen to be *equivalent to*

246

(2.5) $P \geq 0$ and $Q \geq 0$

(all entries ≥ 0 in Varga's notation [5]), whereas *conservation of energy and the exact taking up and re-distribution of energy generated by the source term and flowing in via the boundary inhomogeneities is guaranteed if we have the redistribution properties*

(2.6) $\sum_{j \in J} p_{jk} = 1$ and $\sum_{j \in J} q_{jk} = 1$ for all $k \in J$.

This is obvious from the interpretation of p_{jk} as the fraction of $h\hat{w}_{k,n}$ thrown from point x_k to point x_j, and an analogous interpretation of q_{jk}. See (2.4)
From (2.6) and the discretized initial conditions

(2.7) $h\hat{w}_{j,0} = \int_{x_{j-1/2}}^{x_{j+1/2}} \gamma(x) g(x) dx$ for $j \in J$

it follows that

(2.8) $S(t_n) = \hat{S}_n$ for all $n+1 \in \mathbf{N}$.

i.e.: *the discretized system (2.4) has the same total energy as the problem* $(G) = \{(1.1), (1.4), (1.8)\}$.

The conditions (2.6) mean that all column sums of the matrices P and Q are 1. With the 1-vector $\eta = (1,1,1,\ldots,1)^T$ they can be written as

(2.6') $\eta^T P = \eta^T$ and $\eta^T Q = \eta^T$,

together with (2.5) meaning that P^T and Q^T are *stochastic matrices*. In the particular case of all $\tilde{f}_n$ vanishing (2.4') formally represents a *Markov chain*.

For further analysis we address our attention to the *standard class of difference schemes*

$$(2.9) \qquad \gamma_j \frac{1}{\tau}(\hat{u}_{j,n+1}-\hat{u}_{j,n}) = \Theta L_h \hat{u}_{j,n+1}+\overline{\Theta}L_h \hat{u}_{j,n}+\tilde{f}_{j,n}$$

where $\Theta \in [0,1]$, $\Theta+\overline{\Theta} = 1$, and L_h is an appropriate spatial discretization operator (to be specified in 3.).

REMARK 2.2. If γ has a jump-discontinuity at a gridpoint x_k (problem (I)) remember

$$\gamma_k = \frac{1}{2}(\gamma(x_k-0)+\gamma(x_k+0)).$$

We introduce the diagonal matrix $\Gamma = \text{diag}(\gamma_{1/2}, \gamma_{3/2},\ldots,\gamma_{m-1/2})$ and the *spatial discretization matrix* H related to the operator L_h. From (2.9), using $\tilde{w}_n = \Gamma\tilde{u}_n$ and $\tilde{u}_n = \Gamma^{-1}\tilde{w}_n$, we deduce

$$(2.10) \qquad \tilde{w}_{n+1}-\tilde{w}_n = \mu H\Gamma^{-1}(\Theta\tilde{w}_{n+1}+\overline{\Theta}\tilde{w}_n)+\tau\tilde{f}_n,$$

hence as in (2.4') $\tilde{w}_{n+1} = P\tilde{w}_n+\tau Q\tilde{f}_n$ with

$$(2.11) \qquad Q = (I-\mu\Theta H\Gamma^{-1})^{-1}, \qquad P = Q(I+\mu\overline{\Theta}H\Gamma^{-1}).$$

THEOREM 2.1. *If all column sums of H are equal to zero then all column sums of Q and of P are equal to 1. In short-hand notation:*

$$\eta^T H = 0 \Rightarrow (2.6') \quad \eta^T Q = \eta^T \quad \text{and} \quad \eta^T P = \eta^T.$$

PROOF. $\eta^T H = 0 \Rightarrow \eta^T H\Gamma^{-1} = 0$. Now $Q = \sigma(H\Gamma^{-1})$ and $P = \rho(H\Gamma^{-1})$ with the rational functions $\sigma(s) = 1/(1-\mu\Theta s)$ and $\rho(s) = \sigma(s)(1+\mu\overline{\Theta}s)$ yielding 1 for

$s = 0$. $H\Gamma^{-1}$ having η^T as left eigenvector with eigen-value 0 therefore implies Q and P having η^T as left eigenvector with eigenvalue 1.

3. CONSTRUCTION OF ENERGY CONSERVING SCHEMES

Because of theorem 2.1 *we try to construct the matrix H so that all its column sums are equal to zero in problems (P) and (R).*

A special treatment is necessary for problem (I) in case of a jump discontinuity of γ *at* $x = \xi = x_k$ *and therefore of non-applicability of (2.9) for* $j = k$. *We want, of course, to have* $h^2 + \tau$ *as order of convergence as is usual for second order parabolic equations. In 4. we shall show that the schemes we are going to construct have this global order of convergence (in the maximum norm) if they have the properties (2.5) and (2.6), namely if they conserve non-negativity and energy just as the energy diffusion process.*

We shall first look for fulfilling

$$(3.1) \qquad \eta^T H = 0$$

from which the redistribution properties (2.6') follow.

This goes straight forward in the *problem (P) of periodicity*. Simply discretize $(au)_{xx}$ and $(bu)_x$ by central difference quotients

$$h^{-2}(a_{j+1}u_{j+1,n} - 2a_j u_{j,n} + a_{j-1}u_{j-1,n}),$$

$$(2h)^{-1}(b_{j+1}u_{j+1,n} - b_{j-1}u_{j-1,n}).$$

We obtain a tridiagonal matrix with additional entries in the lower left and in the upper right corner. Taking indices modulo m if $\notin J$ rows $j-1, j, j+1$ of H look as follows:

column / row	$j-2$	$j-1$	j	$j+1$	$j+2$
$j-1$	$a_{j-2} - \frac{h}{2}b_{j-2}$	$-2a_{j-1}$	$a_j + \frac{h}{2}b_j$		
j		$a_{j-1} - \frac{h}{2}b_{j-1}$	$-2a_j$	$a_{j+1} + \frac{h}{2}b_{j+1}$	
$j+1$			$a_j - \frac{h}{2}b_j$	$-2a_{j+1}$	$a_{j+2} + \frac{h}{2}b_{j+2}$

It can be seen that all columns add up to 0 as required.

Now let us construct H for *problem* (R) *of reflecting walls*. Considering the leftmost column and the rightmost column of the matrix constructed for problem (P), namely

$$
\begin{bmatrix}
-2a_{1/2} \\
a_{1/2} - \frac{h}{2}b_{1/2} \\
\cdot \\
\cdot \\
\cdot \\
a_{1/2} + \frac{h}{2}b_{1/2}
\end{bmatrix}
\quad \text{and} \quad
\begin{bmatrix}
a_{m-1/2} - \frac{h}{2}b_{m-1/2} \\
\cdot \\
\cdot \\
\cdot \\
a_{m-1/2} + \frac{h}{2}b_{m-1/2} \\
-2a_{m-1/2}
\end{bmatrix}
$$

- the dots are here standing for zero entries - we see that *we* have to *delete the lower left entry* $a_{1/2} + \frac{h}{2}b_{1/2}$ *and the upper right entry* $a_{m-1/2} - \frac{h}{2}b_{m-1/2}$. After deleting the corresponding column sums are no longer zero. Therefore *we replace the upper left entry* $-2a_{1/2}$ *by*

$-a_{1/2} + \frac{h}{2}b_{1/2}$ *and the lower right entry* $-2a_{m-1/2}$ *by*
$-a_{m-1/2} - \frac{h}{2}b_{m-1/2}$. The convergence analysis in 4. will reveal, *this remarkable trick is just the right thing to do.*

For treating the *interface problem (I)* we put $\tilde{a}^- = a(\xi-0)$, $\tilde{a}^+ = a(\xi+0)$, analogously for b^-, b^+, γ^-, γ^+. We assume the interface point ξ to be a grid point: $\xi = x_r$ for interior index, say $r \in \{7/2,\ 5/2,\ldots,m-7/2\}$ for ease of presentation. We can use (2.9) for all indices $j \in J$ with $j \neq r$.

Introducing a matrix M (still to be specified) *instead of* $H\Gamma^{-1}$ *we want to retain* (2.4'), *but with*

$$(3.2) \qquad Q = (I-\mu\Theta M)^{-1}, \qquad P = Q(I+\mu\overline{\Theta}M)$$

replacing (2.11). Use of (2.9) for $j = r-2,\ r-1,\ r+1,\ r+2$ gives the following section of M with still unde–termined entries m_{rk} in row r:

column \ row	$r-2$	$r-1$	r	$r+1$	$r+2$
$r-2$	$\dfrac{-2a_{r-2}}{\gamma_{r-2}}$	$\dfrac{a_{r-1}+\lambda b_{r-1}}{\gamma_{r-1}}$			
$r-1$	$\dfrac{a_{r-2}-\lambda b_{r-2}}{\gamma_{r-2}}$	$\dfrac{-2a_{r-1}}{\gamma_{r-1}}$	$\dfrac{a^- +\lambda b^-}{\gamma_r}$		
r		$m_{r,r-1}$	$m_{r,r}$	$m_{r,r+1}$	
$r+1$			$\dfrac{a^+ -\lambda b^+}{\gamma_r}$	$\dfrac{-2a_{r+1}}{\gamma_{r+1}}$	$\dfrac{a_{r+2}+\lambda b_{r+2}}{\gamma_{r+2}}$
$r+2$				$\dfrac{a_{r+1}-\lambda b_{r+1}}{\gamma_{r+1}}$	$\dfrac{-2a_{r+2}}{\gamma_{r+2}}$

Here $\lambda = h/2$. Rows with index $\leq r-2$ and $\geq r+2$ are the same as in $H\Gamma^{-1}$ with H as in (P) or (R), respectively.

Instead of Theorem 2.1 we now use the analogous implication

$$(3.3) \qquad \eta^T M = 0 \Rightarrow (2.6') \quad \eta^T Q = \eta^T \quad \text{and} \quad \eta^T P = \eta^T.$$

How to fill in row r of matrix M? We use the same *trick* as in problem (R), namely fixing the still unknown entries in row r, so as to get all column sums zero:

$$m_{r,r-1} = \frac{a_{r-1} - \lambda b_{r-1}}{\gamma_{r-1}}, \quad m_{r,r} = \frac{-a^- - \lambda b^-}{\gamma_r} + \frac{-a^+ + \lambda b}{\gamma_r},$$

$$m_{r,r+1} = \frac{a_{r+1} + \lambda b_{r+1}}{\gamma_{r+1}}$$

again with $\lambda = h/2$.

REMARK 3.1. *A trick, twice applied, is a method.*

The *general interface problem* (G) is now treated as follows: Take initial values as in (2.7) and – if there is no additional interior interface condition – proceed further with the scheme (2.10) using the matrix H constructed for problem (R). If there is an additional interface condition

$$(3.4) \qquad u(\xi-0,\ t) = u(\xi+0,\ t),\ (Bu)(\xi-0,\ t) = (Bu)(\xi+0,\ t)$$

for $t > 0$ and a point $\xi = x_r \in (0,1)$ replace $H\Gamma^{-1}$ by the matrix M constructed for problem (I).

Up to now we have constructed the matrices H and M so that the schemes have the redistribution properties

(2.6). To find sufficient conditions for (2.5), *conservation of non-negativity*, we look at the representations (2.11) and (3.2) for Q and P. We have, generally, as in (3.2),

$$(3.5) \qquad Q = (I-\mu\Theta M)^{-1}, \quad P = Q(I+\mu\overline{\Theta}M), \text{ where } M = H\Gamma^{-1}$$

or is directly constructed, respectively.

To obtain the most stringent conditions, we consider the problem (G) with additional interface condition (3.4) and investigate the corresponding matrix M for sufficient conditions implying $Q \geq 0$ and $P \geq 0$. The appropriate method is to use the *theory of non-negative matrices and of M-matrices*, see V a r g a [5] and C o l l a t z [1]. From C o l l a t z [1], page 297, we quote the following theorem:

LEMMA 3.1. *If A is a real quadratic matrix with all off-diagonal entries ≤ 0 and all row-sums > 0 then A^{-1} exists and is ≥ 0.*

We apply this lemma to $A = I-\mu\Theta M$. Note that in case $M = H\Gamma^{-1}$ we simply have to multiply each column j of H with $1/\gamma_j$ to obtain M. We seek conditions for $I-\mu\Theta M$ to have an inverse $Q \geq 0$ and subsequently conditions for $I+\mu\overline{\Theta}M \geq 0$, so that $P \geq 0$ as product of two non-negative matrices. For abbreviation put

$$(3.6) \qquad \overline{a}(x) = a(x)/\gamma(x), \quad \overline{b}(x) = b(x)/\gamma(x)$$

with corresponding meaning of $\overline{a}_j$, $\overline{a}^+$, etc.

ANALYSIS OF $A = I-\mu\Theta M$.

The *off-diagonal entries of M* being

$$\bar{a}_j + \frac{h}{2}\bar{b}_j, \quad \bar{a}_j - \frac{h}{2}\bar{b}_j \quad \text{for} \quad j \neq r,$$

$$\bar{a}^- + \frac{h}{2}\bar{b}^-, \quad \bar{a}^+ - \frac{h}{2}\bar{b}^+ \quad \text{for} \quad j = r$$

we conclude all off-diagonal entries of the matrix A and also those of the matrix $A\Gamma$ to be ≤ 0 if

$$(3.7) \qquad 0 \leq h \leq 2 \, \min\{\frac{a(x)}{|b(x)|} \,|\, 0 \leq x \leq 1\}$$

Now note that Γ is a diagonal matrix with positive diagonal entries, whence it follows that A is an M-matrix if and only if $A\Gamma$ is an M-matrix. So it remains to inverstigate the row sums of $A\Gamma = I - \mu\Theta H$. Because of the smoothness assumptions we have agreed on in 1. we find all our row sums to be $\gamma_j + O(h)$, so they all are >0 if only h is small enough.

Sufficient upper estimates for h can be given in terms of maxima of

$$|a(x)|, \quad |a'(x)|, \quad |a''(x)|, \quad |b(x)|, \quad |b'(x)|.$$

We deduce: $Q \geq 0$ *if* h *is small enough.*

ANALYSIS OF $I+\mu\overline{\Theta}M$

The *off-diagonal entries* are those of $\mu\overline{\Theta}M$, they are all ≥ 0 if (3.7) is fulfilled.

The *diagonal entries* are

$$1+\mu\overline{\Theta}(-\overline{a}_{1/2}+\tfrac{h}{2}\,\overline{b}_{1/2}) \qquad \text{for} \quad j = 1/2,$$

$$1+\mu\overline{\Theta}(-\overline{a}_{m-1/2}-\tfrac{h}{2}\,\overline{b}_{m-1/2}) \quad \text{for} \quad j = m-1/2,$$

$$1-2\mu\overline{\Theta}\,\overline{a}_{j} \qquad\qquad\qquad \text{for} \quad j\notin\{1/2,r,m-1/2\},$$

$$1+\mu\overline{\Theta}\{-(\overline{a}^{-}+\overline{a}^{+})+\tfrac{h}{2}(\overline{b}^{+}-\overline{b}^{-})\} \quad \text{for} \quad j = r.$$

With the bounds a^{*}, $\check{\gamma}$, b^{*} of (1.3) they all are ≥ 0 if the condition

$$(3.8) \qquad 0 < \mu < \check{\gamma}/(2\overline{\Theta}a^{*}+hb^{*})$$

for the step-length parameter $\mu = \tau/h^{2}$ is fulfilled.

We deduce: $I+\mu\overline{\Theta}M \geq 0$ *if (3.7) and (3.8) are fulfilled.*

RESULT. *If h is sufficiently small and the condition (3.8) is met then $Q \geq 0$ and $P \geq 0$.*

COMMENT TO (3.8). For the fully implicit scheme ($\Theta = 1$, $\overline{\Theta} = 1-\Theta = 0$) (3.8) means $0 < \mu < \check{\gamma}/(hb^{*})$, a very mild restriction on the time-step τ. If $0 < \Theta \leq 1$ the second sign "<" can be replaced by the sign "≤".

REMARK 3.2 Instead of taking the operators L and B as (1.2) and (1.6), one may like to take them as in

(1.2') and (1.6') (see Remark 1.1). The whole theory and method of 2. and 3. can be carried over to treatment of problems with these modified operators. The modification required is now to use the natural discretization

$$h^{-2}\{a_{j-1/2}\hat{u}_{j-1,n} - (a_{j-1/2} + a_{j+1/2})\hat{u}_{j,\,n} +$$

$$+\,a_{j+1/2}\hat{u}_{j+1,\,n}\}$$

for $(a(x)u_x)_x$ at $x = x_j$, $t = t_n$. $(\beta(x)u)_x$ is to be discretized just as previously $(b(x)u)_x$ by the central difference quotient using x_{j-1} and x_{j+1}. It is an easy exercise to construct the matrices H and M corresponding to this discretization. Boundary and interface conditions are treated by the trick of filling in unknown entries so as to obtain column sums zero. See Remark 3.1.

4. CONSISTENCY AND CONVERGENCE

We sketch the ideas of a proof of convergence in an arbitrary bounded rectangle $[0,1] \times [0,t^*]$ for problem (G) with an interior interface condition

$$u(\xi-0,\ t) = (\xi+0,\ t),\ (Bu)(\xi-0,\ t) = (Bu)(\xi+0,\ t).$$

The proof is by *majorizing error comparison functions* and makes essential use of *inverse isotonicity* (monotonic kind, conservation of non-negativity). See Gorenflo [2] and [3].

With $H\Gamma^{-1}$ replaced by M in (2.10) and with

$$w_n = (w(x_{1/2},\ t_n),\ w(x_{3/2},\ t_n),\ldots$$

$$\ldots,w(x_{m-1/2},\ t_n))^T$$

we define the consistency vector

$$(4.1) \qquad \varepsilon_n = \frac{1}{\tau}(w_{n+1} - w_n) - h^{-2} M(\Theta w_{n+1} + \overline{\Theta} w_n) - \tilde{f}_n$$

whose j-th competent is

$$O(h + \tau/h) \quad \text{for} \quad j \in \{1/2, \ r, \ m-1/2\} \quad \text{and}$$

$$O(h^2 + \tau) \quad \text{for the other indices} \quad j \in J.$$

Subtracting

$$(4.2) \qquad 0 = \frac{1}{\tau}(\tilde{w}_{n+1} - \tilde{w}_n) - h^{-2} M(\Theta \tilde{w}_{n+1} + \overline{\Theta} \tilde{w}_n) - \tilde{f}_n$$

from (4.1) we find

$$(4.3) \qquad \frac{1}{\tau}(\delta_{n+1} - \delta_n) - h^{-2} M(\Theta \delta_{n+1} + \overline{\Theta} \delta_n) = \varepsilon_n$$

for the error vectors $\delta_n = w_n - \tilde{w}_n$. For the initial errors we have $\delta_{j,0} = O(h^2)$ if g is sufficiently smooth also across $x = \xi$.

We assume h sufficiently small and (3.8) satisfied so that we have conservation of non-negativity or, in other words, isotonic dependence of the vectors δ_n on the vectors ε_n and on the initial errors $\delta_{j,0}$.

We now have to look for a positive function $y_{j,n}$ such that component-wise

$$(4.4) \qquad y_0 \geq ch^2 \eta$$

$$(4.5) \qquad \frac{1}{\tau}(y_{n+1} - y_n) - h^{-2} M(\Theta y_{n+1} + \overline{\Theta} y_n) \geq z_n$$

where

- 17 -

$$z_n = c(h+\tau/h,\ h^2+\tau\ ,\ldots,h+\tau/h,$$

(4.6)

$$h^2+\tau,\ldots,h^2+\tau,\ h+\tau/h)^T.$$

Again $\eta = (1,1,\ldots,1)^T$, and the constant $c > 0$ is chosen so large that the corresponding initial and truncation errors are majorized in modulus. From the majorizing principle (see [3] for example) $y_{j,n} \geq |\delta_{j,n}|$ for all relevant indices.

A suitable comparison function is given by

$$(4.7) \qquad y_{j,n} = K(h^2+\tau)\exp(Ct_n)\cosh(\Omega p(x_j))$$

with ω fixed so that $0 < \xi-\omega < \xi < \xi+\omega < 1$ and

$$p(x) = \begin{cases} x-\xi+\omega & \text{for} \quad 0 \leq x \leq \xi \\[2ex] x-\xi-\omega & \text{for} \quad \xi \leq x \leq 1. \end{cases}$$

The positive constants K, C, Ω can be determined so that for h sufficiently small (4.4) and (4.5) are satisfied. By $y_{j,n} = O(h^2+\tau)$ *we have proved convergence of order* $h^2+\tau$ *as* $h \to 0$ *and* $\tau = \mu h^2 \to 0$. See [2] for a convergence proof for discretization of a parabolic interface problem where the same comparison function is used.

5. A NUMERICAL CASE STUDY

For the problem (G), without additional interface condition, a test case was run on the Control Data computer of Freie Universität Berlin. I am indebted to Dipl. - Math. W. Bayer for doing the programming work. The data were as follows:

$$\gamma(x) = 1+x, \quad a(x) = 1+x+x^2+x^3, \quad b(x) = x^2+x^3,$$

$$f(x, t) = -(1+x^2)\sin t - (4+2x+12x^2+4x^3)(2+\cos t),$$

$$g(x) = 3(1+x^2)/(1+x), \quad \varphi(t) \equiv 0, \quad \psi(t) = 10(2+\cos t).$$

The exact solution is known: $w(x, t) = (1+x^2)(2+\cos t)$.

Computation was with $\Theta = 1$, $\mu = 30$, $h = 1/30$. We show numerical results for $x = 0.75$, $t = 30$, and compare $\hat{w}(0.75, 30)$ not only with $w(0.75, 30)$, but also with the mean value $\overline{w}(0.75, 30)$ which is more appropriate in the light of the construction principle of the scheme. Note:

$$\overline{w}_{j,n} = \overline{w}(x_j, t_n) = h^{-1}\int_{x_j-h/2}^{x_j+h/2} w(x, t_n)\,dx.$$

The table of results (we omit the arguments) is as follows:

w	$\overline{w}$	$\hat{w}$	$w-\hat{w}$	$\overline{w}-\hat{w}$
3.36602	3.36622	3.40844	-0.04242	-0.04222

As expected we have $|\overline{w}-\hat{w}| < |w-\hat{w}|$.

In order to demonstrate the usefulness of exactly catching and conserving all energy we did the computation once more, but with point values (instead of mean values) in the vectors $\tilde{f}_n$ and with $\hat{u}_{j,0} = g(x_j)$. We obtained:

w	$\hat{w}$	$w-\hat{w}$
3.36602	3.51030	-0.14428

NOTE: Now $|w-\hat{w}|$ is much bigger than previously. Test computations not displayed here confirmed for both types of schemes that h^2 is the order of convergence if μ is kept fixed. In order to compare errors at coinciding points we used step length $h = 1/10$ and $h = 1/30$ and obtained error quotients $\approx 9 = 3^2$ in the points of the coarser grid.

The test problem presented here corresponds to that with $\gamma(x) \equiv 1$ treated in [4]; see 6. for the required transformation.

6. CONCLUDING REMARKS

For the operators L and B given in (1.2) and (1.6) it is possible to reduce the discretization of this paper for non-constant $\gamma(x)$ directly to the case of $\gamma(x) \equiv 1$ treated in Gorenflo [4]. Simply write

$$w_t = \overline{L}w + f$$

and take $\overline{a}(x) = a(x)/\gamma(x), \qquad \overline{b}(x) = b(x)/\gamma(x),$
and

$$(\overline{L}w)(x,\ t) = (\overline{a}(x)w(x,\ t))_{xx} + (\overline{b}(x)w(x,\ t))_x$$

$$(\overline{B}w)(x,\ t) = (\overline{a}(x)w(x,\ t))_x + \overline{b}(x)w(x,\ t),$$

The inhomogeneous boundary conditions of problem (B) now are

$$-(\overline{B}w)(0,\ t) = \varphi(t),\quad (\overline{B}w)(1,\ t) = \psi(t).$$

The discretizations of [4] then yield the discretization of this paper after back-transcription.

It is, however, advantageous to have available the more general treatment of this paper because this possibility of reduction is a property of the specific operators L and B. The reduction cannot so easily be carried out in the case of the operators L and B defined in (1.2') and (1.6').

The arising complications can be overcome, but the resulting difference schemes now are different from each other, the one resulting from reduction being more complicated and requiring function evaluations also of the derivative of $1/\gamma(x)$.

For generalizations and applications to non-linear problems, to indeterminate two-point boundary value problems, and to the conservative longitudinal line method we refer the reader to [4]. It should be an easy exercise to generalize the mass-conserving line method of [4] to an energy-conserving line method (i.e. to generalize from $\gamma(x) \equiv 1$ to $\gamma(x)$ non-constant).

It should be remarked that the schemes presented can also be viewed from the standpoint of finite elements. This amounts to approximate $w(x, t)$ by piecewise constant functions, namely $\hat{w}(x, t_n) = \hat{w}_{j,n}$ in $x_j - \frac{h}{2} < x < x_j + \frac{h}{2}$. The finite-element-method with continuous piecewise linear functions leads to another type of energy-conserving difference schemes, namely to such ones with grid-points, $0, h, 2h, \ldots, mh$.

In this case a modified interpretation of amounts of energy concentrated at grid points is required at the boundary points but one which naturally follows form the finite element ansatz. The details are intended to be published in a forth-coming paper.

REFERENCES

[1]   L. Collatz, *Funktionalanalysis und Numerische Mathematik*, Springer-Verlag, Berlin, 1964.

[2] R. Gorenflo, Differenzenschemata monotoner Art für
     lineare parabolische Randwertaufgaben, *ZAMM*
     51(1971), 595-610.

[3] R. Gorenflo, Über S. Gerschgorins Methode der
     Fehlerabschätzung bei Differenzenverfahren, *Lecture
     Notes in Mathematics* 333(1973), 128-143.

[4] R. Gorenflo, Conservative difference schemes for
     diffusion problems,   *ISNM* 39 (1978), 101-124,
     Birkhäuser-Verlag, Basel.

[5] R.S. Varga, *Matrix Iterative Analysis* , Prentice
     Hall, Englewood Cliffs, N.J., 1962.

[6] W. Walter, *Differential and Integral Inequalities*
     Springer-Verlag, Berlin, 1970.

R. Gorenflo
Fachbereich Mathematik
Freie Universität Berlin
Arnimallee 2-6
D-1000 Berlin 33

ARITHMETIC MODELING IN APPLIED MATHEMATICS
D. GREENSPAN

1. INTRODUCTION

One of the primary aims of applied mathematics is
the modeling of natural phenomena. We do this in order
to understand more clearly the forces of nature, with
the hope of controlling them. Sophisticated mathematical
models were made possible first by the development of the
calculus, whose power is derived from the limit concept.
Today, new types of models, called discrete models, have
emerged, and these are made possible by the power to do
arithmetic and execute basic logical decisions with
exceptional speed on modern, digital computers. It is
this new type of modeling which we will examine in this
paper.

2. GRAVITY

It is always difficult to know how to begin
correctly, so let us develop some intuition first by
studying the following experiment with a force with which

we are all familiar, namely, gravity.

If a particle P of mass m situated at height h above ground, is dropped from a position of rest, one can measure its height x above ground every Δt seconds as it falls. For example, if one has a camere whose shutter time is Δt, then one can take a sequence of pictures at the times $t_k = k\Delta t$, $k=0,1,2,\ldots,$ and from the knowledge of h determine the heights $x_k = x(t_k)$ directly from the photographs by elementary ratio and proportion. Suppose, then, that this has been done, say, for $\Delta t = 1$, and that, to the nearest foot, one finds

$$x_0 = 400, \quad x_1 = 384, \quad x_2 = 336, \quad x_3 = 256,$$

$$x_4 = 144, \quad x_5 = 0.$$

These data are recorded in column A of Table 1. By rewriting x_0, x_1, x_2, x_3, x_4 and x_5 as

$$x_0 = 400-0, \quad x_1 = 400-16, \quad x_2 = 400-64 ,$$

$$x_3 = 400-144, \quad x_4 = 400-256, \quad x_5 = 400-400,$$

(which express the height above ground as the difference of the initial height and the distance fallen) and by factoring, one readily finds the interesting relationships

$$x_0 = 400-16(0)^2, \quad x_1 = 400-16(1)^2,$$

$$x_2 = 400-16(2)^2, \quad x_3 = 400-16(3)^2,$$

$$x_4 = 400-16(4)^2, \quad x_5 = 400-16(5)^2,$$

which can be written concisely as

$$(2.1) \qquad x_k = 400 - 16(t_k)^2; \qquad k = 0, 1, 2, 3, 4, 5.$$

In the traditional manner, one would now interpolate from (2.1) to obtain the continuous formula

$$(2.2) \qquad x = 400 - 16t^2, \qquad 0 \le t \le 5,$$

from which, by differentiation, one would find

$$(2.3) \qquad v(t) = x'(t) = -32t, \qquad 0 \le t \le 5$$

$$(2.4) \qquad a(t) = v'(t) = -32, \qquad 0 \le t \le 5.$$

The particle's velocities $v_0 = v(0)$, $v_1 = v(1)$, $v_2 = v(2)$, $v_3 = v(3)$, $v_4 = v(4)$, and $v_5 = v(5)$, at the times when the corresponding heights x_0, x_1, x_2, x_3, x_4, and x_5 have been recorded, are now determined directly from (2.3), and are recorded in column B of Table I. The particle's accelerations at these times are determined from (2.4) and are recorded in column C of Table I.

Note that formulas (2.3) and (2.4), and the interesting conclusion that the acceleration due to gravity is constant, with the value -32, have all been *deduced* from the given distance measurements x_0, x_1, x_2, x_3, x_4, and x_5.

Let us show now that *all* the above conclusions could have been deduced without ever having introduced the concepts and methodology of the calculus. To do so, let us define the particle's velocity $v_k = v(t_k)$, $k = 0, 1, 2, 3, 4, 5$, as an *average* (rather than *instantaneous*) rate of change of height with respect to time by the arithmetic formula

$$(2.5) \qquad \frac{v_{k+1}+v_k}{2} = \frac{x_{k+1}-x_k}{\Delta t} \ ; \qquad k=0,1,2,3,4.$$

Since averaging procedures are both common and useful
in the analysis of experimental data, the left-hand side
of (2.5) is perfectly reasonable.

TABLE I

Time	A Measured height	B Velocity by calculus	C Acceleration by calculus	D Velocity by arithmetic	E Acceleration by arithmetic
$t_0=0$	$x_0=400$	$v_0=0$	$a_0=-32$	$v_0=0$	$a_0=-32$
$t_1=1$	$x_1=384$	$v_1=-32$	$a_1=-32$	$v_1=-32$	$a_1=-32$
$t_2=2$	$x_2=336$	$v_2=-64$	$a_2=-32$	$v_2=-64$	$a_2=-32$
$t_3=3$	$x_3=256$	$v_3=-96$	$a_3=-32$	$v_3=-96$	$a_3=-32$
$t_4=4$	$x_4=144$	$v_4=-128$	$a_4=-32$	$v_4=-128$	$a_4=-32$
$t_5=5$	$x_5=0$	$v_5=-160$	$a_5=-32$	$v_5=-160$	

Next, for computational convenience, let us rewrite
(2.5) in the form

$$(2.6) \qquad v_{k+1} = -v_k + 2(x_{k+1}-x_k)/(\Delta t); \quad k=0,1,2,3,4.$$

Assuming that $v_0=0$ when a particle is dropped from a
position of rest, one finds from (2.6) that

$$v_1 = -v_0 + 2(x_1-x_0)/(\Delta t) = 0+2(384-400)/1 = -32$$

$$v_2 = -v_1 + 2(x_2 - x_1)/(\Delta t) = 32 + 2(336-384)/1 = -64$$

$$v_3 = -v_2 + 2(x_3 - x_2)/(\Delta t) = 64 + 2(256-336)/1 = -96$$

$$v_4 = -v_3 + 2(x_4 - x_3)/(\Delta t) = 96 + 2(144-256)/1 = -128$$

$$v_5 = -v_4 + 2(x_5 - x_4)/(\Delta t) = 128 + 2(0-144)/1 = -160,$$

which are *identical* with the results of column B in Table I, and are recorded in column D.

Next, since x_0 and v_0, but not a_0, are known initially, let us define a_k as the *average* (rather than *instantaneous*) rate of change of velocity with respect to time by the *arithmetic* formula

$$(2.7) \qquad a_k = \frac{v_{k+1} - v_k}{\Delta t} \; ; \qquad k=0,1,2,3,4.$$

From the values v_k just generated, one finds from (2.7) that $a_0 = a_1 = a_2 = a_3 = a_4 = -32$, which are identical with entries in column C of Table I, and are recorded in column E. Formula (2.7) does not allow a determination of a_5 because this would require knowing v_6. Nevertheless, the entries do indicate quite clearly that the acceleration due to gravity is constant, with the value -32.

Now, just because our arithmetic formulas (2.5) and (2.7) have given the same results as (2.3) and (2.4) does not mean that we have, as yet, a formulation which is of physical significance. Indeed, the physical significance of Newtonian mechanics is characterized by the laws of *conservation* of energy, linear momentum, and angular momentum, and by *symmetry*, that is, by the invariance of its laws of motion under fundamental coordinate transformations [1]. Surprisingly enough, our approach

to gravity will also yield conservation and symmetry. We
will, however, confine attention here only to the
conservation of energy, not only for simplicity, but
because of the intimate relationship between energy
conservation and computational stability [25].

For completeness, recall now the fundamental
Newtonian dynamical equation:

$$(2.8) \qquad F = ma,$$

the classical formula for kinetic energy K:

$$(2.9) \qquad K = \frac{1}{2} mv^2,$$

and, for a falling body with $a = -32$, the formula for
potential energy V:

$$(2.10) \qquad V = 32mx.$$

The classical energy conservation law then states that
if K_0 and V_0 are the kinetic and potential energies,
respectively, at time $t_0 = 0$, while K_n and V_n are the
kinetic and potential energies, respectively, at time
$t_n > t_0$, then

$$(2.11) \qquad K_n + V_n \equiv K_0 + V_0,$$

for all $t_n > t_0$.

Now, the data in column A of Table I were obtained
from photographs at the distinct times $t_k = k\Delta t$. For
this reason, we will concentrate only on these times, so
that (2.8)-(2.10) need be considered only as follows:

$$(2.12) \qquad F_k = ma_k; \qquad k=0,1,2,\ldots$$

$$（2.13）\quad K_k = \frac{1}{2}\,m(v_k)^2; \qquad k=0,1,2,\ldots$$

$$（2.14）\quad V_k = 32mx_k; \qquad k=0,1,2,\ldots$$

Next, define the work W_n, $n=1,2,3,\ldots,$ by

$$（2.15）\quad W_n = \sum_{i=0}^{n-1}(x_{i+1}-x_i)F_i.$$

Then, by (2.5), (2.7) and (2.12)

$$W_n = m\sum_{i=0}^{n-1}(x_{i+1}-x_i)\left(\frac{v_{i+1}-v_i}{\Delta t}\right)$$

$$= \frac{m}{2}\sum_{i=0}^{n-1}(v_{i+1}+v_i)(v_{i+1}-v_i)$$

$$= \frac{m}{2}\,v_n^2 - \frac{m}{2}\,v_0^2,$$

so that

$$（2.16）\quad W_n \equiv K_n - K_0, \qquad n=1,2,3,\ldots$$

which, incidentally, is independent of the structure of F. On the other hand, since $a_k \equiv -32$, one has from (2.12) and (2.15) that

$$W_n = -32m\sum_{i=0}^{n-1}(x_{i+1}-x_i) = -32mx_n + 32mx_0,$$

so that, from (2.14)

$$（2.17）\quad W_n = -V_n + V_0, \qquad n=1,2,3,\ldots.$$

Finally, elimination of W_n between (2.16) and (2.17) yields

$$（2.18）\quad K_n + V_n \equiv K_0 + V_0, \qquad n=1,2,3,\ldots$$

in complete analogy with (2.11). Moreover, since K_0 and V_0 are determined from the initial conditions x_0 and v_0, it follows from (2.9), (2.10), (2.13), and (2.14) that $K_0 + V_0$ is the same in both (2.11) and (2.18), so that our strictly arithmetic approach conserves *exactly* the same total energy, *independently* of Δt, as does classical Newtonian theory.

It is also worth noting that in the derivations of (2.1) and (2.17), the telescopic sums

$$\sum_{i=0}^{n-1} (v_{i+1}^2 - v_i^2) \equiv v_n^2 - v_0^2$$

$$\sum_{i=1}^{n-1} (x_{i+1} - x_i) = x_n - x_0$$

play the same roles in the derivations of (2.16) and (2.17) as does integration in the classical continuous derivations.

We want next to extend the ideas of this section to more complex forces than gravity and, for the purpose of modeling, we will want to consider the basic types of forces which actually are present in dynamical interactions. These are of two types, the long range forces, like gravity and gravitation, and the short range forces, like atomic and molecular attraction and repulsion. Recall, then, that from the classical point of view, two molecules attract like, say $1/r^8$, while they repel like, say, $1/r^{12}$, so that molecular repulsion is a much stronger force than is molecular attraction.

3. EXTENSIONS

Recently, arithmetic, conservative formulas have been found for all long and short range forces, and these will now be summarized. In each case the proof of

conservation of energy follows in the same spirit as
that of Section 2, while the proofs of the other conservation laws and of symmetry are to be found in the references ([2]-[9], [21]-[24]).

Consider first the planar motion of a single particle under the influence of gravitation. For this purp-
ose, if $\Delta t > 0$ and $t_k = k\Delta t$, $k=0,1,2,\ldots$, let particle P of mass m be located at $\vec{r}_k = (x_k, y_k)$, have
velocity $\vec{v}_k = (v_{k,x}, v_{k,y})$, and have acceleration $\vec{a}_k =$
$= (a_{k,x}, a_{k,y})$ at time t_k. In analogy with (2.5) and
(2.7), let

$$(3.1) \qquad \frac{\vec{v}_{k+1} + \vec{v}_k}{2} = \frac{\vec{r}_{k+1} - \vec{r}_k}{\Delta t}, \qquad k=0,1,2,\ldots$$

$$(3.2) \qquad \vec{a}_k = \frac{\vec{v}_{k+1} - \vec{v}_k}{\Delta t}, \qquad k=0,1,2,\ldots .$$

To relate force and acceleration at each time t_k, we
assume a discrete Newtonian dynamical equation

$$(3.3) \qquad \vec{F}_k = m\vec{a}_k ,$$

where

$$(3.4) \qquad \vec{F}_k = (F_{k,x}, F_{k,y}).$$

Suppose now that a massive object, like the sun,
whose mass is M, is positioned at the origin of the
XY coordinate system and is assumed to have no motion.
Then, in analogy with the continuous, conservative gravitational force on P, which has components

$$F_x = - \frac{GMm}{r^2} \frac{x}{r}, \qquad F_y = - \frac{GMm}{r^2} \frac{y}{r},$$

where G is a constant, the arithmetic and conservative gravitational force on P has components ([2], [4], [6]):

$$(3.5) \quad F_{k,x} = - \frac{GMm}{r_k r_{k+1}} \cdot \frac{\frac{x_{k+1}+x_k}{2}}{\frac{r_{k+1}+r_k}{2}} = - \frac{GMm(x_{k+1}+x_k)}{r_k r_{k+1}(r_k+r_{k+1})} ,$$

$$(3.6) \quad F_{k,y} = - \frac{GMm(y_{k+1}+y_k)}{r_k r_{k+1}(r_k+r_{k+1})} ,$$

where

$$(3.7) \quad r_k^2 = x_k^2 + y_k^2, \qquad k=0,1,2,\ldots .$$

Now, gravitation is a $1/r^2$ law. Suppose, as in classical molecular mechanics, one would desire an arithmetic and conservative formulation of a $1/r^p$, $p \geq 2$, law of attraction. Then, in this case, (3.5) and (3.6) need be modified only as follows ([4], [8]):

$$(3.9) \quad F_{k,x} = - \frac{GMm[\sum_{j=0}^{p-2} (r_k^j r_{k+1}^{p-j-2})] (x_{k+1}+x_k)}{r_k^{p-1} r_{k+1}^{p-1}(r_{k+1}+r_k)} , \quad G \geq 0 ,$$

while $F_{k,y}$ is the same as $F_{k,x}$ except that x and y are exchanged. In the particular case where $p=2$, (3.9) reduces to (3.5).

In classical molecular mechanics, however, particles attract like $1/r^p$ only when they are relatively far apart. When they are close, they repel like $1/r^q$, $q > p$ [1]. To simulate both these effects, simultaneously, it follows directly from (3.9) that the conservative formulas are

$$(3.10) \quad F_{k,x} = - \frac{GMm[\sum_{j=0}^{p-2} (r_k^j r_{k+1}^{p-j-2})] (x_{k+1}+x_k)}{r_k^{p-1} r_{k+1}^{p-1} (r_{k+1}+r_k)} +$$

$$+ \frac{HMm[\sum_{j=0}^{q-2} (r_k^j r_{k+1}^{q-j-2})] (x_{k+1}+x_k)}{r_k^{q-1} r_{k+1}^{q-1} (r_{k+1}+r_k)} , \quad G \geq 0, \ H \geq 0,$$

while $F_{k,y}$ is the same as $F_{k,x}$ except that x and y are exchanged.

Finally, with regard to the motion of a single particle, it is of interest to note that *all* the arithmetic conservative formulas developed thus far are special cases of the following general formula [21]. For any Newtonian potential $\Phi(r)$, let

$$(3.11) \quad \vec{F}_k = - \frac{\Phi(r_{k+1})-\Phi(r_k)}{r_{k+1}-r_k} \cdot \frac{\vec{r}_{k+1}+\vec{r}_k}{r_{k+1}+r_k} .$$

Arithmetic formula (3.11) conserves exactly the same energy, linear momentum and angular momentum as does its continuous, limiting counterpart

$$(3.12) \quad \vec{F} = -(\frac{\partial \Phi}{\partial r}) \frac{\vec{r}}{r} .$$

Since we have now explored rather completely the motion of a single particle, the next extension is to a system of particles $P_1, P_2, \ldots, P_n$. To do this, let particle P_i of mass m_i be at $\vec{r}_{i,k} = (x_{i,k}, y_{i,k})$, have velocity $\vec{v}_{i,k} = (v_{i,k,x}, v_{i,k,y})$ and have acceleration $\vec{a}_{i,k} = (a_{i,k,x}, a_{i,k,y})$ at time t_k. Position, velocity and acceleration are assumed to be related by

$$(3.13) \qquad \frac{\vec{v}_{i,k+1} + \vec{v}_{i,k}}{2} = \frac{\vec{r}_{i,k+1} - \vec{r}_{i,k}}{\Delta t}$$

$$(3.14) \qquad \vec{a}_{i,k} = \frac{\vec{v}_{i,k+1} - \vec{v}_{i,k}}{\Delta t} \ .$$

If $\vec{F}_{i,k} = (F_{i,k,x}, F_{i,k,y})$ is the force acting on P_i at time t_k, then force and acceleration are assumed to be related by

$$(3.15) \qquad \vec{F}_{i,k} = m_i \vec{a}_{i,k} \ .$$

If, in particular, we assume that all particles interact with all other particles with attraction like $1/r^p$ and repulsion like $1/r^q$, then the arithmetic, conservative force on each P_i, $i=1,2,\ldots,n$, is given, in analogy with (3.10), by [4]:

$$(3.16) \qquad \vec{F}_{i,k} = m_i \sum_{\substack{j=1 \\ j \neq i}}^{n} \left\{ m_j \left\{ - \frac{G[\sum_{\xi=0}^{p-2} (r_{ij,k}^{\xi}\, r_{ij,k+1}^{p-\xi-2})]}{r_{ij,k}^{p-1}\, r_{ij,k+1}^{p-1} (r_{ij,k} + r_{ij,k+1})} \right.\right.$$

$$\left. + \frac{H[\sum_{\xi=0}^{q-2} (r_{ij,k}^{\xi}\, r_{ij,k+1}^{q-\xi-2})]}{r_{ij,k}^{q-1}\, r_{ij,k+1}^{q-1} (r_{ij,k} + r_{ij,k+1})} \right\} \times$$

$$\times\ (\vec{r}_{i,k+1} + \vec{r}_{i,k} - \vec{r}_{j,k+1} - \vec{r}_{j,k}) \bigg\} ,$$

where $G > 0$, $H > 0$, $q > p \geq 2$, and $r_{ij,k}$ is the distance between P_i and P_j at t_k.

It should be noted, in particular, that only for simple forces, like gravity, do the continuous and discrete approaches yield *exactly* the same dynamical behaviour. In general ([23], [24]), the two approaches yield results which differ by terms of order $(\Delta t)^3$.

Recently, new numerical formulas have been developed
([23], [24]) which increase this order of magnitude
difference to any prescribed exponent, but, for these,
the conservation of angular momentum of systems which
have more than one particle is still to be proved [24].

4. DISCRETE MODELS

The arithmetic approach developed thus far lends
itself naturally and consistently to discrete, or part-
icle-type, models of complex physical phenomena. These
have been developed in both the conservative, implicit
fashion ([4], [6]-[9], [22]) and the less expensive,
nonconservative, explicit fashion ([11], [16]-[20]).
Viable discrete models have been developed for vibrating
strings [4]; heat conduction and convection ([4], [7],
[11], [17], [20]); free surface, laminar and turbulent
fluid flows ([9],[11], [16]-[19]); shock wave generation
[4]; interface problems ([17], [18], [20]); and elastic
vibration ([4], [7]). For illustrative purposes, we will
summarize next a variety of computer simulations using
discrete models. Whenever possible, the derived physical
insights and advantages will be described.

Figure 4.1 shows the conservative, elastic vibration of a flexible bar from position of tension [7].
What emerges clearly is that the bar does not swing
smoothly, but flutters up, due to waves which travel
through the bar as part of its gross upward motion.
Engineers have been aware, for some time, of such waves
on the surfaces of vibrating materials.

Figure 4.2 shows how particles of a liquid emerge
from a nozzle at relatively low speeds [9]. Gravity is
not included in this conservative model, although it
can be, because we will be concerned only with the

(g) t=t₁₂₀
(h) t=t₁₄₀
(i) t=t₁₆₀
(j) t=t₁₈₀
(k) t=t₂₀₀
(l) t=t₂₂₀

(a) t=t₀
(b) t=t₂₀
(c) t=t₄₀
(d) t=t₆₀
(e) t=t₈₀
(f) t=t₁₀₀

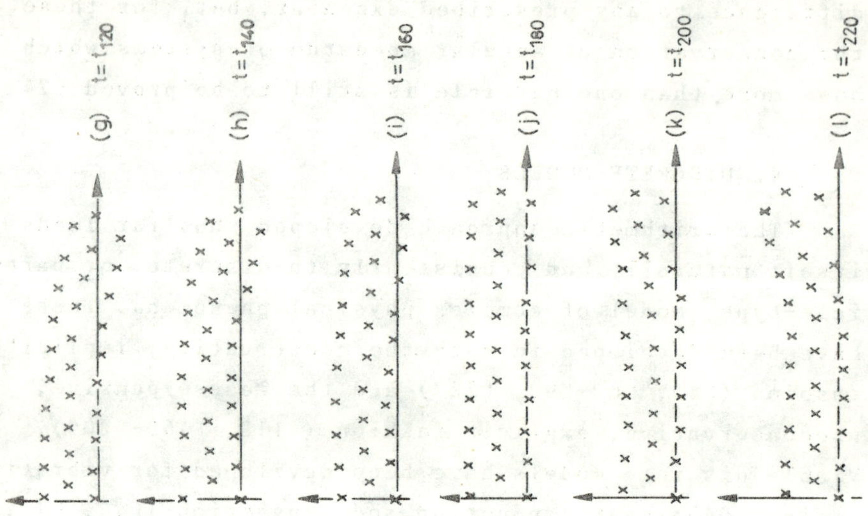

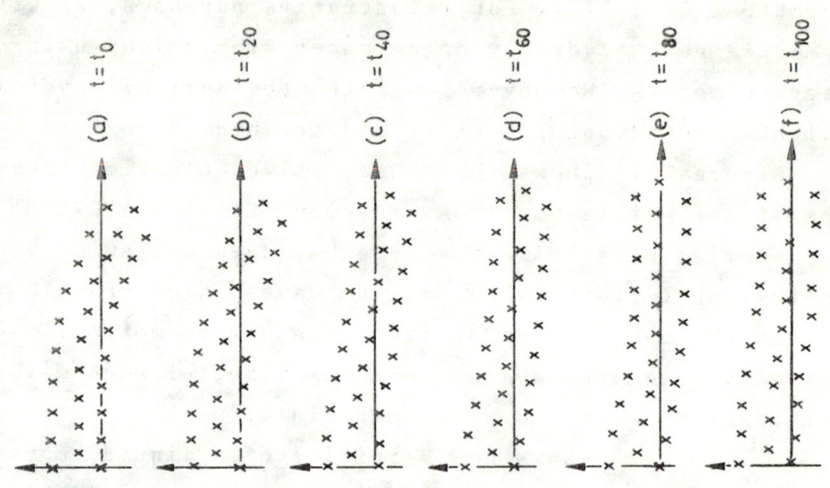

FIGURE 4.1

initial emission phase. The flow is what is usually
called laminar. As the particle velocities are increased
moderately, the rows of particles maintain their relative
positions, as shown in Figure 4.3, but the flow is
becoming relatively chaotic. The disturbance arises from
the increase in velocities, since faster moving particles
can come closer to each other than can more slowly moving
ones, and greater repulsive forces thereby result. Finally, in Figure 4.4, the velocities have been increased
to the point where the rows no longer maintain their
relative positions, and the motion is called turbulent.
Indeed, if a vortex is defined as a set of particles
which are rotating together in a clockwise, or a counter-
clockwise, fashion, as shown in Figure 4.5, then what
we have called turbulent flow exhibits, with time, the
phenomenon of many vortices appearing and disappearing
quickly. Again, this phenomenon is well known to fluid
engineers. It is also rather interesting to note that
the model allows one to transform from laminar flow to
turbulent flow merely by increasing velocities. In
contemporary continuum mechanics, not only is this not
possible, but there is, as yet, no viable model of
realistic turbulence [26].

If one does not have sufficient resources for the
implicit, conservative modeling described thus far,
then one can still formulate and study discrete models
by using explicit formulas, but, of course, one no
longer has exact conservation [4]. Some of these will be
described next.

Figure 4.6 (a)-(d) shows how shock waves can be
generated [4]. In (a) is shown a gas in a long tube. The
gas particles are distributed relatively uniformly. In
(b), a plunger has been inserted into the tube. When the
plunger is moved slowly down the tube, as is shown in

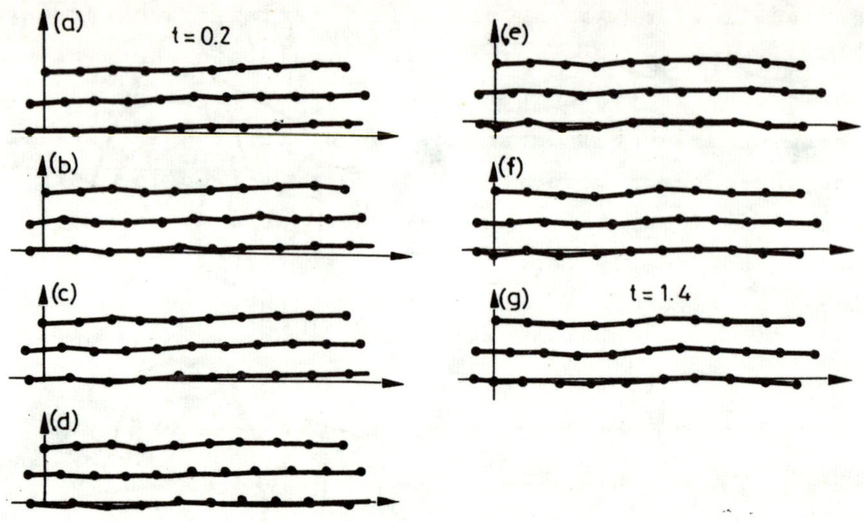

FIGURE 4.2

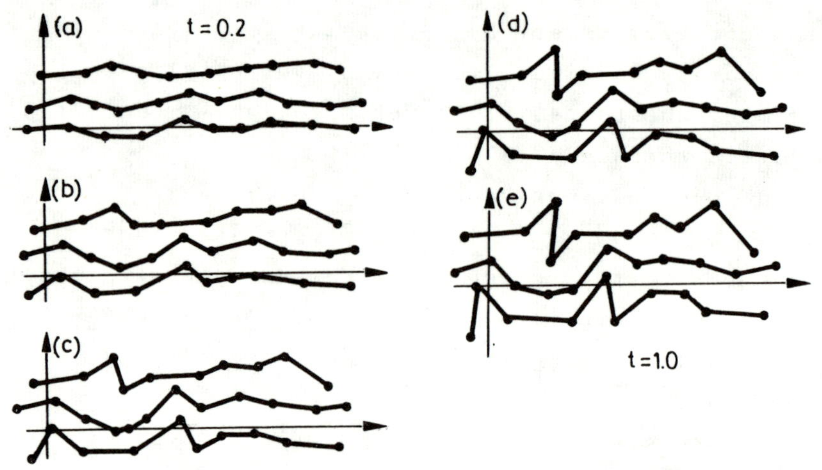

FIGURE 4.3

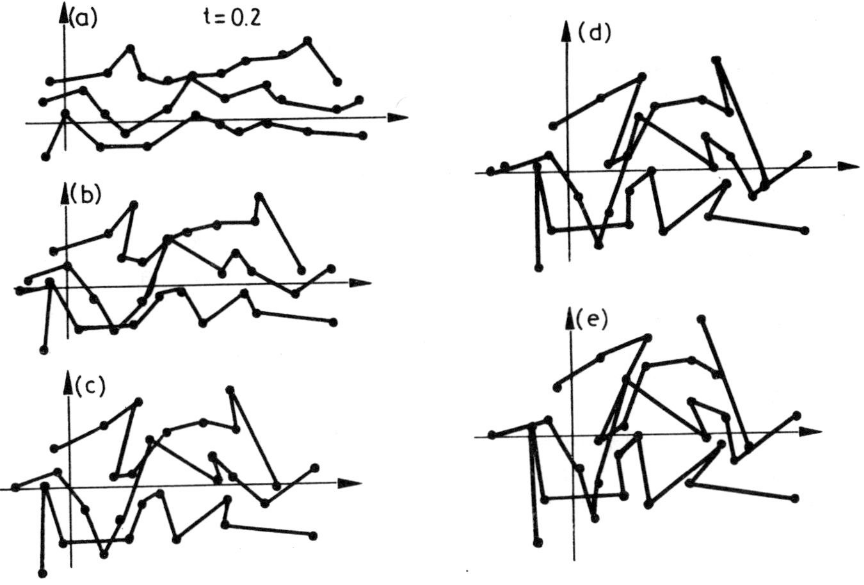

FIGURE 4.4

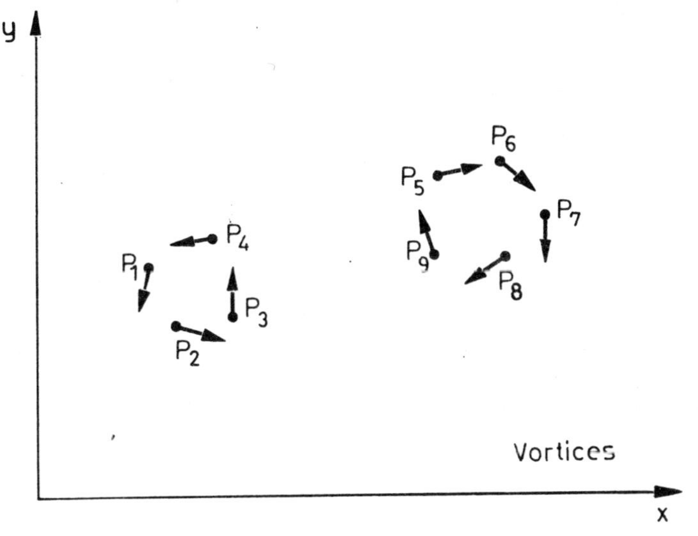

FIGURE 4.5

279

(c), the gas simply reorganizes itself into a new, but
relatively uniform, distribution. However, when the
plunger is moved down the tube at a very high speed, then
gas particles do not have the time to reorganize, and
they pack up on the face of the plunger, as shown in (d).
The gas now separates into two portions, one that has,
approximately, the initial density, and one that is
highly dense and is impacted on the face of the plunger.
The boundary between these two portions is called a
shock wave and the computer generation of "realistic"
shock wave formation is shown in Figure 4.7. We have
used the term "realistic" because this model also includes the heating of the walls of the tube, a phenomenon of
fundamental importance which is usually too difficult
to incorporate into continuous models.

Figure 4.8 shows the vibration of a string under a
nonlinear elastic force [4]. It is important to notice
the emergence of small trailing waves, which are known
to exist, physically, and rarely appear in continuous
models.

Figure 4.9 shows a stable ocean-bay particle model
[19]. An earthquake is simulated by the upward motion of
the right-hand portion of the ocean floor, as shown in
Figure 4.10. The resultant compression of particles
yields large repulsive forces, which generate an upward
moving compression wave, as shown in Figure 4.11, and
which results in ocean waves which later break into the
bay, as shown in Figure 4.12. The mechanism described in
this model for ocean wave motion appears in no continuous
models, because these models assume water to be
incompressible. Note that the model combines gravity and
short range forces.

Figure 4.13 shows a rotating fluid with particles
of three different masses represented, proportionately,

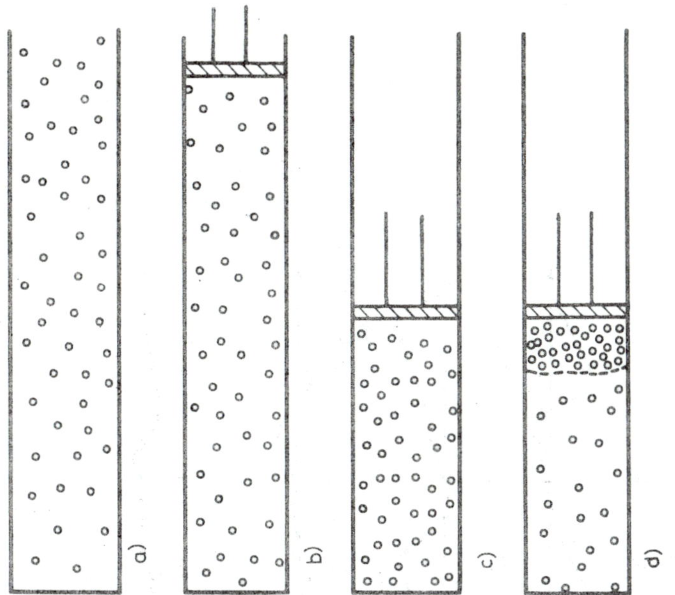

FIGURE 4.6

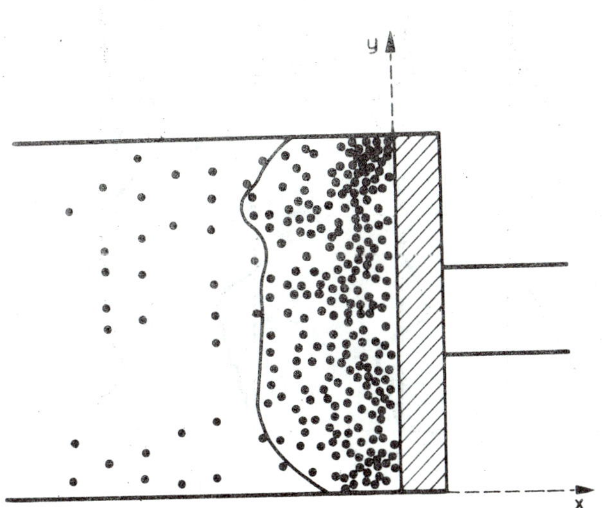

FIGURE 4.7

t= 0.00
0.05
0.10
0.15
0.20
0.25
0.30
0.35
0.40
0.45
0.50
0.55
0.60
0.65
0.70
0.75
1
- 0.3
0.3
-0.6
0.4
-0.7
0.4
-0.3

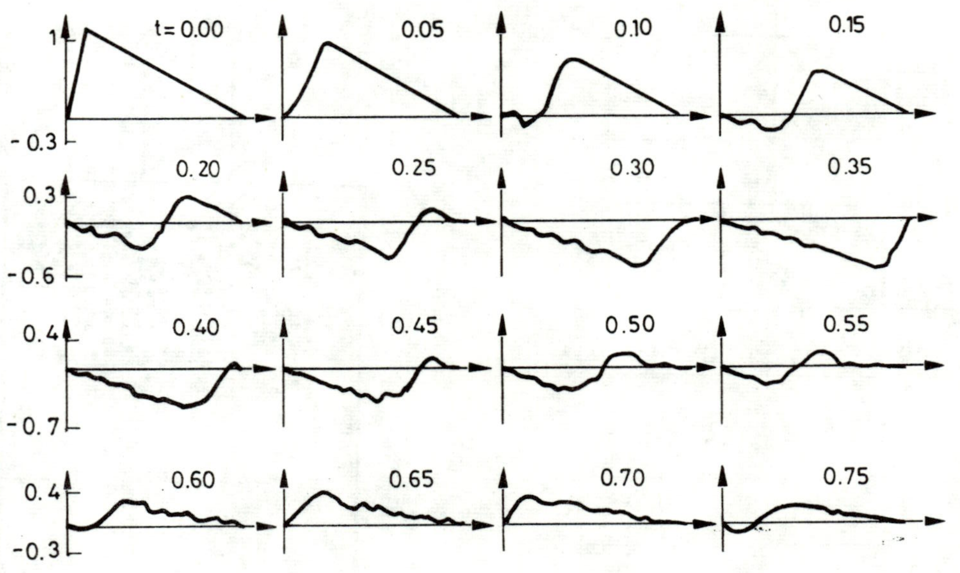

FIGURE 4.8

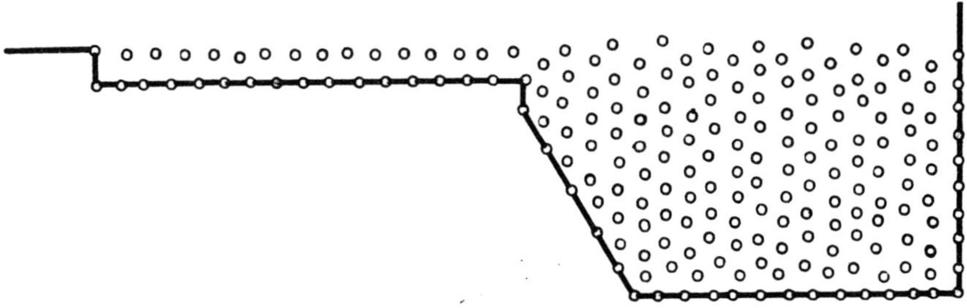

FIGURE 4.9

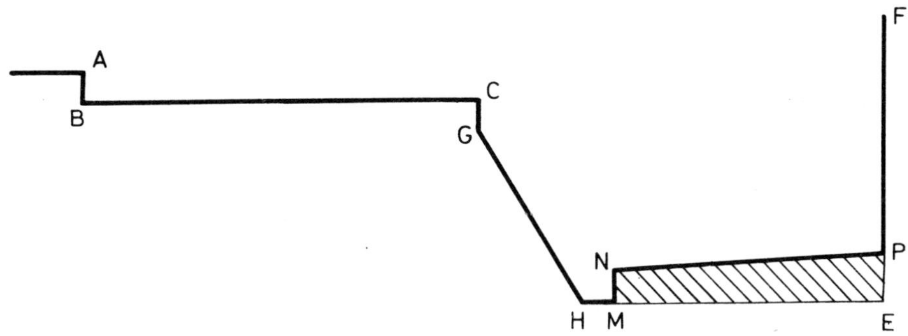

FIGURE 4.10

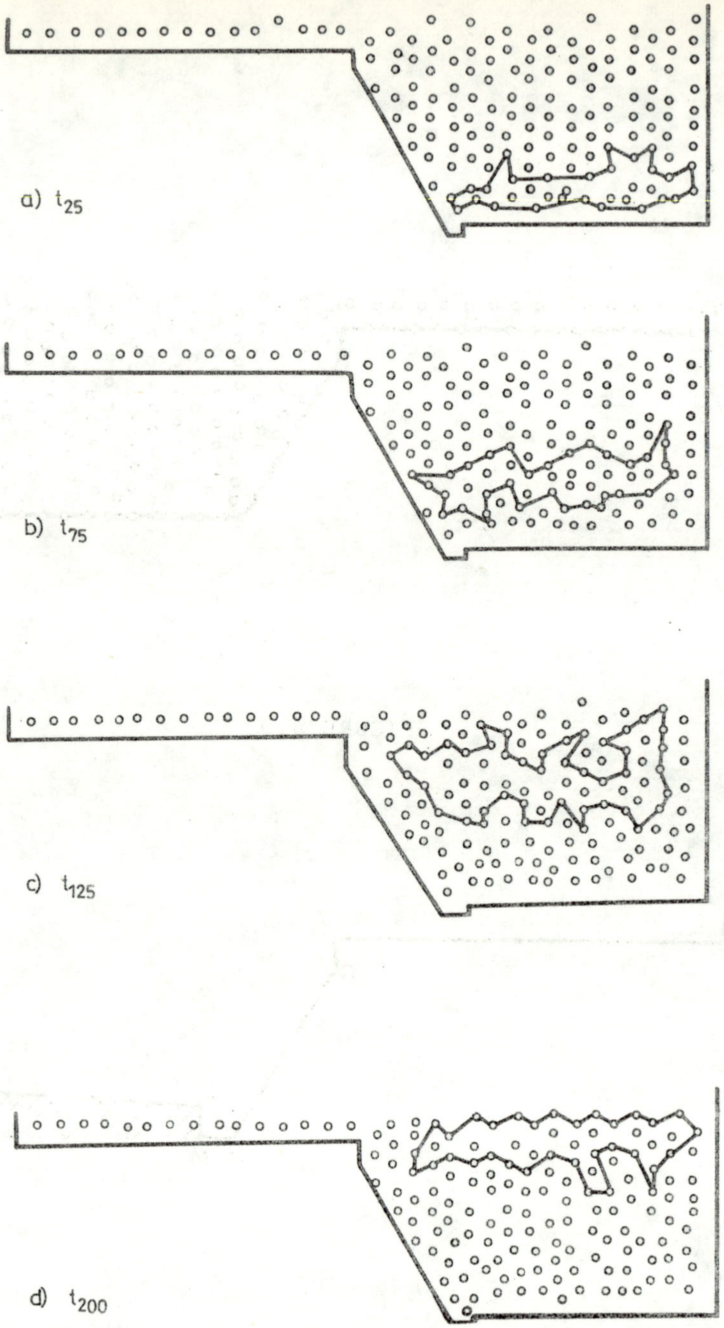

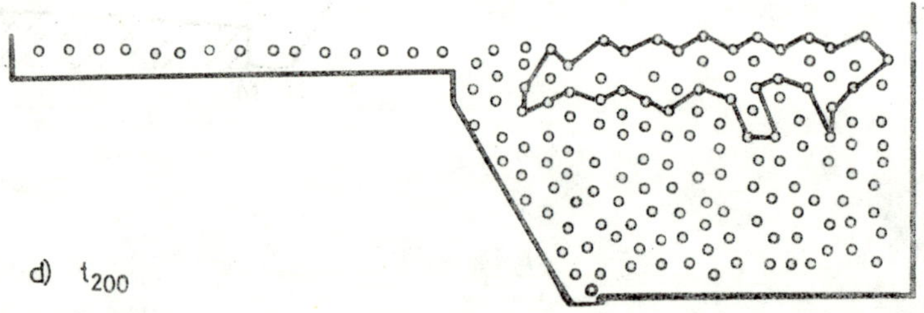

FIGURE 4.11

284

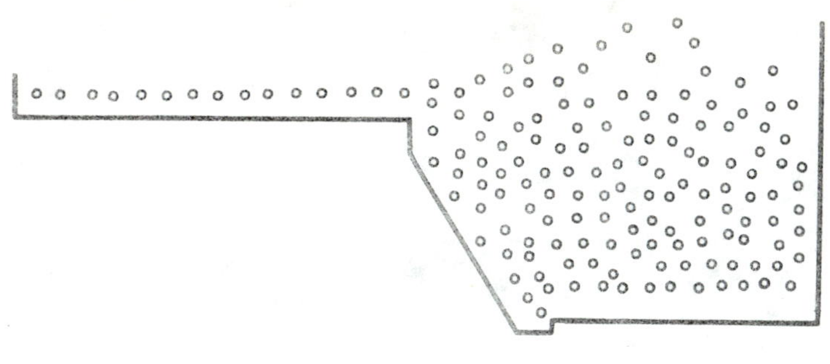

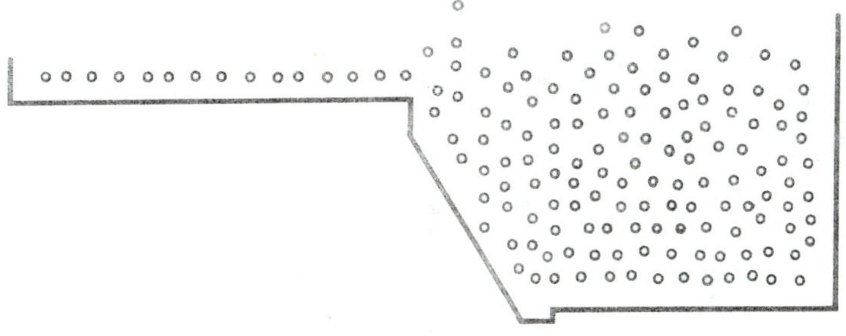

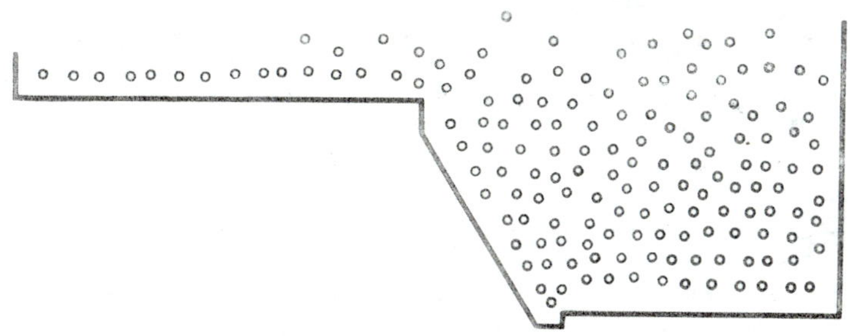

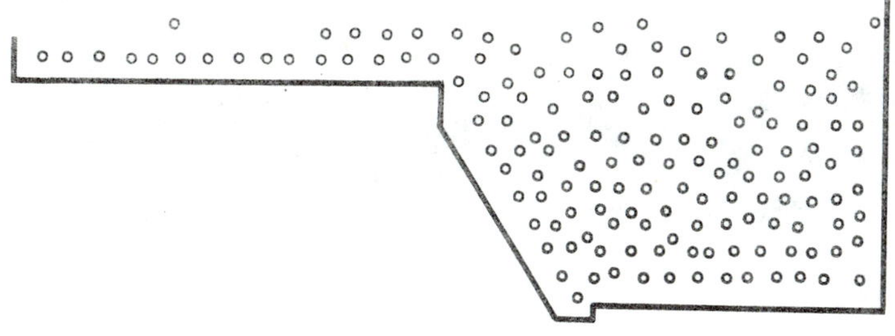

FIGURE 4.12

285

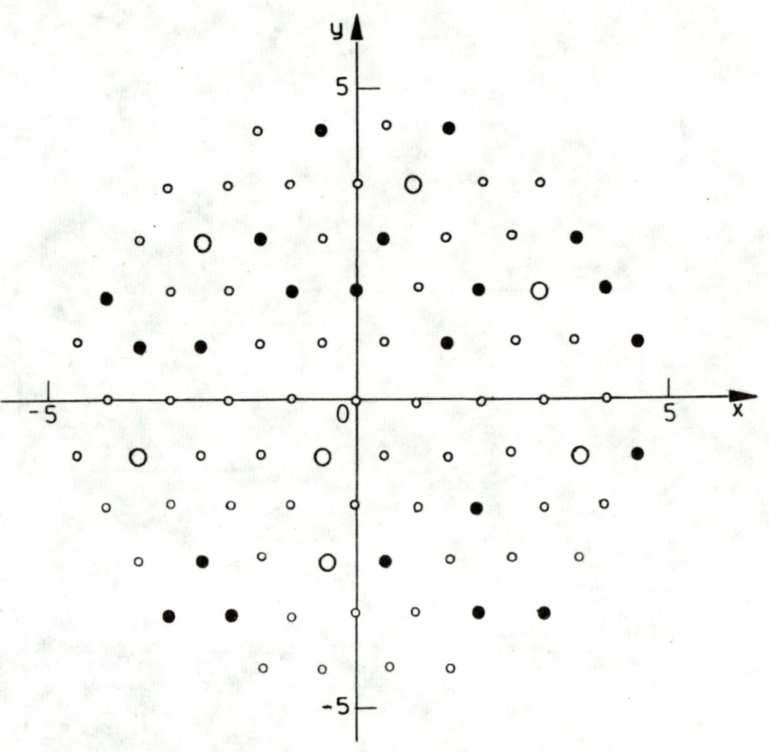

FIGURE 4.13

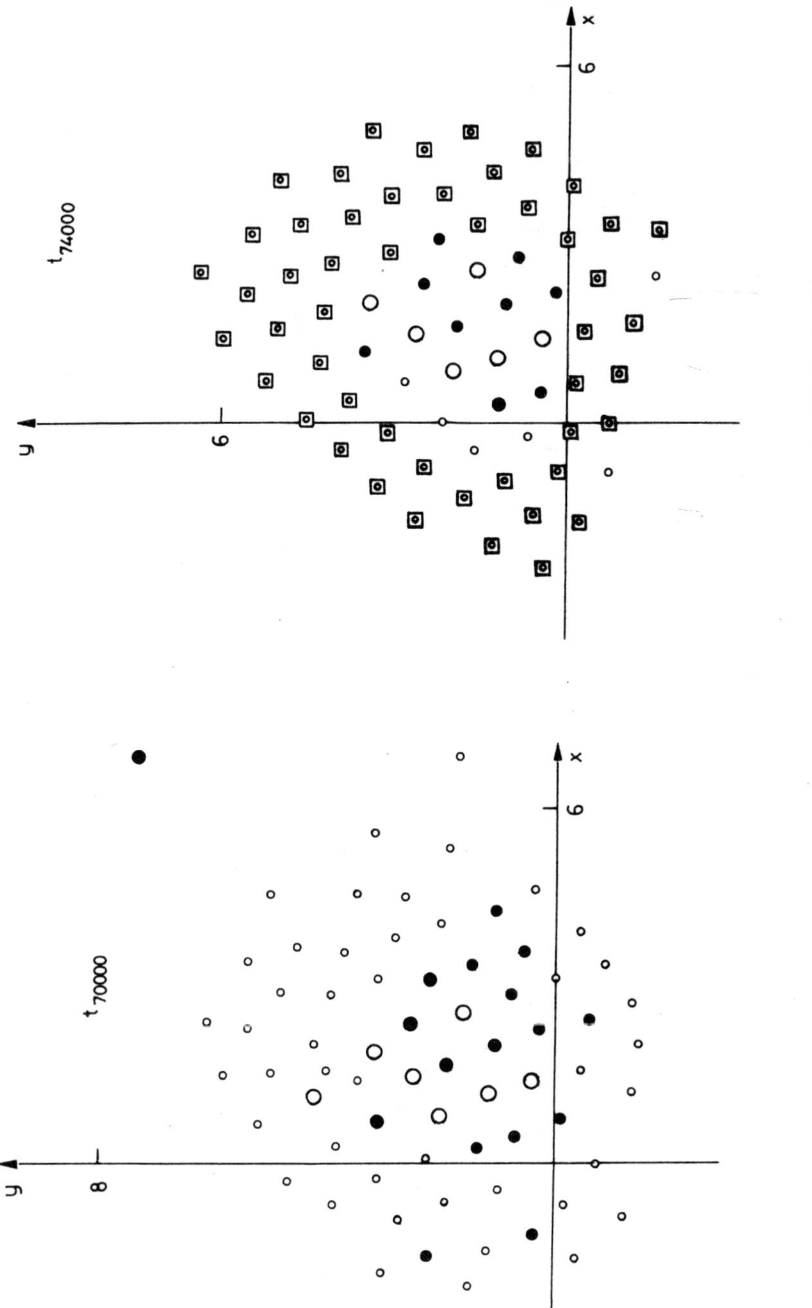

287

by circles of three different radii [18]. Figure 4.14
reveals that, with time, the particles reorganize themselves with the heaviest at the center and the lightest
around the outside. Introducing radiative heat loss at
this point results, as shown in Figure 4.15, in the
formation of a crust, shown by means of squares, around
the outside, while the inner particles remain hot and
fluid. This naive simulation of planetary evolution is
now being refined and extended. Note that the model
combines gravitation and short range forces.

5. SPECIAL RELATIVITY

Finally, let us show that an arithmetic basis
exists also for the other broadly accepted, deterministic
theory of mechanics, that is, special relativity. For
simplicity, we will do this in one dimension only and
for variety will emphasize the symmetry property rather
than energy conservation, which was emphasized for
Newtonian mechanics. The extension to more than one
dimension, the establishment of the conservation laws,
and the proof that $E = mc^2$ all follow from the basic
numerical formulas to be given and from arithmetic
analgoues of various relativistic concepts [12]-[15].

Consider, then, two Euclidean coordinate systems
XYZ and $X'Y'Z'$ which, at some initial time, coincide.
Let the $X'Y'Z'$ system, called the rocket frame, be
in constant uniform motion relative to the XYZ frame,
called the lab frame. Assume that the constant relative
speed is u and that the axes Y and Y' are always
parallel, as are the axes Z and Z'. For $\Delta t > 0$, an
observer in the lab frame makes observations at the
distinct times $t_k = k\Delta t$, $k=0,1,2,\ldots$. Using an
identical, synchronized clock, an observer in the rocket
frame makes observations at the times t'_k , where t'_k

on the rocket clock corresponds to t_k on the lab clock.

Now, if particle P is at (x_k, y_k, z_k) in the lab frame at time t_k, while it is at (x_k', y_k', z_k') in the rocket frame at time t_k', then these variables are related by the linear, algebraic, Lorentz transformation ([27], [28]):

$$(5.1) \qquad x_k' = c(x_k - ut_k)/(c^2 - u^2)^{1/2}$$

$$(5.2) \qquad y_k' = y_k$$

$$(5.3) \qquad z_k' = z_k$$

$$(5.4) \qquad t_k' = (c^2 t_k - ux_k)/[c(c^2 - u^2)^{1/2}],$$

where c is the speed of light.

In the lab frame, let particle P be in motion in the X-direction. Then, at time t_k, P's velocity $v(t_k) = v_k$ and acceleration $a(t_k) = a_k$ are defined by

$$(5.5) \qquad v_k = \Delta x_k/\Delta t_k$$

$$(5.6) \qquad a_k = \Delta v_k/\Delta t_k,$$

where the forward difference operator Δ is defined in usual way by $\Delta f(t_k) = f(t_{k+1}) - f(t_k)$. In the rocket frame, at time t_k', one defines v_k' and a_k' by

$$(5.7) \qquad v_k' = \Delta x_k'/\Delta t_k'$$

$$(5.8) \qquad a_k' = \Delta v_k'/\Delta t_k' .$$

In order to find the relationship between v_k and v_k', and between a_k and a_k', note that (5.1)-(5.4) imply

(5.9) $\Delta x_k' = c(\Delta x_k - u\Delta t_k)/(c^2 - u^2)^{1/2}$

(5.10) $\Delta y_k' = \Delta y_k$

(5.11) $\Delta z_k' = \Delta z_k$

(5.12) $\Delta t_k' = (c^2 \Delta t_k - u\Delta x_k)/[c(c^2 - u^2)^{1/2}]$.

Thus, (5.7), (5.9) and (5.12) imply

(5.13) $v_k' = [c^2(v_k - u)]/(c^2 - uv_k)$,

while (5.8), (5.12), and (5.25) imply

(5.14) $a_k' = \dfrac{c^3(c^2 - u^2)^{3/2}}{(c^2 - uv_k)^2(c^2 - uv_{k+1})} \, a_k$.

Next, we assume that the mass m of particle P depends on its velocity in the following way. In the lab frame, the mass m_k of P at time t_k is assumed to satisfy the relationship

(5.15) $m_k = cm_0/(c^2 - v_k^2)^{1/2}$,

while its mass m_k' at the corresponding time t_k' in the rocket frame is assumed to satisfy

(5.16) $m_k' = cm_0/(c^2 - v_k'^2)^{1/2}$.

In (5.15) and (5.16), m_0 is the rest mass of P and both $|v_k|$ and $|v_k'|$ are assumed to be smaller than c.

We now come to the problem of interest. From the

dynamical point of view, the actual motion of a particle in, say, the lab frame can be determined from (5.5) and (5.6) once an equation which relates force and acceleration is given. We will take this equation to be

$$(5.17) \quad F_k = \frac{c^2 m_k}{[\,(c^2 - v_k^2)(c^2 - v_{k+1}^2)\,]^{1/2}} \frac{\Delta v_k}{\Delta t_k} \ .$$

However, by the principle of relativity [28], in the rocket frame, one must have symmetry, that is:

$$(5.18) \quad F_k' = \frac{c^2 m_k'}{[\,(c^2 - v_k'^2)(c^2 - v_{k+1}'^2)\,]^{1/2}} \frac{\Delta v_k'}{\Delta t_k'} \ .$$

But this is valid only if the right-hand side of (5.18) maps into the right-hand side of (5.17) under the Lorentz transformation. Fortunately, this is correct, since

$$\frac{c^2 m_k'}{[\,(c^2 - v_k'^2)(c^2 - v_{k+1}'^2)\,]^{1/2}} \frac{\Delta v_k'}{\Delta t_k'}$$

$$= \frac{c^3 m_0}{(c^2 - v_k'^2)(c^2 - v_{k+1}'^2)^{1/2}} a_k'$$

$$= \frac{c^3 m_0}{(c^2 - v_k^2)(c^2 - v_{k+1}^2)^{1/2}} a_k$$

$$= \frac{c^2 m_k}{[\,(c^2 - v_k^2)(c^2 - v_{k+1}^2)\,]^{1/2}} \frac{\Delta v_k}{\Delta t_k} \ .$$

Note that taking limits in (5.17) yields the particular form

$$F = \frac{c^2 m}{c^2 - v^2} \frac{dv}{dt}$$

of the classical Einstein equation

$$F = \frac{d}{dt}\,(mv),$$

where

$$m = cm_0/(c^2 - v^2)^{1/2}.$$

Note also that the use of (5.17) in the lab frame and of
(5.18) in the rocket frame implies that the results are
related by the Lorentz transformation [15]. That is,
if one were to install identical computers in the lab
and rocket frames and use the force laws (5.17) and
(5.18), in the respective frames, then all resultant
computations would be related by the Lorentz transformation.

CONCLUDING REMARKS

In this paper we have shown how to reformulate both
Newtonian and special relativistic mechanics using only
arithmetic. Such dramatic simplification is reasonable
and possible only because of the availability of modern
digital computers. The possible additional advantages
which can be derived are varied and extensive. For
example, from the educational point of view, the current
availability of inexpensive computers means that excit-
ing mathematics and physics, considered previously to
be "advanced", can be presented now at a most elementary
level. Indeed, there is not one model described in
Section 4 which cannot be presented after a first course
in trigonometry. From the scientific point of view, the
availability of discrete models provides researchers with
additional tools in their study of natural phenomena,

and the importance of these new cannot be underestimated
at a time when both sub-atomic and cosmic physics are
revealing that Nature is far from the simplistic entity
it was once thought to be.

Also from the scientific point of view, it may seem
questionable that we have emphasized Newtonian models
almost to the exclusion of relativistic and quantum
mechanical models. However, at present there is good
reason for this emphasis, for neither relativity nor
quantum mechanics apply practically to the study of
dynamical interaction, relativity because it precludes
the n-body problem for $n \geq 3$, and quantum mechanics
because it requires a prohibitive number of dimensions
for $n \geq 4$. Thus, Newtonian mechanics, not relativity,
provides the only practical tools for the determination
of astrophysical trajectories, while atomic and molecular
trajectories are often determined now by "quasi" quantum
mechanical methods, which utilize, in part, Newtonian
mechanics.

REFERENCES

[1] R.P. Feynman – R.B. Leighton – M. Sands, *The Feynman
     Lectures on Physics*, vol. I, Addison-Wesley,
     Reading, Mass., 1963.

[2] D. Greenspan, Discrete Newtonian gravitation and
     the n-body problem, *Utilitas Math.*, 2(1972), 105-
     -126.

[3] D. Greenspan, An energy conserving, stable
     discretization of the harmonic oscillator, *Bull.
     Poly. Inst. Iasi*, 17(22), Section 1(1972), 205-209.

[4] D. Greenspan, *Discrete Models*, Addison-Wesley,
     Reading, Mass., 1973.

[5]   D. Greenspan, Symmetry in discrete mechanics, *Foundations of Physics*, 3(1973), 247-253.

[6]   D. Greenspan, Discrete Newtonian gravitation and the three-body problem, *Foundations of Physics*, 4(1974), 299-310.

[7]   D. Greenspan, Discrete bars, conductive heat transfer, and elasticity, *Computers and Structures*, 4(1974), 243-251.

[8]   D. Greenspan, A physically consistent, discrete n-body model, *Bull. Amer. Math, Soc.*, 80(1974), 553-555.

[9]   D. Greenspan, An arithmetic, particle theory of fluid dynamics, *Comp. Meth. Appl. Mech. Eng.*, 3(1974), 293-303.

[10] D. Greenspan, *Discrete Numerical Methods in Physics and Engineering*, Academic, N.Y., 1974.

[11] D. Greenspan, Computer studies of a von Neuman type fluid, Tech. Rpt. 261, Dept. Comp. Sci., Univ. Wisconsin, Madison, 1975, to appear in *Appl. Math. and Comp.*

[12] D. Greenspan, Computer Newtonian and special relativistic mechanics, *Proceedings of the Second USA-Japan Computer Conference*. Amer. Fed. Inf. Proc. Soc., Montvale, N.J., 1975, 88-91.

[13] D. Greenspan, The arithmetic basis of special relativity - I, *Tech. Rpt.* 232, Dept. Comp. Sci., Univ. Wisconsin, Madison, 1975.

[14] D. Greenspan, The arithmetic basis of special relativity - II, *Tech. Rpt.* 240, Dept. Comp. Sci., Univ. Wisconsin, Madison, 1975.

[15] D. Greenspan, Lorentz invariant computations, *Tech. Rpt.* 251, Dept. Comp. Sci., Univ. Wisconsin, Madison, 1975, to appear in *Int. Jour. Theor. Physics* in combination with [13] and [14].

[16] D. Greenspan, Cavity flow of a particle fluid, *Tech. Rpt.* 265, Dept. Comp. Sci., Univ. Wisconsin, Madison, 1976.

[17] D. Greenspan, A particle model of the Stefan problem, *Comp. Meth. Appl. Mech. and Eng.*, 13 (1978), 95-104.

[18] D. Greenspan, Computer studies of interactions of particles with differing masses, *Jour. Comp. Appl. Math.*, 3 (1977), 145-154.

[19] D. Greenspan - M. Cranmer - J. Collier, A particle model of ocean waves generated by earthquakes, *Tech. Rpt.* 277, Dept. Comp. Sci., Univ. Wisconsin, Madison, 1976.

[20] D. Greenspan - M. Rosati, Computer generation of particle solids, *Tech. Rpt.* 273, Dept. Comp. Sci., Univ. Wisconsin, Madison, 1976, to appear in *Computers and Structures*.

[21] R.A. LaBudde - D. Greenspan, Discrete mechanics - a general treatment, *Jour. Comp. Physics*, 15 (1974), 134-167.

[22] R.A. LaBudde - D. Greenspan, Discrete mechanics for nonseparable potentials with application to the LEPS form, *Tech. Rpt.* 210, Dept. Comp. Sci., Univ. Wisconsin, Madison, 1974.

[23] R.A. LaBudde - D. Greenspan, Energy and momentum conserving methods of arbitrary oder for the numerical integration of equations of motion - I, *Numerische Math.*, 25 (1976), 323-346.

[24] R.A. LaBudde - D. Greenspan, Energy and momentum
     conserving methods of arbitrary order for the
     numerical integration of equations of motion - II,
     Numerische Math., 26 (1976), 1-16.

[25] R.D. Richtmyer - K.W. Morton, *Difference Methods
     for Initial-Value Problems*, 2nd ed., Wiley, N.Y.,
     1967.

[26] P.G. Saffman, Lectures on homogeneous turbulence,
     in *Topics in Nonlinear Physics*, Springer-Verlag,
     N.Y., 1968. 485-614.

[27] J.L. Synge, *Relativity: The Special Theory*, North
     Holland, Amsterdam, 1965.

[28] E.F. Taylor - J.A. Wheeler, *Spacetime Physics*, San
     Francisco, 1966.

D. Greenspan
Mathematics Department
University of Texas
Arlington, Texas
USA 76019

ON A COMPUTATIONAL METHOD OF THE TIME-OPTIMAL
CONTROL PROBLEM

ÉVA GYURKOVICS – T. VÖRÖS

1. INTRODUCTION

This paper is concerned with the computation of
time-optimal controls of systems described by a set of
linear differential equations. It is well-known that the
optimal control can be found applying P o n t r i a g i n's
Maximum Principle if a certain solution of the related
adjoint differential equation is known. However, the
initial value for this solution is not given explicitly.
Since the early 1960's, different iterative procedures
for computing this initial value and the optimal time
have been proposed. Here we deal with a method developed
by B a b u n a s h v i l i, M a t s h a i d z e,
K h a r a t i s h v i l i and Z i s k a r i d z e [1].
Section 3. contains a brief review of the method. In
Sec. 4. we present a proof of the convergence of the
method. In Sec. 5 the problems of the approximation
are discussed and the convergence of the difference-
-approximation is proved. In Sec. 6 we refer to a com-

puter program, developed on the basis of this method.

2. STATEMENT OF THE PROBLEM

Consider the time-optimal control of a system described by the differential equation

$$(1) \qquad \frac{dx}{dt} = A(t)x + B(t)u + f(t),$$

where x is the n-dimensional state vector, u is the r-dimensional vector control function, $A(t)$, $B(t)$ are $n \times n$, $n \times r$ matrix functions, respectively, and $f(t)$ is an n-dimensional vector function. $A(t)$, $B(t)$, $f(t)$ are assumed to be piecewise continuous for $t \geq t_0$ on any finite interval.

A control function $u(t)$ is said to be admissible, if it is measurable over any finite interval and takes its values from a given convex, compact set $\Omega \subset R^r$. Let U denote the set of all admissible control functions.

If there are given an initial state

$$(2) \qquad x(t_0) = x_0 \qquad (\neq 0)$$

and a control function $u(t) \in U$, $t_0 \leq t \leq t_1$, then Eq. (1) has a unique solution $x(t)$. The time-optimal control problem is to find an admissible control which turns $x(t)$ into 0 in the shortest possible time t. If an optimal control and an optimal time exist, they will be denoted by $v(t)$ and $t[v]$. (We shall assume that the time-optimal control problem has a solution.)

3. COMPUTATION OF THE OPTIMAL TIME

The following equivalent statement of the time--optimal control problem is well-known: for $t \geq t_0$

find the minimal time $t = t[v]$ for which

(3) $\qquad z(t) \in X(t)$,

where $z(t)$, $X(t)$ are given by the formulae:

$$z(t) = -(x_0 + \int_{t_0}^{t} \Phi^{-1}(s) f(s) ds),$$

(4)

$$X(t) = R(t, U), \quad R(t, u) = \int_{t_0}^{t} \Phi^{-1}(s) B(s) u(s) ds$$

and $\Phi(t)$ is the matrix solution of $d\Phi/dt = A(t)\Phi$, $\Phi(t_0) = I$.

It is known that $X(t)$ is compact and convex in R^n for every $t \geq t_0$ and depends continuously on t.

Let $p(\chi, t)$ be a function defined by

$$p(\chi, t) = \chi X X(t) - \chi z(t), \text{ for } t \geq t_0, \quad \chi \in R^n,$$

where

$$\chi X X(t) = \max_{x \in X(t)} \chi x.$$

It can be easily seen that a necessary and sufficient condition for $z(t')$ to belong to $X(t')$ for some t' is:

$$p(\chi, t') \geq 0 \text{ for every } \chi \in R^n.$$

Function $p(\chi, t)$ is continuous, hence the set

$$V_\chi = \{t: p(\chi, t) < 0\}$$

is open for any $\chi \in R^n$ and it can be composed as follows:

$$V_\chi = \bigcup_{n=1}^\infty I_n, \qquad I_n = (a_n, b_n).$$

Obviously $V_{\alpha\chi} = V_\chi$, if $\alpha > 0$ and real.

Let $\tau \geq t_0$, and let N_τ denote the set of indices m for which

$$a_m < \tau, \quad b_m > t_0, \quad \text{if} \quad \tau > t_0, \quad \text{and}$$

$$a_m \leq \tau, \quad b_m > t_0, \quad \text{if} \quad \tau = t_0.$$

Let $\pi(\chi, \tau)$ be defined for $\chi \in G$ (G – the unit sphere of R^n) and $\tau \geq t_0$ as

$$\pi(\chi, \tau) = \begin{cases} \sup_{m \in N_\tau} b_m, & \text{if} \quad N_\tau \neq \phi \\ \tau - p(\chi, \tau), & \text{if} \quad N_\tau = \phi. \end{cases}$$

The role of the function $\pi(\chi, \tau)$ is shown by the following fact [1].

If it is known that $z(t) \notin X(t)$, $t_0 \leq t < \tau_0$ and for some $\chi \in G$

$$\tau_1 = \pi(\chi, \tau_0) > \tau_0,$$

then

$$z(t) \notin X(t), \qquad t_0 \leq t < \tau_1.$$

Thus, knowing an approximation t_k to $t[v]$ ($t_k < t[v]$), one can find a better approximation t_{k+1} by maximization of $\pi(\chi, t_k)$ with respect to $\chi \in G$. Taking t_0 as starting-point and finding succeeding approximations as mentioned above, a sequence $\{t_n\}_{n=0}^\infty$ is obtained which tends to the optimal time $t[v]$.

This method has the disadvantage that it requires the maximization of a function of n variables in each step. Reference [1] proposes a change in the rule of choosing τ and computing $\pi(\chi,\tau)$ by means of which the result can be found by a single maximization. Practically, the function $\pi(\chi,\tau)$ is computed in each step in the course of maximization for the value of τ which equals the largest value of π found until that time.

4. CONVERGENCE

One has to prove that the procedure of maximization performed by the modified rule of computing function π, gives a sequence that converges to the optimum.

Let $\{\chi_i\}_{i=0}^{\infty}$, $\chi_i \in G$ and t_0 be given. Assign the sequence $\{\tau_i\}_{i=0}^{\infty}$ to them as

$$\tau_0 = t_0, \qquad\qquad t_1 = \pi(\chi_0,\tau_0),$$

$$\tau_1 = \max\{t_1,\tau_0\}, \qquad t_2 = \pi(\chi_1,\tau_1),$$

(5)

$$\cdots\cdots\cdots\qquad\cdots\cdots\cdots$$

$$\tau_n = \max\{t_n,\tau_{n-1}\}, \qquad t_{n+1} = \pi(\chi_n,\tau_n),$$

$$\cdots\cdots\cdots\qquad\cdots\cdots\cdots$$

The sequence $\{\tau_i\}_{i=0}^{\infty}$ is a nondecreasing and bounded with respect to $\{\chi_i\}_{i=0}^{\infty}$ as a consequence of the assumption that there exists a solution of the problem in Sec. 2. Therefore $\lim_{i\to\infty} \tau_i$ exists and it is less then some C for any $\{\chi_i\}_{i=0}^{\infty}$.

DEFINITION. The $\{\chi_i^*\}_{i=0}^{\infty}$ is said to be a maximizing sequence if for the corresponding $\{\tau_i^*\}_{i=0}^{\infty}$ the equality

$$\tau^* = \lim_{i \to \infty} \tau_i^* = \sup_{\{\chi_j\}_{j=0}^{\infty} \subset G} (\lim_{i \to \infty} \tau_i) = \tau^0$$

holds.

LEMMA 1. *There exists a maximizing sequence.*

PROOF. Let

$$(6) \qquad \{\chi_i^{(k)}\}_{i=0}^{\infty} \ , \qquad k=0,1,\dots$$

be a sequence for which the corresponding sequences $\{\tau_i^{(k)}\}_{i=0}^{\infty}$, give limits $\tau^{(k)}$ for $k=0,1,\dots$, tending to τ^0 as $k \to \infty$. Let $\{\chi_i^*\}_{i=0}^{\infty}$ denote a sequence formed from (6) in the following manner:

$$
\begin{array}{cccc}
\chi_0^{(0)} \to & \chi_1^{(0)} & \chi_2^{(0)} & \cdots \\
& \swarrow \quad \nearrow & & \\
\chi_0^{(1)} & \chi_1^{(1)} & \chi_2^{(1)} & \cdots \\
\downarrow \quad \nearrow & & & \\
\chi_0^{(2)} & \chi_1^{(2)} & \chi_2^{(2)} & \cdots
\end{array}
$$

Obviously $\lim_{i \to \infty} \tau_i^* \leq \tau^0$. On the other hand, $\{\tau_i^*\}_{i=0}^{\infty}$ has a subsequence $\{\tau_{i_k}^*\}_{k=0}^{\infty}$ for which $\tau_{i_k}^* \geq \overline{\tau}_k$, where $\overline{\tau}_k = \max\{\tau_i^{(j)} : i+j=k\}$. This follows by induction from the fact, that for any χ and $\tau' < \tau''$, $\pi(\chi,\tau') \leq \pi(\chi,\tau'')$. Since $\overline{\tau}_k \to \tau_0$ if $k \to \infty$, one has

$$\tau^0 = \lim_{k \to \infty} \overline{\tau}_k \leq \lim_{k \to \infty} \tau_{i_k}^* = \lim_{i \to \infty} \tau_i^* \ ,$$

thus $\tau^0 = \tau^*$, i.e. $\{\chi_i^*\}_{i=0}^{\infty}$ is a maximizing sequence.

THEOREM 1. *Let τ^* be determined by a maximizing sequence. Then $\tau^* = t[v]$ is the optimal time of the control problem.*

PROOF. Suppose the contrary. Then only the inequality $\tau^* < t[v]$ can hold, hence $z(t) \notin X(t)$ for $t \in [t_0, \tau^*]$. Therefore there exists a χ^0, for which

$$-\delta = p(\chi^0, \tau^*) < 0$$

with some positive δ. Since $p(\chi, t)$ is continuous, there exists a neighbourhood W of (χ^0, τ^*) $(W = W_{\chi^0} \times (\tau', \tau''), \ \tau^* \in (\tau', \tau''))$ such that for any $(\chi, \tau) \in W$

$$|p(\chi, \tau) - p(\chi^0, \tau^*)| < \frac{\delta}{2} ,$$

i.e. $p(\chi, \tau) < -\frac{\delta}{2}$. Furthermore, there exists an n_0, such that $\tau_n^* \in (\tau', \tau'')$ for $n \geq n_0$. Consider any $\overline{\chi} \in W_{\chi^0}$, and let

$$\widetilde{\chi}_i = \begin{cases} \chi_i^*, & i < n_0, \\ \overline{\chi}, & i \geq n_0. \end{cases}$$

From $\pi(\overline{\chi}, \tau_{n_0}) > \tau'' > \tau^* = \tau^0$ it follows that $\lim_{t \to \infty} \widetilde{\tau}_i > \tau^0$ for $\{\widetilde{\chi}_i\}_{i=0}^{\infty}$, which contradicts the definition of τ^0.

Following Ref. [1], the system described by Eq. (1) will be called normal (nondegenerated) if for any $\chi \in R^n$, $\chi \neq 0$ the control function $u_0(t)$ is uniquely determined by Pontriagin's Maximum Principle [2]

$$(7) \qquad \max_{u \in \Omega} \chi \Phi^{-1}(t) B(t) u = \chi \Phi^{-1}(t) B(t) u_0(t)$$

almost everywhere over any interval of finite length.
Obviously $\{\chi_i^*\}_{i=0}^{\infty}$ has a subsequence $\{\chi_{i_k}^*\}_{k=0}^{\infty}$, for which

$$(8) \qquad p(\chi_{i_k}^*, \tau_{i_k+1}^*) = 0.$$

Since G is compact in R^n, there is a convergent subsequence in $\{\chi_{i_k}^*\}_{k=0}^{\infty}$. Denote it by $\{\chi_{i_k}^*\}_{k=0}^{\infty}$, too, and let $\chi_{i_k}^* \to \chi^*$.

THEOREM 2. *Suppose that system (1) is normal. Then*
(a) the control function $u_0^*(t)$ determined from (7) with
   $\chi = \chi^*$ over $[t_0, t[v]]$ is optimal;
(b) if $u_{i_k}(t)$ is determined from (7) with $\chi = \chi_{i_k}^*$
   over $[t_0, \tau_{i_k+1}^*]$, and

$$u_{i_k}^*(t) = \begin{cases} u_{i_k}(t), & t_0 \leq t \leq \tau_{i_k+1}^*, \\ \overline{u}, & \tau_{i_k+1}^* < t \leq t[v], \end{cases}$$

where $\overline{u} \in \Omega$, then $u_{i_k}^*(t) \to u_0^*(t)$, weakly in
$L^2(t_0, t[v])$ as $k \to \infty$;
(c) the solutions $x_{i_k}(t)$ of Eq. (1) with $u(t) =$
   $= u_{i_k}^*(t)$ tend uniformly on $[t_0, t[v]]$ to $x^*(t)$.

PROOF. (a) In consequence of the choice of $\{\chi_{i_k}^*\}_{k=0}^{\infty}$ and the continuity of $p(\chi, t)$ it is evident that $p(\chi^*, t[v]) = 0$, which implies that

$$\chi^* z(t[v]) = \int_{t_0}^{t[v]} \chi^* \Phi^{-1}(s) B(s) u_0^*(s) ds.$$

An optimal control $v(t)$ exists and it satisfies the

relation

$$\chi^* z(t[v]) = \int_{t_0}^{t[v]} \chi^* \Phi^{-1}(s)B(s)v(s)ds.$$

From the normality of the system and from the inequality

$$\chi^* \Phi^{-1}(t)B(t)u_0^*(t) = \max_{u \in \Omega} \chi^* \Phi^{-1}(t)B(t)u \geq$$

$$\geq \chi^* \Phi^{-1}(t)B(t)v(t)$$

one concludes that

$$u_0^*(t) \equiv v(t)$$

must hold a.e. in $[t_0, t[v]]$.

(b) The set of admissible controls over $[t_0, t[v]]$ is weakly compact [4]. Therefore from $\{u_{i_k}^*(t)\}_{k=0}^{\infty}$ one can choose a weakly convergent subsequence; denote it also by $\{u_{i_k}^*(t)\}_{k=0}^{\infty}$ and the corresponding subsequences of $\{\chi_{i_k}^*\}$, $\{\tau_{i_k}^*\}$, in the same way. Let $u_{i_k}^*(t) \to u^*(t)$, $k \to \infty$, where $\to$ denotes the weak convergence on $[t_0, t[v]]$. One has to prove that $u^*(t) = u_0^*(t)$ a.e. in $[t_0, t[v]]$. Making use of the definition of $p(\chi, t)$ and (8), one gets

$$\chi_{i_k}^* z(\tau_{i_k+1}^*) = \int_{t_0}^{\tau_{i_k+1}^*} \chi_{i_k}^* \Phi^{-1}(s)B(s)u_{i_k}^*(s)ds =$$

$$(9) \qquad = \int_{t_0}^{t[v]} (\chi_{i_k}^* - \chi^*)\Phi^{-1}(s)B(s)u_{i_k}^*(s)ds - \int_{\tau_{i_k+1}^*}^{t[v]} \chi_{i_k}^* \Phi^{-1}(s)B(s)\bar{u}ds +$$

$$+ \int_{t_0}^{t[v]} \chi^* \Phi^{-1}(s)B(s)u_{i_k}^*(s)ds.$$

The left-hand side of (9) tends to $\chi^* z(t[v])$, the first two terms on the right-hand side tend to zero, and the third term to $\int_{t_0}^{t[v]} \chi^* \Phi^{-1}(s) B(s) u^*(s) ds$. Therefore it follows in the same way as in (a) that $u^*(t) = u_0^*(t)$ a.e. in $[t_0, t[v]]$. On the other hand, $u_0^*(t)$ is uniquely determined, thus the entire original sequence $\{u_{i_k}^*(t)\}_{k=0}^{\infty}$ converges weakly to $u_0^*(t)$.

(c) It is known [3] that if a sequence of controls converges weakly to the optimal control then the sequence of the corresponding trajectories tend uniformly to the optimal trajectory, which completes the proof of Theorem 2.

REMARK. It can be easily seen that $p(\chi, t)$ can be calculated by solving the system of differential equations

$$\frac{d\psi}{dt} = -\psi A(t),$$

(10)

$$\frac{dm}{dt} = \psi f(t) + \max_{u \in \Omega} (\psi B(t) u),$$

with the initial values $\psi(t_0) = \chi$, $m(t_0) = \chi x_o$. Then the obtained $m(t)$ equals $p(\chi, t)$.

5. THE APPROXIMATION OF THE METHOD

In computational realizations, all the functions mentioned above are known only approximatively. In view of the fact that $\pi(\chi, \tau)$ must be computed by finding certain zeros of the function $p(\chi, t)$ that is known with some error, it is clear, that the convergence of the approximation of the method is not self-evident.

First of all, it should be observed that for any fixed T the following estimation holds:

$$|p(\chi,t)-p(\chi,s)| \leq L(T)|t-s|, \quad \chi\in G, \quad t,s\in[t_0,T],$$

where $L(T) = c_1 e^{c_2(T-t_0)}$, $c_1,c_2 > 0$. This follows immediately from (10) and the conditions on $A(t)$, $B(t)$, $f(t)$, U.

Similarly as in Ref. [5], it can be proved that choosing a fine enough subdivision of any finite interval $[t_0,T]$ the approximate value p_χ^n of $p(\chi,t_n)$ can be obtained by Euler's method with the required accuracy. Consider $\varepsilon > 0$, $\chi\in G$, and a subdivision B for which $|p(\chi,t_n)-p_\chi^n| < \varepsilon$. If $p_\chi^n < -2\varepsilon$, then $z(t_n)\notin X(t_n)$. Let a sequence $\{\delta_i\}_{i=0}^\infty$ be given, where $\delta_i > 0$, $\delta_{i+1} < \delta_i/4$, $i=0,1,\dots$. For any $\{\chi_i\}_{i=0}^\infty$ there exists a sequence of subdivisions $\{B_i\}_{i=0}^\infty$ which satisfies the following conditions:

(A) $t_{n,i} \to \infty$, $n \to \infty$, where $t_{n,i}$ denotes the n-th mesh-point of B_i;

(B) $(t_{n,i}-t_{n-1,i})L(t_{n,i}) \leq \dfrac{\delta_i}{4}$;

(C) all points of discontinuity of $A(t)$, $B(t)$ and $f(t)$ belong to the set of mesh-points of any B_i;

(D) $|p(\chi_i,t_{n,i})-p_{\chi_i}^n| < \dfrac{\delta_i}{4}$, if $t_{n,i}\in[t_0,T]$,

where T is sufficiently large.

Introduce the following notations:

$$I_1^{\chi_i} = \{j : p_{\chi_i}^j < -\delta_i\}, \quad I_2^{\chi_i} = \{j : p_{\chi_i}^j \geq -\delta_i\},$$

$$
N'_{t_{n,i}} = \begin{cases} \left\{ j : \begin{array}{l} k, \quad 0 \le k \le n, \\ k,k+1,\ldots,j \in I_1^{\chi_i}, \ j+1 \in I_2^{\chi_i} \end{array} \right\}, & n \ne 0, \\[3mm] \left\{ j : \ 0,1,\ldots,j \in I_1^{\chi_i}, \ j+1 \in I_2^{\chi_i} \right\}, & n = 0, \end{cases}
$$

furthermore let

$$
\pi_{\chi_i}^{t_{n,i}} = \begin{cases} \displaystyle\sup_{j \in N'_{t_{n,i}}} t_{j+1,i} & \text{if } N'_{t_{n,i}} \ne \phi, \\[3mm] t_{n,i} - p_{\chi_i}^n, & \text{if } N'_{t_{n,i}} = \phi. \end{cases}
$$

Concerning this construction, the following lemma will be needed.

LEMMA 2. *Suppose that* $z(t) \notin X(t)$, $t_0 \le t < t_{n,i}$ *and for* χ_i

$$
(11) \qquad t_{m,i} = \pi_{\chi_i}^{t_{n,i}} > t_{n,i}
$$

holds. Then $z(t) \notin X(t)$, $t_0 \le t < t_{m,i}$.

PROOF. From (11) it follows that $N'_{t_{n,i}} \ne \phi$, i.e. there exist indices k and m, such that $0 \le k \le n < m$ and $k, k+1,\ldots,m-1 \in I_1^{\chi_i}$. In consequence of the definition of $I_1^{\chi_i}$ and the assumptions (B) and (D), this means that

$$
p(\chi_i,t) < 0, \qquad t_0 \le t < t_{m,i}.
$$

Therefore the statement of this lemma is valid.

On the basis of $\pi_{\chi_i}^{t_{n,i}}$, a sequence $\{\tau'_i\}_{i=0}^{\infty}$ can be defined in the same way as in (5). Let us assume that

τ'_i belongs to the mesh-points of B_i. (If it is not
the case, let τ'_i be joined to B_i.)

The sequences $\{\tau'_i\}_{i=0}^{\infty}$ are uniformly bounded, thus
one can talk about maximizing sequences.

To show the existence of the maximizing sequence
for this construction, one has to prove that $\pi_{\chi_i}^{t_{n,i}}$
keeps its monotone character in the following sense. If
$\chi = \chi_i = \chi_k$, $i < k$ and $\tau'_i < \tau'_k$, then $\tau'_{i+1} \leq \tau'_{k+1}$.
The cases of $\pi_{\chi_i}^{\tau'_i} \leq \tau'_i = \tau'_{i+1}$ and $\pi_{\chi_k}^{\tau'_k} \leq \tau'_k = \tau'_{k+1}$ or
$\pi_{\chi_i}^{\tau'_i} \leq \tau'_i = \tau'_{i+1}$ and $\tau'_{k+1} = \pi_{\chi_k}^{\tau'_k} > \tau'_k$ are obvious.
Suppose that $\tau'_{i+1} = \pi_{\chi_i}^{\tau'_i} > \tau'_i$, $\pi_{\chi_k}^{\tau'_k} \leq \tau'_k = \tau'_{k+1}$ and
$\tau'_{i+1} > \tau'_{k+1}$. For B_i there exists a subinterval
$I = [t_{n,i}, t_{n+1,i}]$ such that $\tau'_k \in I$ and $p_{\chi_i}^{t_{n,i}} < -\delta_i$. In
consequence of (B) and (D) it follows that

$$\left| p(\chi_i, \tau'_k) - p_{\chi_i}^{t_{n,i}} \right| < \frac{\delta_i}{2}, \text{ i.e. } p(\chi_i, \tau'_k) < -\frac{\delta_i}{2}.$$

On the other hand, $p_{\chi_k}^{\tau'_k} > -\delta_k > -\frac{\delta_i}{4}$ because of the
assumption that $\pi_{\chi_k}^{\tau'_k} \leq \tau'_k$. Thus $p_{\chi_k}^{\tau'_k} - p(\chi_k, \tau'_k) > \frac{\delta_i}{4} > \delta_k$
follows, which contradicts to (D), i.e. $\tau'_{k+1} \geq \tau'_{i+1}$.
The case of $\tau'_{i+1} = \pi_{\chi_i}^{\tau'_i} > \tau'_i$ and $\tau'_{k+1} = \pi_{\chi_k}^{\tau'_k} > \tau'_k$ can
be handled analogously.

The proof of the existence of the maximizing
sequence can be performed in the same way as in Lemma 1.
An equivalent of Theorem 1 is valid, which is formulated
in the next theorem.

THEOREM 3. *Let $\tau^{*\prime}$ be determined by a maximizing
sequence which is constructed by the aid of the function*

$\pi_{\chi_i}^{t_{n,i}}$. *Then* $\tau^{*\prime} = t[v]$, *where* $t[v]$ *is the optimal*

time of the control problem.

PROOF. It is fully analogous to that of Theorem 1, hence it is omitted.

Define a sequence of controls by

$$(12) \qquad u_i^*(t) = \begin{cases} u_i^n, & t \in [\, t_{n,i}, t_{n+1,i} \,) \subset [\, t_0, \tau_{i+1}^* \,), \\ \overline{u}, & t \in [\, \tau_{i+1}^*, t[v] \,], \end{cases}$$

where $u_i^n \in \Omega$ for which

$$\max_{u \in \Omega} \psi_{\chi_i^*B_{n,i}}^n u = \psi_{\chi_i^*B_{n,i}}^n u_i^n , \quad \overline{u} \in \Omega \quad \text{arbitrary.}$$

Because of the compactness of G, a subsequence $\{\chi_{i_k}^*\}_{k=0}^{\infty}$ convergent to some $\chi^{*\prime}$ can be chosen from $\{\chi_i^*\}_{i=0}^{\infty}$ and from it another subsequence, which will be denoted also by $\{\chi_{i_k}^*\}_{k=0}^{\infty}$ and for which

$$(13) \qquad p_{\chi_{i_k}^*}^{\tau_{i_k}^{*\prime}} < -\delta_{i_k}, \qquad p_{\chi_{i_k}^*}^{\tau_{i_k}^{*\prime}+1} \geq -\delta_{i_k}.$$

THEOREM 4. *Suppose that system* (1) *is normal. Then*
(a) the control function $u_0^*(t)$ *determined from* (7)
with $\chi = \chi^{*\prime}$ *over* $[\, t_0, t[v] \,]$ *is optimal,*
(b) the sequence $\{u_{i_k}^*(t)\}_{k=0}^{\infty}$ *given in* (12) *with*
$\{\chi_{i_k}^*\}_{k=0}^{\infty}$ *converges weakly in* $L^2(t_0, t[v])$ *to* $u_0^*(t)$
as $k \to \infty,$
(c) if $x_{i_k}(t)$ *denotes the approximation, by Euler's*
method, of the solution of Eq. (1) *taken with* $u_{i_k}^*(t),$

then $\{x_{i_k}(t)\}_{k=0}^{\infty}$ tends uniformly to the optimal trajectory.

PROOF. (a) Because of (D) and (13), the inequalities

$$p(\chi_{i_k}^{*},\tau_{i_k}^{*\prime}) < -\frac{3}{4}\delta_{i_k}, \qquad p(\chi_{i_k}^{*},\tau_{i_k}^{*\prime}+1) > -\frac{5}{4}\delta_{i_k}$$

hold. Since $\chi_{i_k}^{*} \to \chi^{*\prime}$, $\tau_{i_k}^{*\prime} \to t[v]$, $\delta_{i_k} \to 0$ and $p(\chi,\tau)$ is continuous, it follows that $p(\chi^{*\prime},t[v]) = 0$. Thus (a) can be shown in the same way as part (a) of Theorem 2.

(b) The control functions $u_{i_k}^{*}(t)$ are admissible, thus in consequence of weak compactness of U one can choose a weakly convergent subsequence. Denote it by $\{u_{i_k}^{*}\}_{k=0}^{\infty}$, too, and let $u_{i_k}^{*}(t) \to u^{*\prime}(t)$, $k \to \infty$. The approximate value of $p_{\chi_{i_k}^{*}}^{t_n}$, computed by Euler's method, can be given as follows:

$$p_{\chi_{i_k}^{*}}^{t_{n,i_k}} = \chi_{i_k}^{*}x_0 + \sum_{j=0}^{n-1}(t_{j+1,i_k}-t_{j,i_k})[\psi_{\chi_{i_k}^{*}}^{j}f^{j} +$$

(14)

$$+ \psi_{\chi_{i_k}^{*}}^{j}B^{j}u_{i_k}^{j}],$$

where $f^{j} = f(t_{j,i_k}+0)$, $B^{j}=B(t_{j,i_k}+0)$.

Let $t_{n,i_k} = \tau_{i_k}^{*\prime}+1$. A simple computation shows that the right-hand side in (14) tends to

$$\chi^{*\prime}x_0 + \int_{t_0}^{t[v]}\psi^{*}(s)f(s)ds + \int_{t_0}^{t[v]}\psi^{*}(s)B(s)u^{*}(s)ds,$$

and the left-hand side tends to

$$p(\chi^{*\prime}, t[v]) = \chi^{*\prime} x_0 + \int\limits_{t_0}^{t[v]} \psi^{*}(s) f(s)\, ds +$$

$$+ \int\limits_{t_0}^{t[v]} \psi^{*}(s) B(s) u_0^{*}(s)\, ds, \quad \text{as} \quad k \to \infty.$$

Therefore statement (b) of this theorem follows in the same way as in Theorem 2.

(c) Using the results of Ref [5] and the convergence of Euler's method, the proof of (c) is trivial.

6. COMPUTATIONAL EXPERIMENTS

On the basis of this method, a system of programs was developed in the Computing Centre for Universities for handling the following cases.

1. The set Ω is a convex polyhedron:

$$\Omega = \{u \in R^r : \sum_{k=1}^{r} \alpha_{i,k} u_k \leq \beta_i, \quad i = 1, 2, \ldots, s\}.$$

In this case the maxima in (7) and (10) can be found by the well-known methods of linear programming.

2. The set Ω is a parallelepiped:

$$\Omega = \{u \in R^n : \alpha_i \leq u_i \leq \beta_i, \quad i = 1, 2, \ldots, r\}.$$

In this case the required maxima can be found by means of a simple transformation taking the absolute values of the appropriate functions in (7) and (10).

In both cases the following possibilities were considered:

(a) $A(t)$, $B(t)$, $f(t)$ are arbitrary piecewise continuous functions;

(b) A, B are constant matrices and $f(t)$ is a piecewise continuous vector function;

(c) A, B are constant matrices and $f(t) \equiv 0$.

3. A, B, f are constant, the eigenvalues of matrix A are known and the set Ω is the same as in 1, or in 2. Then the solution of (10), i.e. the function $p(\chi,t)$ can be given explicitly by means of Hermite's interpolational polynomials. This yields a program that works much more quickly than in the previous cases.

A more detailed description of the algorithm and the programs can be found in [6].

REFERENCES

[1] T.G. Babunashvili - Z.A. Mathshaidze - G.L. Kharatishvili - K.S. Ziskaridze, Computation of time-optimal control of linear system (In Russian), *Trudy Inst. Prikl. Mat. T.G.U.*, 3(1972), 207-238.

[2] L.S. Pontriagin - V.G. Boltianskii - R.V. Gamkrelidze - E.F. Mishchenko, *The Mathematical Theory of Optimal Processes*, John Wiley and Sons Inc., 1962.

[3] F.P. Vasilyev, *Lectures on the Numerical Methods of Extremal Problems*, (In Russian), Moscow State University, 1974.

[4] E.B. Lee - L. Marcus, *Foundations of Optimal Control Theory*, John Wiley and Sons Inc., 1967.

[5] B.M. Budak - F.P. Vasilyev, *Some Computational Aspects of the Optimal Control Problem* (In Russian) Moscow State University, 1975.

[6] E. Gyurkovics - T. Vörös, Some computational algorithms of linear time-optimal control systems (In Russian), *Series ESzK T.20.*, Budapest, 1977.

Éva Gyurkovics - T. Vörös
Computing Centre for Universities
Budapest, Dimitrov tér 8.
H-1098
Hungary

ACCELERATED OVERRELAXATION METHOD

A. HADJIDIMOS

1. INTRODUCTION

For the numerical solution of linear systems numerous direct as well as indirect methods exist. Among the indirect or iterative methods the Successive-Overrelaxation (SOR) and related methods play a very important role and are the most popular ones. These methods are fully covered in the excellent books by V a r g a [3], by W a c h s p r e s s [4], and in the most recent one by Y o u n g [5].

The purpose of this paper is to present a two-parameter generalization of the SOR method and also the first basic results concerning this method which has been called Accelerated Overrelaxation (AOR) method. As will be seen the well known methods of Jacobi, of Gauss-Seidel, of Simultaneous Overrelaxation and of Successice Overrelaxation can be derived, as special cases, from the AOR method.

2. DERIVATION OF THE AOR METHOD

We consider a system of N linear equation with N unknows written in matrix form

(2.1) $Ax = b,$

where the matrix A has nonvanishing diagonal elements and $\det(A) \neq 0$. These two assumptions will be considered throughout this paper even when they are not mentioned. We also consider the following splitting of A

(2.2) $A \equiv D - A_L - A_U ,$

where D is a diagonal matrix and A_L and A_U are strictly lower and upper triangular matrices respectively.

For the numerical solution of equation (2.1) we propose to use the most general linear stationary iterative scheme whose matrix coefficients are linear functions of the components of A and the coefficient of the new iterate is an at most lower triangular matrix. This scheme must be of the form

$$(\alpha_1 D + \alpha_2 A_L) x^{(n+1)} = (\alpha_3 D + \alpha_4 A_L + \alpha_5 A_U) x^{(n)} + \alpha_6 b \mid n =$$

$$= 0, 1, 2, \ldots,$$

where $\alpha_i \mid i = 1(1)6$ are constants to be determined $(\alpha_1 \neq 0)$ and $x^{(0)}$ an arbitrary initial approximation to the solution x of (2.1). By dividing through by α_1 we obtain

$$(2.3) \quad (D + \alpha_2' A_L) x^{(n+1)} = (\alpha_3' D + \alpha_4' A_L + \alpha_5' A_U) x^{(n)} + \alpha_6' b \mid n =$$

$$= 0, 1, 2, \ldots,$$

where we have set $\alpha_i' = \alpha_i/\alpha_1 | i = 2(1)6$. Sufficient conditions for scheme (2.3) to be consistent with equation (2.1) are

$$(2.4) \quad (1-\alpha_3')D+(\alpha_2'-\alpha_4')A_L-\alpha_5'A_U \equiv \alpha_6'A, \quad \alpha_6' \neq 0.$$

In view of (2.2), the first relationship of (2.4) gives

$$1-\alpha_3' = \alpha_6', \quad \alpha_2'-\alpha_4' = -\alpha_6' \quad \text{and} \quad -\alpha_5' = -\alpha_6'.$$

The above set of equations has the following two-parameter solution

$$\alpha_2' = -r, \quad \alpha_3' = 1-\omega, \quad \alpha_4' = \omega-r, \quad \alpha_5' = \omega \quad \text{and}$$

$$\alpha_6' = \omega,$$

where r and $\omega \neq 0$ are any two fixed parameters. Consequently (2.3) becomes

$$(2.5) \quad (I-rL)x^{(n+1)} = [(1-\omega)I+(\omega-r)L+\omega U]x^{(n)}+\omega c \,|\, n =$$
$$= 0,1,2,\ldots,$$

where we have set

$$(2.6) \quad L = D^{-1}A_L, \quad U = D^{-1}A_U, \quad c = D^{-1}b$$

and I the identity matrix of order N. Method (2.5) is what is called from now on Accelerated Overrelaxation (AOR) method or $M_{r,\omega}$ -method. We observe that for specific values of the parameters r and $\omega M_{r,\omega}$ -method reduces to well known methods. Thus

$M_{0,1}$ -method is the Jacobi method

$M_{1,1}$-method is the Gauss-Seidel method

$M_{0,\omega}$-method is the Simultaneuous Overrelaxation method and

$M_{\omega,\omega}$-method is the Successice Overralaxation method.

From now on we shall call r the acceleration parameter and ω the overrelaxation parameter and shall use the notations $L_{r,\omega}$ for the iterative matirx of scheme (2.5) given by

$$L_{r,\omega} = (I-rL)^{-1}[(1-\omega)I+(\omega+r)L+\omega U]$$

and $\rho(L_{r,\omega})$ for the spectral radius of $L_{r,\omega}$.

It should be noted that, except for the case $r = 0$, the AOR method is essentially the Extrapolated (E)SOR method with overrelaxation parameter r and extrapolation one $s = \omega/r$, for it is easy to show that

$$L_{r,\omega} = sL_{r,r} + (1-s)I.$$

Thus if ν is an eigenvalue of $L_{r,r}$ $(r \neq 0)$ and λ the corresponding one of $L_{r,\omega}$ we have that

$$(2.7) \qquad \lambda = s\nu+(1-s).$$

In what follows we shall try to give, under various assumptions on the original matrix A, the restrictions imposed on the parameters r and ω so that the $M_{r,\omega}$-method converges. The proofs for the corresponding theorems are given in [2].

3. IRREDUCIBLE MATRICES WITH WEAK DIAGONAL DOMINANCE

If A is an irreducible matrix with weak diagonal dominance then it will be nonsingular with nonvanishing diagonal elements. In such a case the following theorem can be proved.

THEOREM. *If A is an irreducible matrix with weak diagonal dominance then the $M_{r,\omega}$ -method converges for all $0 \leq r \leq 1$ and $0 < \omega \leq 1$.*

Considering now the $M_{r,\omega}$ -method corresponding to the pairs $(r, \omega) = (0, 1)$, $(1, 1)$, $(0, \omega)$ and (ω, ω) we can obtain, as a corollary of the previous theorem, the following well known statement.

COROLLARY. *If A is an irreducible matrix with weak diagonal dominance then the methods of Jacobi, of Gauss--Seidel, of Simultaneous Overrelaxation and of Successive Overrelaxation, the last two methods for $0 < \omega \leq 1$, converge.*

4. L-MATRICES

If A is a L-matrix that is a matrix whose elements $a_{ij} | i, j = 1(1)N$ satisfy the relationships

$$a_{ii} > 0 | i = 1(1)N \quad \text{and} \quad a_{ij} \leq 0 | i \neq j, \; i, \; j = 1(1)N$$

then the following theorem concerning the AOR method can be proved.

THEOREM. *If A is a L-matrix then for all r and ω such that $0 \leq r \leq \omega \leq 1$ $(\omega \neq 0)$ the $M_{r,\omega}$ -method* *converges if and only if the $M_{0,1}$ -method converges.*

5. CONSISTENTLY ORDERED MATRICES

In this section we assume that matrix A is a consistently ordered one that is a matrix for which the expression $\det(\alpha A_L + \alpha^{-1} A_U - \beta D)$ is independent of α for $\alpha \neq 0$ and for all β. As can be easily found out the analysis of this section also applies in the case where A is a matrix which has property A.

Before we begin our analysis we give three lemmas which will be very useful in the sequel.

LEMMA 1. *If A is a consistently ordered matrix with nonvanishing digaonal elements then if $\mu \neq 0$ is an eigenvalue of $L_{0,1}$ of multiplicity p then $-\mu$ is also an eigenvalue of $L_{0,1}$ of multiplicity p.*

LEMMA 2. *If A is a consistently ordered matrix with nonvanishing digaonal elements then if μ is an eigenvalue of $L_{0,1}$ and ν satisfies*

$$(5.1) \qquad (\nu - 1 + r)^2 = r^2 \mu^2 \nu$$

then ν is an eigenvalue of $L_{r,r}$ and vica versa.

LEMMA 3. *If β and γ are real then both roots of the quadratic equation $\lambda^2 - \beta\lambda + \gamma = 0$ are less than one in modulus if and only if*

$$(5.2) \qquad |\gamma| < 1, \quad |\beta| < 1 + \gamma.$$

Having stated the three lemmas above we give in the sequel three theorems concerning the AOR method.

THEOREM 1. *If A is a consistently ordered matrix
with nonvanishing diagonal elements then if μ is an eigenvalue
of $L_{0,1}$ and λ satisfies*

$$(5.3) \qquad (\lambda-1+\omega)^2 = \omega\mu^2[r(\lambda-1)+\omega]$$

then λ is an eigenvalue of $L_{r,\omega}$ and vica versa.

THEOREM 2, *If A is a consistency ordered matrix
with nonvanishing diagonal elements and if $L_{0,1}$ has
real eigenvalues $\mu_i|i = 1(1)N$, with $\underline{\mu} = \min_i|\mu_i|$ and
$\overline{\mu} = \max_i |\mu_i|$, then the $M_{r,\omega}$ -method converges if and
only if the $M_{0,1}$ -method coverges and the parameters
ω and r take values from the intervals I_ω and I_r
respectively defined as follows for $\underline{\mu} \neq 0$:*

$$I_\omega \equiv (-2/(1-\underline{\mu}^2)^{1/2}, 0) \quad and \quad I_r \equiv (\beta(\underline{\mu}^2), \alpha(\overline{\mu}^2))$$

$$or \quad I_\omega \equiv (0,2] \quad and \quad I_r \equiv (\alpha(\overline{\mu}^2), \beta(\overline{\mu}^2)) \quad or$$

$$I_\omega \equiv [2,2/(1-\underline{\mu}^2)^{1/2}) \quad and \quad I_r \equiv (\alpha(\overline{\mu}^2), \beta(\underline{\mu}^2))$$

while

for $\underline{\mu} = 0$:

$$I_\omega \equiv (0,2) \quad and \quad I_r \equiv (\alpha(\overline{\mu}^2), \beta(\overline{\mu}^2)),$$

where

$$\alpha(z) \equiv \frac{1}{\omega z} (\tfrac{1}{2}\omega^2 z - \tfrac{1}{2}\omega^2 + 2\omega - 2) \quad and \quad \beta(z) \equiv \frac{1}{z}(\omega z - \omega + 2).$$

THEOREM 3. *If A is a consistently ordered matrix
with nonvanishing diagonal elements and if $L_{0,1}$ has
real eigenvalues $\mu_i|i = 1(1)N$ such that $0 < \mu = \underline{\mu} =$*

$$= \min_i |\mu_i| = \bar{\mu} = \max_i |\mu_i| < 1 \quad then \ for \quad (r,\omega) =$$
$$= (2(1+(1-\mu^2)^{1/2})\mu^2, \quad -1/(1-\mu^2)^{1/2}) \quad or \quad (2/(1+(1-\mu^2)^{1/2}),$$
$$1/(1-\mu^2)^{1/2}), \quad \rho(L_{r,\omega}) = 0.$$

NOTE: Since, as we have just seen that, when $\mu = \bar{\mu}$, $\rho(L_{r,\omega})$ can be made zero, a value far better than the corresponding one for the optimum SOR which is $\rho(L_{\omega,\omega})$ $= (\mu/(1+(1-\mu^2)^{1/2}))^2$, it is strongly suggested that because of the continuity and at least in cases where μ is very close to $\bar{\mu}$ we shall be able to find an optimum AOR method which will be better than the corresponding optimum SOR one.

It should be added that Dr. Avdelas and the author have investigated further the rpevious suggestion and have found out so far that if $1-\mu^2 \leq (1-\bar{\mu}^2)^{1/2}$ then for $(r,\omega) = (2/(1-\bar{\mu}^2)^{1/2}, \quad (1-\underline{\mu}^2-(1-\bar{\mu}^2)^{1/2})/ /(1-\mu^2)(1-(1-\bar{\mu}^2)^{1/2}))$, the corrrespoinding AOR method converges faster than the optimum SOR one. As soon as this investigation has been completed we shall give a complete set of the optimum results in forthcoming paper [1].

6. FINAL REMARKS

As has been seen the AOR or ESOR (for $r \neq 0$) method can be proved to be a very simple and powerful technique for solving linear system of equations. Its powerfulness compared with the other well known methods (e.g. the SOR method) lies in the fact that two parameters instead of usually at most one are present. Full exploitation of the presence of these two parameters will provide us with methods which will converge faster than any other method of the same type. The determination of optimum acceleration and overrelaxation parameters is a matter which needs further investigation.

REFERENCES

[1] G. Avdelas - A. Hadjidimos, Optimum Accelerated
     Overrelaxation Method in a Special Case, T.R. No.9,
     Dept. Math., Univ.of Ioannina,Ioannina, Greece, 1978.

[2] A. Hadjidimos, Accelerated Overrelaxation Methods,
     Mathematics of Computation, 32 (1978), 149-157.

[3] R.S. Varga, *Matrix Iterative Analysis*, Prentice Hall,
     Englewood Cliffs, New Jersey, 1962.

[4] E.L. Wachpress, *Iterative Solution of Elliptic Sys-
     tems and Applications to the Newtron Diffusion
     Equations of Reactor Physics*, Prentice Hall, Engle-
     wood Cliffs, New Jersey, 1966.

[5] D.M. Young, *Iterative Solution of Large Linear
     Systems*, Academic Press, New York and London, 1971.

A. Hadjidimos
Department of Mathematics,
University of Ioannina,
Ioannina,
Greece

COLLOQUIA MATHEMATICA SOCIETATIS JÁNOS BOLYAI
22. NUMERICAL METHODS, KESZTHELY (HUNGARY), 1977.

A RECURSIVE COMPUTATION OF THE DETERMINANT OF A BANDTRIANGULAR MATRIX

L. HALADA

1. INTRODUCTION

A simple and well-known 3-term recursive relation holds for the computation of all leading principal minors of a tridiagonal matrix. Similar recursive forms for the computation of the leading principal minors of a pentadiagonal matrix have been already derived by some authors, too.

The first algorithm of this kind has been described in [1]. According to the author it could be arranged into a very complicated 7-term recursive relation. The algorithm described in [2] is already simpler and it is composed of some 7-term and two 3-term recursive relations. The last algorithm has been proposed for a pentadiagonal matrix including only ones in the last over diagonal. It has been published in the paper [3] and it is composed of a twofold computation of a simple 5-term recursive relation.

In this paper a new recursive algorithm for the

computation of leading principal minors and determinant of a bandtriangular matrix, particularly a band matrix with arbitrary bandwidth, is proposed.

For the case of a pentadiagonal matrix, this algorithm requires less arithmetical operations than the previously mentioned ones.

THE BASIC RESULT

Let A be a real n by n matrix. Denote its columns $C_1, \ldots, C_n$ and let $C_i = (a_{1i}, \ldots, a_{ni})^T$, $i = n+1, \ldots, 2n$ be arbitrary non-zero n-dimensional vectors.

DEFINITION 1. We shall say that the matrix $A^{(k_1, \ldots, k_r)}$ is the r-times modified matrix A if it originates from matrix A by omitting columns $C_{k_1}, \ldots, \ldots, C_{k_r}$ and by adjoining columns $C_{n+1}, \ldots, C_{n+r}$ in the given sequence, after the last column of matrix A, where $1 \le r \le n$ and k_i, $i = 1, \ldots, r$ are mutually different integers satisfying $1 \le k_i \le n$.

The following relations obviously hold:

$$\left[A^{(k_1)} \right]^{(k_2)} = A^{(k_1, \, k_2)},$$

$$A^{(k_1, \ldots, k_r)} = A^{(k_{i_1}, \ldots, k_{i_r})},$$

where $(i_1, \ldots, i_r)$ is an arbitrary permutation of the numbers $(1, \ldots, r)$.

Let us further denote the leading principal minors of matrices A and $A^{(k_1, \ldots, k_r)}$ with M_j and $M_j^{(k_1, \ldots, k_r)}$, respectively. Then the following assertion holds:

THEOREM. *Let A be a real n by n matrix and k,q integers satisfying $1 \leq k < q \leq n$. Then if $M_{j-1}^{(k,q)} \neq 0$, we have*

$$(1) \qquad M_j = \frac{M_{j-1}^{(k)} M_j^{(q)} - M_{j-1}^{(q)} M_j^{(k)}}{M_{j-1}^{(k,q)}} \qquad j=q,q+1,\ldots,n.$$

PROOF. Let k, q be numbers satisfying the condition of the Theorem and let $j \geq q$. Let the square matrix B of order $(2j-1)$ be defined as

$$B = \begin{bmatrix} S_1 & \cdots & S_k & \cdots & S_q & \cdots & S_j \\ 0 & \cdots & 0 \; S_k & 0\ldots0 & S_q & 0\ldots0 \\ 0 & \cdots & 0 \; a_{jk} & 0\ldots0 & a_{jq} & 0\ldots0 \end{bmatrix}$$

$$\begin{matrix} 0 & \ldots\ldots\ldots\ldots\ldots\ldots\ldots\ldots\ldots\ldots\ldots & 0 & S_{j+1} \\ S_1 & \cdots \; S_{k-1} \; S_{k+1} \; \cdots \; S_{q-1} \; S_{q+1} \; \cdots \; S_j & S_{j+1} \\ a_{j1} & a_{jk-1} \; a_{jk+1} \cdots \; a_{jq-1} a_{jq+1} \cdots a_{jj} & a_{jj+1} \end{matrix} \Bigg] ,$$

where S_i represents the $(j-1)$-dimensional vector including the first $j-1$ elements of the vector c_i, $i=1,\ldots,n$ and 0 stands for the zero $(j-1)$-dimensional vector in the 1st and 2nd row.

Then $\det(B) = 0$. To prove it add the $(j+p)$-th column of B to the p-th column for $p=1,\ldots,k-1$; the $(j+p)$-th column to the $(p+1)$-st column for $p=k,\ldots,q-2$; the $(j+p)$-th column to the $(p+2)$-nd column for $p=q-1,\ldots$ $\ldots,j-1$ and subtract the m-th row from the $(m+j-1)$-th row for $m=1,\ldots,j-1$. After these re-arrangements when obtaining the determinant by expanding via the Laplace theorem with respect to its first $(j-1)$ rows, all relevant algebraical adjugates are equal to zero.

Now expand the original B analogically. Because of the special form of B, three terms are non-zero only in the resulting sum of products. Thus we get the equality ($|A|$ denotes the determinant of matrix A):

$$|s_1 \cdots s_{k-1} \cdots s_{k+1} \cdots s_j| \cdot \begin{vmatrix} s_1 & \cdots & s_{q-1} & s_{q+1} & \cdots & s_j & s_{j+1} \\ a_{j1} & \cdots & a_{jq-1} & a_{jq+1} & \cdots & a_{jj} & a_{jj+1} \end{vmatrix} (-1)^{j^2-1} +$$

$$+ |s_1 \cdots s_{q-1} s_{q+1} \cdots s_j| \cdot \begin{bmatrix} s_1 & \cdots & s_{k-1} & s_{k+1} & \cdots & s_j & s_{j+1} \\ a_{j1} & \cdots & a_{jk-1} & a_{jk+1} & \cdots & a_{jj} & a_{jj+1} \end{bmatrix} (-1)^{j^2-2} +$$

$$+ |s_1 \cdots s_{k-1} s_{k+1} \cdots s_{q-1} s_{q+1} \cdots s_j s_{j+1}| \cdot$$

$$\cdot \begin{vmatrix} s_1 & \cdots & s_j \\ a_{j1} & \cdots & a_{jj} \end{vmatrix} (-1)^{j^2-2j-4} = 0,$$

i.e., $M_{j-1}^{(k)} M_j^{(q)} (-1)^{j^2-1} + M_{j-1}^{(q)} M_j^{(k)} (-1)^{j^2-2} + M_{j-1}^{(k,q)} M_j (-1)^{j^2-2j-4} = 0,$

that yields relation (1) for both even and odd number j.

COROLLARY. *For the special case of* $k=1$ *and* $q=2$ *the relation*

$$(2) \qquad M_j = \frac{M_{j-1}^{(1)} M_j^{(2)} - M_{j-1}^{(2)} M_j^{(1)}}{M_{j-1}^{(1,2)}} \qquad j = 2,3,\ldots,n$$

holds.

Further on, relation (2) will be utilized recursively to calculate the leading principal minors (from now on LPM) of the next class of matrices.

DEFINITION 2. The real n by n matrix A is called a t-bandtriangular matrix (t-BT matrix) if

1. $a_{ij} = 0$ for $j-i > t$

2. $a_{ij} \neq 0$ for $j-i=t$, $i=1,\ldots,n-t$; $j=i+t,\ldots,n$.

Note that each band matrix not having a zero element in the last over-diagonal is a bandtriangular matri⌄

3. THE APPLICATION OF THE THEOREM

Let A be a t-BT matrix with $t > 1$. The hitherto arbitrary non-zero vectors C_i, $i=n+1,\ldots,2n$ are further specified as $C_i = E_{i-t}$, $i=n+1,\ldots,n+t$, where $E_j = (0,\ldots,0,\alpha_j,0,\ldots,0)^T$ with $\alpha_j = 1$.

If (2) is applied to matrix A with such a choice of vectors C_i, $i=n+1,\ldots,n+t$, we get its LPM M_j with the aid of the LPM of matrices

$$A^{(1)} = [C_2,\ldots,C_{n+1}],$$

$$A^{(2)} = [C_1, C_3,\ldots,C_{n+1}] \quad \text{and}$$

$$A^{(1,2)} = [C_3,\ldots,C_{n+2}],$$

these being $(t-1)$ and $(t-2)$-BT matrices, respectively. Thus, and it is important, the width of the over-diagonal band of the matrices needed to calculate M_j has been narrowed.

This process of a successive restriction of the overdiagonal bandwidth can, of course, be continued. If (2) is applied for matrices $A^{(1)}$, $A^{(2)}$ (or their LPM $M_j^{(1)}$, $M_j^{(2)}$, respectively) we get the LPM of the already mentioned matrix $A^{(1,2)}$ and to the LPM of matrices

$A^{(1,3)}$, $A^{(2,3)}$, $A^{(1,2,3)}$ these being already $(t-2)$
and $(t-3)$-BT matrices, respectively.

It is obvious that after a successive $(t-1)$-
-multiple recursive application of (2) we attain to
1-BT matrices, i.e., to the Hessenberg matrices
$A^{(1,\ldots,i-1,i+1,\ldots,t)}$, $i=1,\ldots,t$ and to the 0-BT
matrix, i.e., to a lower triangular matrix $A^{(1,\ldots,t)}$.

These matrices are already sufficiently simple.
Their LPM may be obtained by a recurrence calculation,
e.g., by eliminating the elements of the first column of
a Hessenberg matrix (see the Appendix) and by taking the
product of the diagonal elements for a triangular matrix.
Being aware of their values, arbitrary M_j, $j=2,\ldots,n$
can be calculated by recursive substitution.

The algorithm ALMIN for calculating the LPM of a
t-BT matrix thus consists of three steps:

Step 1: Calculate $M_j^{(1,\ldots,t)}$, $j=1,\ldots,n$;

Step 2: For $i=1,\ldots,t$, calculate $M_j^{(1,\ldots,i-1,i+1,\ldots,t)}$,
$\qquad j=1,\ldots,n$;

Step 3: For $r=t-2,\ldots,1,0$, calculate recursively

$$(3)\qquad M_j^{(k_1,\ldots,k_r)} = \frac{M_{j-1}^{(1,\ldots,r+1)} M_j^{(k_1,\ldots,k_r,r+2)} - M_{j-1}^{(k_1,\ldots,k_r,r+2)} M_j^{(1,\ldots,r+1)}}{M_{j-1}^{(1,\ldots,r+2)}}$$
$$j=2,\ldots,n$$

for all $(k_1,\ldots,k_r)\in C_{r+1}^r$, where C_{r+1}^r stands for the
set of all combinations of the r-th class from $r+1$
elements $\{1,2,\ldots,r+1\}$. If $r=0$ then $M_j^{(k_1,\ldots,k_r)} = M_j$.

The algorithm ALDET for calculating the determinant of a $t-BT$ matrix:

Step 1: Calculate $M_j^{(1,\ldots,t)}$, $j=n-t+1,\ldots,n$;

Step 2: For $i=1,\ldots,t$, calculate $M_j^{(1,\ldots,i-1,i+1,\ldots,t)}$, $j=n-t+1,\ldots,n$;

Step 3: Step 3 from ALMIN for $j=n-r,\ldots,n$.

It is obvious that relation (3) is applicable only if the denominator values for given indexes r and j differ from zero.

REMARKS

1. The number of arithmetical operations of algorithms for the computation of the determinant and the LPM of a pentadiagonal matrix are given in the following table. The algorithm in [3] is not compared since it has been proposed for a special form of a pentadiagonal matrix.

Algorithm given by	Computation of					
	M_n			M_j, $j=1,\ldots,n$		
	$\pm$	$\times$	$:$	$\pm$	$\times$	$:$
relation in [1]	$7n-16$	$17n-43$	$n-3$	$7n-16$	$17n-43$	$n-3$
relation in [2]	$7n-14$	$15n-31$	$-$	$7n-14$	$15n-31$	$-$
relation (3)	$6n-7$	$11n-25$	$n-2$	$7n-9$	$14n-32$	$n-2$

2. In our opinion the fastest parallel method of computation of the determinant of a full matrix A may

be obtained from the fast parallel matrix inversion
algorithm [4], in which the last unknown of a triangular
linear system is equal to the determinant of A. For the
determinant of a $t-BT$ matrix this method requires
$\frac{3}{2}(\log_2 n)^2 + \frac{7}{2}\log_2 n + 1$ parallel steps of computation
with $O(n^4/\log_2 n)$ processors. A parallel version of
ALDET requires $\frac{1}{2}(\log_2 n)^2 + \frac{3}{2}\log_2 n + 3t - 1$ parallel
steps with $O(n^3)$ processors. Thus, if

$$t < \frac{1}{3}[(\log_2 n)^2 + 2\log_2 n + 2]$$

then the parallel version of ALDET for the determinant
of a $t-BT$ matrix requires less parallel steps than the
parallel algorithm [4].

ACKNOWLEDGEMENT. My thanks are due to J. Mikloško,
for his critical comments.

APPENDIX

Let us have a Hessenberg matrix

$$H = \begin{bmatrix} b_{11} & b_{12} & & & & \\ b_{21} & b_{22} & b_{23} & & & \\ \vdots & & \ddots & \ddots & & \\ & & & \ddots & \ddots & b_{n-1,n} \\ b_{n1} & \cdots & \cdots & \cdots & \cdots & b_{nn} \end{bmatrix}$$

and let us define

$$R_i = b_{i1} + \sum_{j=2}^{i} b_{ij} x_j, \qquad i = 1, \ldots, n$$

$$x_i = -\frac{1}{b_{i-1,i}} R_{i-1}, \qquad i = 2, \ldots, n.$$

Then for the LPM P_j of the matrix H

$$P_j = (-1)^{j+1} R_j \prod_{i=1}^{j-1} b_{i,i+1}, \quad j=1,\ldots,n$$

holds.

REFERENCES

[1] R.A. Sweet, A recurrence relation for the determinant
of a penta-diagonal matrix, *CACM*, 12(1969), 330-332.

[2] D.J. Evans, A recursive algorithm for determining
the eigenvalues of a quindiagonal matrix, *Computer
Journal*, 18(1975), 321-325.

[3] J. Mikloško, A recursive computation of the determinant of a pentadiagonal matrix, *Journal Comp.
Appl. Math.*, 1(1975), 73-78.

[4] L. Csanky, Fast parallel matrix inversion algorithms
16th *Ann. Symp. on Foundations of Computer Science
(SWAT)*, Berkeley, October, 1975.

L. Halada
Institute of Cybernetics
Slovak Academy of Sciences
809 31 Bratislava
Dúbravska 1
Czechoslovakia

ÜBER GAUSS' BEITRÄGE ZUR NUMERISCHEN MATHEMATIK
H. HEINRICH

Bei der Wahl des Themas erschien es mir angebracht, auch an dieser Stelle dessen zu gedenken, daß sich in diesem Jahr 1977 der Geburstag des grossen GAUSS zum 200-sten Male gejährt hat. Zwei Umstände haben mich darin bestärkt:

1./ der Umstand, dass sich diese Tagung mit Fragen der *Numerischen Mathematik* befasst und wir CARL FRIEDRICH GAUSS bedeutende Beiträge auch zu diesem Teilgebiet der Mathematik verdanken;

2./ der Umstand, dass zu GAUSS' besten Freunden seit der gemeinsamen Studienzeit in Göttingen der Ungar FARKAS (Wolfgang) V. BOLYAI gehört hat, der Vater des durch die geniale Schöpfung einer nichteuklidischen Geometrie berühmt gewordenen JÁNOS (Johann) BOLYAI, den die VR Ungarn zu Recht dadurch ehrt, dass sie ihrer Mathematischen Gesellschaft seinen Namen gegeben hat. Der Briefwechsel zwischen GAUSS und dem Vater BOLYAI, der sich von der gemeinsamen Studienzeit im Jahre 1797 trotz der bald erfolgten und nie mehr überwundenen räumlichen Trennung bis zum Jahre 1853 d.h. bis nahe an das Lebensende der beiden

Männer erstreckt hat, gehört zu den ergreifendsten Dokumenten einer tief wurzelnden Seelenfreundschaft zweier geistig und sittlich hochstehenden Geister. Und welches Wunder war es, dass die Gedanken über die Grundlagen der Geometrie, die bereits in der Gesprächen der beiden jungen Studenten in Göttingen eine bedeutende Rolle spielten, in dem Werke des Sohnes BOLYAI ihre Vollendung gefunden haben, über die GAUSS am 14. Februar 1832 in einem berühmt gewordenen Brief an GERLING schreiben konnte:

> "Noch bemerke ich, dass ich dieser Tage eine kleine Schrift aus Ungarn über die Nicht-Euklidische Geometrie erhalten habe, worin ich alle meine eignen Ideen und Resultate wiederfinde, mit grosser Eleganz entwickelt, obwohl in einer für jemand, dem die Sache fremd ist, wegen der Konzentrierung etwas schwer zu folgenden Form. Der Verfasser ist ein *sehr* junger österreichischer Offizier, Sohn eines Jugendfreundes von mir, mit dem ich 1798 mich oft über die Sache unterhalten hatte, wiewohl damals meine Ideen noch viel weiter von der Ausbildung und Reife entfernt waren, die sie durch das eigne Nachdenken dieses jungen Mannes erhalten haben. Ich halte diesen jungen Geometer V. BOLYAI für ein Genie erster Größe."

Wenige Tage später, am 6. März 1832, schrieb GAUSS auch an seinen Freund, den Vater BOLYAI, über diese Arbeit seines Sohnes:

> "Wenn ich damit anfange, 'daß ich solche nicht loben darf': so wirst Du einen Augenblick stutzen: aber ich kann nicht anders; sie loben, hiesse mich selbst loben, denn der ganze Inhalt der Schrift, der Weg, den dein Sohn eingeschlagen hat, und die Resultate, zu denen er geführt ist, kommen fast durchgehends mit meinen eigenen, zum Teil schon seit 30-35 Jahren angestellten Meditationen überein. In der Tat

bin ich dadurch auf das Äusserste überrascht.

Mein Vorsatz war, von meiner eignen Arbeit, von der übrigens bis jetzt wenig zu Papier gebracht war, bei meinen Lebzeiten gar nichts bekannt werden zu lassen. Die meisten Menschen haben gar nicht den rechten Sinn für das, worauf es dabei ankommt, und ich habe nur wenige Menschen gefunden, die das was ich ihnen mitteilte, mit besonderem Interesse aufnahmen. Um das zu können, muss man erst recht lebendig gefühlt haben, was eigentlich fehlt, und darüber sind die meisten Menschen ganz unklar. Dagegen war meine Absicht, mit der Zeit Alles so zu Papier zu bringen, dass es wenigstens mit mir dereinst nicht unterginge.

Sehr bin ich also überrascht, dass diese Bemühung mir nun erspart werden kann und höchst erfreulich ist es mir, dass gerade der Sohn meines alten Freundes es ist, der mir auf eine so merkwürdige Art zuvorgekommen ist."

Ich laufe offenbar Gefahr, von meinem eigentlichen Thema abzukommen. Aber ich hätte diese brieflichen Äusserungen von GAUSS nicht so ausführlich zitiert, wenn sie nicht so ausserordentlich charakteristisch für ihn, für seine Denk- und seine Arbeitsweise wären, wie sie sich auch bei seinen BEITRÄGEN ZUR NUMERISCHEN MATHEMATIK zeigt.

Bei der Beschäftigung mit diesen Beträgen stellt es sich heraus, daß sich hier ein noch weithin offenes Feld vor einem ausbreitet. Zwar begegnet man in Monographien und Lehrbüchern der Numerischen Mathematik auch heute auf Schritt und Tritt dem Namen GAUSS, aber im Gegensatz zu den Teilgebieten der Mathematik, die wir nach der seit der Mitte des vorigen Jahrhunderts entstandenen Tradition der sogenannten reinen Mathematik zurech-

nen, haben bisher seine ideenreichen und tiefgründigen
Beiträge zur Numerischen Mathematik, zu denen er *in
engster Verbundenheit mit der Praxis* als Astronom und
Geodät gelangt ist, offensichtlich noch keine zusammen-
fassende Darstellung unter einem einheitlichen Gesichts-
punkt erfahren. Ich bin überzeugt, daß sich in dieser
Hinsicht für die künftige GAUSS-Forschung noch eine
reizvolle Aufgabe ergibt.

Wenn wir heute von *Numerischer Mathematik* sprechen,
so meinen wir damit jenes Teilgebiet der Mathematik, das
sich der gesellschaftlich unbedingt notwendigen und im
Zuge des technisch-wissenschaftlichen Fortschritts immer
dringlicher gewordenen Aufgabe angenommen hat, für die
zahllosen, aus den verschiedensten Objektbereichen stam-
menden Probleme, deren Lösung die Angabe *zahlenmäßiger*
Resultate verlangt, mathematisch fundierte und leistungs-
fähige *Verfahren* bereitzustellen, mit denen man eben
diese Ergebnisse zuverlässig gewinnen kann. Wir stellen
heute an solche Verfahren *hohe wissenschaftliche Ansprüche.* Ich will versuchen, ihnen eine Formulierung zu
geben, die auf die Zeit von GAUSS ebenso paßt wie auf
unsere heutige.

1. Die Verfahren sollen *ökonomisch* arbeiten, d.h.
möglichst wenig Zeit- und Arbeitsaufwand erfordern.

2. Sie sollen das Ergebnis mit einer *dem Problem
angepaßten*, d.h. weder übertriebenen noch zu geringen
Genauigkeit liefern.

3. Die mit ihnen erreichte bzw. erreichbare *Genauigkeit* soll zuverlässig und in möglichst engen Grenzen
abschätzbar sein.

4. Die Grenzen, innerhald deren ein Verfahren *durchführbar* ist, müssen bekannt sein und beachtet werden.

5. Die Verfahren sollen *numerisch stabil* sein.

6. Jedes numerisch zu behandelnde *Problem* soll in

einer dem Rechner möglichst leicht zugänglichen Form
aufbereitet sein. Dazu gehört

7. daß das anzuwendende Verfahren in einer *algo-
rithmischen* Form dargestellt wird, die den durchzuführenden
Rechenprozeß unmittelbar widerspiegelt.

8. Die Verfahren sind den jeweils geeignetsten ver-
fügbaren *Rechenhilfsmitteln* anzupassen.

9. Sie sollen schließlich - entsprechend der *Multivalenz* der mathematischen Modelle, die der Beschreibung
quantitativer Zusammenhänge dienen, - vielseitig anwend-
bar sein und für ganze Klassen gleichartiger Aufgaben zur
Verfügung stehen.

Diese Ansprüche bereits erkannt und in seinen zahlreichen und bedeutenden Beiträgen zur Numerischen Mathematik formuliert und in genialer Weise befriedigt zu haben, ist das unbestreitbare Verdienst von CARL FRIEDRICH
GAUSS. Er ist dadurch zum unübertroffenen Lehrmeister der
Numerischen Mathematik geworden. Seine beiträge betreffen:

die *Ausgleichsrechnung* und die ihr zugrunde liegende
Fehlerquadratmethode bis hin zu Ansätzen zur Lösung von
Optimierungsproblemen;

die theoretischen und praktischen Fragen der *Interpolation* - einschließlich der trigonometrischen - und der
numerischen Quadratur;

in engem Zusammenhang mit der Interpolation die Berechnung umfangreicher *Tafeln* für als nützlich erkannte
und ad hoc eingeführte *Hilfsfunktionen;*

Reihen - und *Kettenbruchentwicklungen* als Hilfsmittel
für *Approximationen* und die dabei auftretenden *Konvergenz-
fragen;*

die *Eliminationsalgorithmen* und *Iterationsverfahren*

zur   numerischen Lösung *linearer Gleichungssysteme;*

Ansätze zur *iterativen numerischen Lösung nichtlinearer Gleichungssysteme;*

nicht zuletzt die von ihm immer wieder als besonders
wichtig hervorgehobenen Fragen der zweckmäßisten

Organisation umfangreicher Rechenprozesse.

Schon diese Aufstellung macht es verständlich, daß es
im Rahmen dieses Vortrages unmöglich ist, in irgendeiner
Hinsicht Vollständigkeit der Ausführungen anzustreben und
sie in wünschenswertem Maße durch Zitate aus GAUSS' Werken,
seinen zahlreichen Briefen und seinem Nachlaß zu ergänzen.

Es ist bekannt, daß GAUSS von Kindesbeinen an ein
ausgezeichneter *Rechner* war. Seine an vielen Stellen zu
findenden Angaben über das von ihm in Stoßzeiten täglich
verarbeitete Zahlenmaterial und über den relativ geringen
dafür benötigten Zeitaufwand grenzen ans Unwahrschein-
liche und sind nur durch eine außergewöhnliche Fähigkeit
zum *Kopfrechnen*, ein enormes *Zahlengedächtnis* und eine
hervorragende Begabung, Rechenprozesse durch Ersinnen
immer wieder *neuer Methoden* abzukürzen und zu beschleuni-
gen, erklärbar. Man vergleiche dazu den Brief, den der
Astronom OLBERS Ende April 1807 an den Astronomen VON ZACH
gerichtet hat und in dem er mit Bewunderung darüber berichtet, daß GAUSS ihm die rechnerische Auswertung der
Beobachtungen des Planetoiden Vesta innerhalb von nur
zehn Stunden geliefert hat [1]. Es wäre aber sicher falsch,
GAUSS einen leidenschaftlichen Rechner zu nennen, der
etwa um des Rechnens willen rechnete. Aus einem Gutachten,
das er am 5. April 1831, über

- Eine in Deutschland erfundene Rechenmaschine, -

eine offenbar wieder in der Versenkung verschwundene Erfindung des Frankfurter Mathematiklehrers J.F. SCHIERECK, abgegeben hat, zeigt deutlich wie aufgeschlossen GAUSS gegenüber der Idee war, umfangreiche Rechnungen mit maschinellen Hilfsmitteln zu bewältigen [2].

Vielleicht hätte ihm e i n e Eigenschaft unserer heutigen Rechner mißfallen: alle Ergebnisse, auch in Fällen, in denen es vermeidbar wäre, als *Näherungswerte in Form abgekürzter Dezimalzahlen* zu liefern. Denn obwohl wir ihm die später für alle Funktionentafeln verbindliche Rundungsvorschrift verdanken, hat er es beispielsweise bei Reihen- und Kettenbruchentwicklungen und in Quadraturformeln in auffälliger Weise vorgezogen, die Koeffizienten exakt als Brüche ganzer Zahlen, und seien sie noch so vielstellig, anzugeben.

Trotzdem wäre aber wohl niemand über die Existenz leistungsfähiger Rechenmaschinen glücklicher gewesen als er. Wie aus zahlreichen Briefen an seine Freunde und Mitarbeiter hervorgeht, hat GAUSS häufig genug das übermäßig viele Rechnenmüssen als eine unerträgliche Last empfunden. Dazu eines von vielen möglichen Zitaten, z.B. [3], [4]. In dem Briefwechsel zwischen GAUSS und OLBERS findet sich die folgende bezeichnende Stelle vom September 1813 [5]:

> "Die Pallas-Störungen habe ich seit einigen Monaten ganz beiseite gelegt, ich hatte die geistlosen Rechnungen zu satt und habe mich seitdem wieder mit einigen Spekulationen der höheren Arithmetik beschäftigt."

Angesichts solcher Selbstzeugnisse müssen wir den Grund für die bei GAUSS zweifellos vorhandene *starke Affinität zur Welt der Zahlen* darin suchen, daß zahlenmäßige Rechenergebnisse für ihn entweder eine *Quelle für neue mathematische Erkenntnisse, Ideen und Vermutungen* oder das *Endziel*, waren, *durch das sich die angestrebte Lösung eines Problems einzig und allein adäquat aus-*drücken ließ. Was den erstgenannten Aspekt angeht, so sei an die Worte erinnert, mit denen FELIX KLEIN in seinen Vorbemerkungen zu dem ersten Abdruck des berühmten GAUSS-schen *Tagebuchs* [6] den Arbeitsstil von GAUSS charak-terisiert hat [7]:

> "Und dabei immer wieder die Eigenart seines
> mathematischen Genius: an der Hand von Zahlen-
> rechnungen die Resultate zu finden, um hinterher
> langsam, in härtester Arbeit die Beweise zu zwin-
> gen."

Sowohl in dem im Band X/2 von GAUSS' Werken erschienenen Aufsatz

PAUL BACHMANN, *Über Gauß' zahlentheoretische Ar-beiten* [8]

als auch in dem Aufsatz

PAUL STÄCKEL, *Gauß als Geometer*, [9]

der ebenda abgedruckt ist und für das Kennenlernen der GAUSSschen Arbeitsweise besonders aufschlußreich ist, ist in solchem Zusammenhang von "*numerischem Beobachtungs-material*" die Rede. Man denke hier beispielsweise nur an die Tagebuchnotiz vom 30. Mai 1799 [10] in der GAUSS in knappster Form von seiner Entdeckung berichtet, daß das arithmetisch-geometrische Mittel zwischen den Zahlen 1 und $\sqrt{2}$ *auf elf Dezimalstellen* mit dem Wert π/ω über-einstimmt, bei dem $\omega/2$ der Wert des elliptischen Integ-rals $\int_0^1 \frac{1}{\sqrt{(1-x^4)}}dx$ ist, das bei der Untersuchung der Lemniskate eine Rolle spielt. GAUSS ist sich, was beson-ders bemerkenswert ist, der Tragweite dieser Entdeckung offenbar sofort, *noch bevor er die exakte Übereinstimmung der beiden Zahlen streng beweisen konnte*, bewußt, denn

er gibt gleich in dieser ersten Notiz seiner Überzeugung
Ausdruck, daß sich damit ein "neues Feld in der Analysis"
eröffne. In der Tat war damit ein Tor zur Entwicklung der
Theorie der elliptischen Integrale und Funktionen aufge-
stoßen, die im Grunde erst 25 Jahre später durch ABEL
und JACOBI erfolgt ist.

Was den zweiten - für uns im Vordergrund stehenden-
-Aspekt angeht, so muß man berücksichtigen, daß GAUSS
vom Ausmaß der aufgewendeten Zeit her gesehen vorwiegend
als *Astronom* und *Geodät* gearbeitet hat. Schließlich war
er ja niemals Professor für Mathematik, sondern von 1807
bis an sein Lebensende Professor für Astronomie und Di-
rektor der Sternwarte in Göttingen. Als Astronom hatte
er nicht nur eine analytische Theorie der Bewegungen der
Himmelskörper und ihrer gegenseitigen Störungen aufzubauen,
sondern diese auch für eine numerische Behandlung so auf-
zubereiten, daß ihre Ergebnisse einen zuverlässigen Ver-
gleich mit den beobachteten astronomischen Werten und
eine zuverlässige Deutung dieser Werte zuließen. An den
überaus verwickelten, umfangreichen und zeitraubenden
Rechnungen, die dazu - zu einem erheblichen Teil von ihm
selbst - zu bewältigen waren, hätte der *Mathematiker*
GAUSS zerbrechen können, wenn es nicht seinem mathematischen Scharfsinn gelungen wäre, immer wieder neue
Methoden zu ersinnen, um den Rechenaufwand auf ein Min-
destmaß zu reduzieren, dem mathematischen Hintergrund
dieser Verfahren nachzugehen, sie mit höheren mathematischen
Hilfsmitteln zu verfeinern, ihre Genauigkeit zu steigern
und durch eine hervorragende Organisation der notwendigen
Rechenprozesse den für sie erforderlichen Zeitaufwand
entscheidend zu verringern. Man findet in GAUSS' Briefen

immer wieder Bemerkungen, die sich auf seine "Methoden"
beziehen, "durch welche die weitläufigsten Rechnungen
sehr zusammengezogen würden" [11]. In einem Brief an
OLBERS vom 10. Mai 1805 [3], in dem es um die Berechnung
der Ceres-Störungen geht, heißt es dazu:

> "Ich habe indessen bereite eine andere Methode
> ausgesonnen, die eben so weit führen kann, als
> jene, aber bei weitem weniger - obwohl künst-
> lichere - Arbeit erfordert. ...*Diese Methode hat
> um so mehr Reiz für mich, da ich dabei von vielen,
> schon vor längerer Zeit angestellten, ziemlich
> tiefen Untersuchungen über eigene Arten von trans-
> cendenten Funktionen einen glücklichen Gebrauch
> machen kann.*"

Ein äußerst instruktives Beispiel für die Gesichts-
punkte, unter denen GAUSS einen verwickelten Rechenprozeß
organisiert und entsprechende detaillierte Anweisungen an
seine Mitarbeiter gibt, findet sich in einem Brief vom
3. September 1805 an den Astronomen BESSEL [12]. Es ist
leider zu umfangreich, um hier in Einzelheiten erörtert
werden zu können. Es betrifft die tabellarische Ermittlung
der numerischen Werte der Koeffizienten $A^o, A', A'', \ldots$
der trigonometrischen Reihe

$$(1) \qquad \frac{1}{2}A^o + A'\cos\varphi + A''\cos 2\varphi + A'''\cos 3\varphi + \ldots$$

in die der reziproke Abstand

$$(2) \qquad (a^2 + a'^2 - 2aa'\cos\varphi)^{-\frac{1}{2}}$$

zweiter Punkte im Raum entwickelt werden kann. (Man beachte
die Jahreszahl 1805, verglichen mit dem Erscheinungsjahr
1822 der Théorie analytique de la chaleur von FOURIER

(1768-1830)!)

Es kommt mir nicht so sehr auf eine detaillierte
Darlegung der mehrfachen Umformungen und Einführungen von
Hilfsgrössen an, die GAUSS vornimmt und die auf Potenz-
reihen- und Kettenbruchentwicklungen führen und bei denen
auch die Tabellierung von Hilfsfunktionen mit der zugehörigen Interpolationsrechnung eine Rolle spielen. Wesent-
licher erscheint es mir zu erfassen, wie er die einzelnen
Schritte motiviert. Ich zitiere dazu einige besonders
charakteristische Stellen aus diesem Briefe. Es heißt
da:

1. "Ich habe mich schon seit einiger Zeit mit
Perturbationsrechnungen beschäftigt, wo ich ...
mir eine eigene Methode ausgedacht habe, bei der
allerdings viel, sehr viel Arbeit ist, der aber
dieß nicht zum Vorwurf gereichen kann, da, meinem
Urtheile nach, alle bisherigen Methoden in einem
solchen Falle ganz unzulänglich sind. Etwas Cha-
rakteristisches bei dieser Methode ist es, daß
die Entwicklung der Coeffizienten ... nicht wie
bei den bisher üblichen Methoden bloß für *einen*
bestimmten Wert von *a, a'*, *sondern für eine große
Menge verschiedener Werthe* erfordert werden ...

Nun bin ich im Besitz von besonderen ... Kunst-
griffen, jene Coeffizienten mit sehr großer
Geschwindigkeit zu bestimmen. Indeß macht die
große Menge doch immer die Arbeit beschwerlich,
und ich habe es daher vor, zumal da für jeden
Planeten *die Arbeit mehrere Male* (mit den succesive
verbesserten Elementen) wieder gemacht werden
müßte, ein für alle Mal eine *Tabelle* zu berechnen,
mit deren Hilfe man dann *mit sehr geringer Arbeit*
die gesuchten Coeffizienten angeben könnte ..."

Weiter unten heißt es:

2. "Gewöhnlich berechnet man A^o und A' aus obigen Reihen und dann daraus vermittelst eben gedachter Formeln A'', A''', A^{iv} etc. *Allein dieses Verfahren ist zu meinem Zwecke unbrauchbar,* denn wenn A^o oder A' nur etwas wenig fehlerhaft sind, z.B. nur um 1 in der 7-ten Decimale, so werden die daraus bei A'', A''' etc. entspringenden Fehler immer grösser, beinahe in geometrischer Progression, dieß ist desto beträchtlicher, je kleiner f ist, und man kann für die entfernteren Coeffizienten, z.B. für A^x ... *ganz und gar falsche Werthe* erhalten.

Diese Bemerkung ist meines Wissens noch nicht gemacht, aber von Wichtigkeit. Ich habe daher an eine ganz andere Methode gedacht ..."

Schliesslich füge ich noch folgende Stelle an:

3. "Übrigens werden Sie selbst wissen, daß eine gleichsam fabrikmässige Arbeit sehr viele Zeit erspart; ich meine, nicht etwa erst für einen Werth von W die ganze Rechnung zu machen und dann für einen andern etc. sondern jede einzelne Operation gleich für alle W der Reihe nach zu machen. Man gewinnt dadurch erstens, daß man nicht soviel in den Tafeln hin und her zu blättern braucht, zweitens, daß etwaige Fehler sich, weingstens wenn sie beträchtlich sind, sogleich in Sprüngen in den Differenzen verraten."

Einer der ersten eklatanten Erfolge des rechenden Astronomen GAUSS waren seine Ergebnisse, die es den Astronomen ermöglichten, am 7. Dezember 1801 den nach nur etwa vierzigtägiger Beobachtung wieder verschwundenen, am Neu-

jahrstag 1801 von dem Italiener PIAZZI in Palermo entdeck-
ten ersten Planetoiden Ceres wieder aufzufinden. Dahinter
steckten als Leistungen einerseits die

*Entwicklung einer analytischen Methode, die
charakteristischen Elemente einer elliptischen
Planetenbahn aus nur drei Ortsbestimmungen zu
berechnen,*

andererseits die Anwendung der

Fehlerquadrathmethode,

die es gestattete, die mehr als drei vorhandenen Beobachtungen *in systematischer, von Willkür freier Weise* zu
einem optimal zuverlässigen Ergebnis zu kombinieren. Diese
Fehlerquadrathmethode oder

Methode der kleinsten Quadrate,

wie sie später auch von GAUSS nach dem Vorbild des fran-
zösischen Mathematikers LEGENDRE (1752-1833) genannt
wird, der sie unabhängig von ihm entwickelt und vor ihm
veröffentlicht hat (1806 [13]), begleitet GAUSS durch
sein ganzes Leben. Nach seinen eigenen Angaben hat er sie
bereits 1794 oder spätestens 1795 konzipiert, um sie bald,
was für ihn als Mathematiker charakteristisch ist,
wahrscheinlichkeitstheoretisch zu begründen, sie zu einer
umfassenden *Theorie der Beobachtungsfehler* auszubauen,
als zugehörige numerische Methode eine *Ausgleichsrechnung*
zu schaffen, diese bei seinen astronomischen und geodä-
tischen Arbeiten mit durchschlagendem Erfolg einzusetzen
und sie seine Schüler und Mitarbeiter zu *lehren,* zuletzt
noch in einer Vorlesung, die er im Wintersemester 1850/51
als Vierundsiebzigjähriger gehalten hat und über die sein
damals zwanzigjähriger Schüler RICHARD DEDEKIND (1831-1916)
noch fünfzig Jahre später in einer Festschrift zum 150-
jährigen Bestehen der Königlichen Gesellschaft der Wissenschaften zu Göttingen eindrucksvoll und begeistert berich-

tet hat [14]. Wenn man sich darüber unterrichten will,
wie GAUSS die Zweckmäßigkeit der Fehlerquadratmethode
durch *praktische* Erwägungen begründet hat, - dazu war er
sich nicht zu schade, obwohl er über mathematischtheore-
tische Begründungen verfügte! -, so kann ich mir keine
bessere Darstellung vorstellen, als sie DEDEKIND in diesem
Bericht über die GAUSSsche Vorlesung gegeben hat.

Die Methode der kleinsten Quadrate ist im Laufe der
Zeit nicht nur - entsprechend ihrer ursprünglichen Aufgabe - zu einem unentbehrlichen, immer mehr vervollkommen-
ten und in ihren numerischen Verfahren den modernen
maschinellen Rechenhilfsmitteln angepaßten *Rüstzeug vieler
technischer und wissenschaftlicher Bereiche* ausgebaut worden, sie ist auch zur *Grundlage statistischer Methoden*
geworden und ist darüber hinaus heute ein fester, zum
täglichen Handwerkszeug gehöriger *Bestandteil der Analysis* insbesondere der *Matrizenanalysis* und der *Approximationstheorie*, wo die Verwendung der Quadratsummen- bzw.
der L_2-Norm und die Aufgabe, diese unter bestimmten Bedingungen zu minimieren, unmittelbar auf das Prinzip der
Fehlerquadratmethode zurückgehen.

Im Zusammenhang mit der Methode der kleinsten Quadrate hätte es zu einem Prioritätsstreit zwischen GAUSS
und LEGENDRE kommen können, wenn GAUSS auf die *Priorität*
der Veröffentlichung Wert gelegt hätte. In den Göttingeschen gelehrten Anzeigen vom 17. Juni 1809 [15] zeigt
GAUSS in einem ausführlich gehaltenen Artikel das Erschei-
nen seines astronomischen Hauptwerkes [16]

- Theoria motus corporum coelestium in sectionibus
conicis solem ambientium -

an. Es heißt dort u.a.:

"Die Grundsätze, welche hier ausgeführt werden,
und welche von dem Verfasser schon seit 14 Jahren

angewandt, und von demselben schon vor geraumer
Zeit mehreren seiner astronomischen Freunde mitge-
theilt waren, führen zu derjenigen Methode, welche
auch LEGENDRE in seinem Werke

   - Nouvelles méthodes pour la détermination des
   comètes-

vor einigen Jahren unter dem Namen

   - Méthode des moindres carrés -

aufgestellt hat; *die Begründung der Methode,* welche
von dem Verfasser gegeben wird, *ist diesem ganz
eigenthümlich.* Eine weitere Ausführung hat man von
demselben in der Folge zu erwarten." [17]

Aus einem Brief, den GAUSS am 5. Mai 1812 in einem
anderen mathematischen Zusammenhang an LAPLACE (1749-1827)
geschrieben hat [18], geht indirekt hervor, daß LEGENDRE
offenbar geglaubt hat, aus diesen Bemerkungen von GAUSS
seinen Anspruch auf die Priorität bezüglich der Methode
der kleinsten Quadrate herauslesen zu müssen. Dieser Anspruch bezieht sich aber offensichtlich nur auf die - wahr-
scheinlichkwitstheoretische - Begründung der Methode. Diese
selbst hält GAUSS für so einfach, daß man "sich wundern
muß, daß man sich ihrer nicht schon hundert Jahre früher
bedient hat". Er wehrt sich nur dagegen, daß man glauben
könne, daß er sie von LEGENDRE übernommen habe, und schreibt
am Ende seines Briefes:

"Ich habe in meinen Aufzeichnungen viele Dinge, bei
denen ich vielleicht die Priorität der Veröffentlichung verlieren könnte, Sei es denn! Ich liebe
es mehr, die Dinge reifen zu lassen."

349

Dem Prinzip der *Multivalenz der Mathematischen Methoden* entspricht es, wenn GAUSS die der Fehlerquadratmethode zugrunde liegende Aufgabe, eine Summe von Quadraten unter bestimmten Nebenbedingungen zum Minimum zu machen, auch in Gesetzen der *Mechanik* vorliegen sieht und so zu dem nach ihm benannten *Prinzip des kleinsten Zwanges* gelangt. Über diese Prinzip gibt es eine Dissertation seines Schülers A. RITTER aus dem Jahre 1853, in der ganz klar die Aufgabe formuliert ist, eine über dem n-dimensionalen Raum definierte Ortsfunktion unter Nebenbedingungen zu minimieren, die in Form von *Ungleichungen* gegeben sind [19]. Daß dieses Problem auf GAUSS selbst zurückgeht, läßt sich indirekt aus der Nachschrift ablesen, die RITTER von der bereits erwähnten Verlesung

− Über die Methode der kleinsten Fehlerquadrate −

von 1850/51 angefertigt hat [20]. Dort behandelt GAUSS die Aufgabe, eine Summe von Quadraten

$$(3) \qquad f(\underline{x}) := \underline{x}^T \underline{x} = \sum_{k=1}^{n} x_k^2 \qquad \underline{x} = (x_1, \ldots, x_n) \in R^n$$

unter der Einschränkung durch lineare Ungleichnungen

$$u_j := \sum_{k=1}^{n} a_{jk} x_k + b_j \geq 0 \; ; \quad j = 1(1)m, \quad m > n; \quad b_j \geq 0$$

d.h.

$$\underline{u} := A\underline{x} + \underline{b} \geq 0 \; ; \quad A = [a_{jk}] \in L(R^n, R^m), \quad \mathrm{Rg}\, A = n \; ;$$

$$(4) \qquad \underline{b} = (b_1, \ldots, b_m)^T \in R^m$$

$$\underline{u} = (u_1, \ldots, u_m)^T \in R^m$$

zu minimieren. In unserer heutigen Sprechweise handelt es sich also um ein echtes *Optimierungsproblem* mit der Quadratsumme (3) als nichtlinearer Zielfunktion und mit den

lineáren Restriktionen (4). Wir hätten es nun kaum mit
GAUSS zu tun, wenn er nicht sogleich - und zwar in der
genannten Vorlesung - einen *Algorithmus* zur Lösung dieses
Problems angegeben hätte. Er setzt dabei voraus, daß die
Ungleichungen (4) vor der Durchführung des Algorithmus so
normiert worden sind, daß

$$(5) \qquad \sum_{k=1}^{n} a_{jk}^2 = 1 \qquad j = 1 \, (1) \, m$$

ist. Auf meine Bitte hin hat sich Herr Dr. KLEINMICHEL
(Dresden) der Mühe unterzogen, diesen Algorithmus mit den
in Betracht kommenden heutigen Vorgehensweisen zu ver—
gleichen. Danach entspricht die auf S. 478 ff. l.c. [20]
beschriebene Vorgehensweise dem Auffinden einer *zulässigen Basisdarstellung* des Systems (4), wie es nicht nur
für die Simplexmethode zur Lösung linearer Optimierungs-
probleme kennzeichnend ist, sondern auch für versschiedene
Algorithmen zur Lösung von Optimierungsproblemen mit
nichtlinearer Zielfunktion und linearen Restriktionen,
z.B. für die reduzierte Gradientenmethode von WOLFE [21]
oder - bei quadratischer Zielfunktion - für das Verfahren
von BEALE [22]. Im einzelnen läßt sie sich wohl folgen-
dermaßen darstellen. Es wird ohne wesentliche Beschrän-
kung der Allgemeinheit angenommen, daß sich die (m,n)-
-Matrix A gemäß der Blockdarstellung

$$(6) \qquad A = \begin{bmatrix} A_1 \\ A_2 \end{bmatrix}$$

in eine reguläre (n,n)-Matrix A_1 und eine $(m-n,n)$-Matrix
A_2 aufteilen läßt. Sind dann

$$(7.1) \qquad \underline{b} = \begin{bmatrix} \underline{b}^1 \\ \underline{b}^2 \end{bmatrix}, \qquad \underline{b}^1 = (b_1, \ldots, b_n)^T, \qquad \underline{b}^2 = (b_{n+1}, \ldots, b_m)^T$$

$$(7.2) \quad \underline{u} = \left[\frac{\underline{u}^1}{\underline{u}^2}\right], \quad \underline{u}^1 = (u_1, \ldots, u_n)^T, \quad \underline{u}^2 = (u_{n+1}, \ldots, u_m)^T$$

entsprechende Aufteilungen der Vektoren $\underline{b}$ und $\underline{u}$, so zer-
fällt die Gleichung (4) in die beiden Gleichungen

$$(8.1) \quad \underline{u}^1 = A_1 \underline{x} + \underline{b}^1,$$

$$(8.2) \quad \underline{u}^2 = A_2 \underline{x} + \underline{b}^2.$$

Aus ihnen folgt sofort

$$(9.1) \quad \underline{x} = A_1^{-1} \underline{u}^1 - A_1^{-1} \underline{b}^1$$

$$(9.2) \quad \underline{u}^2 = B\underline{u}^1 + \underline{c}$$

mit

$$(9.3) \quad B := A_2 A_1^{-1}, \quad \underline{c} = \underline{b}^2 - B\underline{b}^1 .$$

Der Übergang von $\left[\frac{\underline{u}^1}{\underline{u}^2}\right]$ zu $\left[\frac{\underline{x}}{\underline{u}^2}\right]$ kann durch ein Austausch-
verfahren bewerkstelligt werden. Bei GAUSS/RITTER findet
sich die Teilgleichung (9.1) als das Gleichungssystem (2)
auf S. 479 l.c., die Teilgleichung (9.2/3) als das
Gleichungssystem (1) auf S. 478 l.c. Die quadratische
Zielfunktion (3) geht, wenn man (9.1) in sie einführt, in
eine Funktion $\varphi(\underline{u}^1)$ über:

$$\varphi(\underline{u}^1) := f(\underline{x}(\underline{u}^1)) = [A_1^{-1}(\underline{u}^1 - \underline{b}^1)]^T [A_1^{-1}(\underline{u}^1 - \underline{b}^1)]$$

$$(10.1)$$

$$= (\underline{u}^1 - \underline{b}^1)^T D (\underline{u}^1 - \underline{b}^1)$$

$$(10.1) \qquad\qquad = \underline{v}^T(\underline{u}^1 - \underline{b}^1) \ ,$$

wobei die (n,n)-Matrix

$$(10.2) \qquad D := (A_1 A_1^T)^{-1} \ ,$$

eine der beiden GAUSSschen Transformierten der Matrix A_1^{-1}, symmetrisch $(D^T = D)$ und voraussetzungsgemäß positiv definit ist und der Vektor $\underline{v}$ durch

$$(10.3) \qquad \underline{v} := D(\underline{u}^1 - \underline{b}^1)$$

definiert ist.

Betrachten wir mit GAUSS/RITTER den Einfluß einer Änderung von $\underline{u}^1$ um $\delta\underline{u}^1$ auf $\underline{x}$, $\underline{u}^2$, $\underline{v}$ und φ, so erhält man für diese Größen, wie aus (9) und (10) folgt, die Änderungen

$$(11.1) \qquad \delta\underline{x} = A_1^{-1}\delta\underline{u}^1$$

$$(11.2) \qquad \delta\underline{u}^2 = B\delta\underline{u}^1$$

$$(11.3) \qquad \delta\underline{v} = D\delta\underline{u}^1$$

$$(11.4) \qquad \delta\varphi = \delta\underline{u}^{1^T} D\delta\underline{u}^1 + 2\underline{v}^T\delta\underline{u}^1$$

Auf dieser Grundlage ist es möglich, *schrittweise* vorzugehen und von einem Startvektor $\underline{u}^1$ aus über eine *Folge* von Änderungen $\delta\underline{u}^1$ und die zugehörigen Änderungen (11) *Folgen* von Vektoren $\underline{u}^1$, $\underline{x}$, $\underline{u}^2$ und $\underline{v}$ und von Werten φ zu erzeugen. Die dabei zu befolgende Strategie betrifft die Wahl der

Änderungen $\delta \underline{u}^1$. Für GAUSS ist es kennzeichnend daß er bei jedem Schritt nur eine Koordinate berücksichtigt und dementsprechend

$$(12.1) \qquad \delta \underline{u}^1 = \tau \underline{e}_k , \qquad k \in \{1,2,\ldots,n\}$$

setzt, wobei $\underline{e}_k$ den k-ten Einheitsvektor bedeutet und τ eine "Schrittweite" ist. Damit ergeben sich die übrigen Änderungen

$$\delta \underline{x} = \tau A_1^{-1} \underline{e}_k , \quad \delta \underline{u}^2 = \tau B \underline{e}_k , \quad \delta \underline{v} = \tau D \underline{e}_k$$

$$(12.2)$$

$$\delta \varphi = \tau^2 d_{kk} + 2 \tau v_k ,$$

wobei

$$(12.3) \qquad d_{kk} := \underline{e}_k^T D \underline{e}_k \quad \text{und} \quad v_k := \underline{v}^T \underline{e}_k$$

das k-te Hauptdiagonalelement der Matrix D bzw. die k-te Koordinate des Vektors $\underline{v}$ sind.

Bei GAUSS/RITTER heißt es nun (1.c. S.478):

"Man sucht zuerst ein System von Werthen der $x_1, x_2, \ldots$, welches n von den Ungleichungen zu 0 macht und die übrigen $m-n$ positiv."

Das bedeutet, daß von der zulässigen Basisdarstellung

$$(13) \qquad \underline{x} = -A_1^{-1} \underline{b}^1 , \quad \underline{u}^2 = \underline{c} \geq \underline{o}$$

ausgegangen wird, die sich aus (9) ergibt, wenn man $\underline{u}^1 = \underline{o}$ setzt. Der Index k wird bei GAUSS/RITTER (s. letzte Formelzeile auf S. 479 l.c. und den zugehörigen Text) so gewählt, daß $\underline{v}_k$ die betragsmäßig größte negative und damit diejenige

Koordinate von $\underline{v}$ ist, die den größten Beitrag zur Vermin-
derung des bezüglich $\underline{u}^1$ linearen Anteils von φ liefert.
Die Schrittweite τ muß so festgelegt werden, daß $\delta\varphi \leq 0$,
$|\delta\varphi|$ möglichst groß und die Nebenbedigung $\underline{u}^2 + \delta\underline{u}^2 \geq 0$ erfüllt wird.

Diese Andeutungen über die von RITTER beschriebene
GAUSSsche Vorgehensweise bei der Lösung des angegebenen
Optimierungsproblems sollten genügen, um den Standpunkt zu
rechtfertigen, daß es angebracht ist, sie in Darstellungen
über mathematische Optimierung als *Vorläuferin heutiger
Verfahren* zu erwähnen.

Sie steht im übrigen in engstem Zusammenhang mit den
übrigen bedeutenden Beiträgen von GAUSS zur *numerischen
linearen Algebra*.

Bei der Ausgleichung astronomischer Beobachtungen
stößt GAUSS auf sogenannte überbestimmte lineare Gleichungssysteme

$$(14) \qquad A\underline{x} + \underline{a} = \underline{o}, \quad A \in L(R^n, R^m), m > n; \quad \underline{x} \in R^n, \underline{a} \in R^m$$

mit spaltenregulärer (m,n)-Koeffizientenmatrix A (Rang
$A = n$). Der gesuchte Vektor $\underline{x}^*$ umfaßt die an den Beobach-
tungswerten anzubringenden differentiellen Korrekturen.
GAUSS sucht ein solches System in dem Sinne "optimal" zu
befriedigen, daß er $\underline{x}^*$ als denjenigen Vektor bestimmt,
für den die Quadratsummennorm des Residuenvektors

$$(15) \qquad \underline{r}(\underline{x}) := A\underline{x} + \underline{a} \in R^m$$

minimal wird:

$$(16) \qquad \Omega(\underline{x}) := \| \underline{r}(\underline{x}) \|^2 = \| A\underline{x} + \underline{a} \|^2 \geq \Omega(\underline{x}^*) \qquad \forall \underline{x} \in R^n.$$

Die notwendige Minimalbedingung führt für $\underline{x}^*$ auf das sogenannte *Normalgleichungssystem*

$$(17) \qquad A^T A \underline{x} + A^T \underline{a} = \underline{o}$$

dessen (n,n)-Koeffizientenmatrix $A^T A$ eine der beiden GAUSSschen Transformierten der Matrix A ist und sich durch Regularität, Symmetrie und positive Definitheit auszeichnet.

In seiner am 25. November 1810 der königlichen Societät der Wissenschaften zu Göttingen als Vorlesung übergebenen, in lateinischer Sprache verfaßten Schrift [23]

— Untersuchung über die elliptischen Elemente der Pallas aus den Oppositionen der Jahre 1803, 1804, 1805, 1807, 1808 und 1809 —

läßt sich GAUSS eingehend über seine Art der numerischen Behandlung der Normalgleichungen aus. Dort findet sich auch der heute allgemein nach ihm benannte *Eliminationsalgorithmus*, mit dessen Hilfe ein vollbesetztes lineares Gleichungssystem in ein sogemanntes gestaffeltes übergeführt werden kann. Es scheint nicht möglich zu sein, in früheren Aufzeichnungen von GAUSS Hinweise auf diesen Algorithmus ausfindig zu machen. Indirekt geht dies auch daraus herver, daß er selbst in einem 1816 verfaßten Nachtrag [24] zum Artikel 183 seiner Theoria motus corporum ... im Hinblick auf sein Eliminationsverfahren auf eben jene Stelle in der Pallas-Schrift als Quelle verweist. Außerdem zeigt ein Bericht über diese Schrift, der am 13. Dezember 1810 in den Göttingeschen gelehrten Anzeigen erschienen ist [25], daß seine Vorgehensweise damals als etwas absolut Neues empfunden worden ist.

GAUSS hat freilich diesen Algorithmus in der besag-
ten Schrift nicht nur dazu benutzt, um die Normalgleichungen zu lösen, sondern es kommt ihm außerdem darauf an,
die Fehlerfunktion durch die Kette von Eliminations-
schritten in

$$\Omega(x) = \Omega(x^*) + \text{eine Summe von Quadraten, deren}$$
Glieder im Minimalfall gleichzeitig
mit den Koordinaten des Residuenvektors Null werden,

umzuformen.

Auf die zahlreichen Varianten und Weiterentwicklun-
gen des GAUSSschen Algorithmus kann hier nicht eingegangen werden. Interessant aber und vielleicht nicht allgemein bekannt ist - was dem Genie GAUSS sicher keinen
Abbruch tut -, daß sich sein Eliminationsverfahren bereits
vor etwa zwei Jahrtausenden unter dem Namen *fang cheng*
in der chinesischen Mathematik findet, und zwar in einer
Gestalt, die der heute üblichen, den Matrizenkalkül ver-
wendenden Darstellungsweise aufs engste verwandt ist. Man
vergleiche dazu, was der sowjetische Mathematik-Histori-
ker JUSCHKEWITSCH in seinem Buch "Mathematik im Mittelalter" [26] über das aus der frühen Han-Zeit, d.h. etwa
aus den beiden letzten Jahrhunderten v.u.Z. stammende, in
einer Ausgabe aus dem Jahre 263 u.Z! erhaltene chinesische
Mathematik-Werk "Mathematik in neun Büchern" ausführt,
und die 1968 erschienene, von K. VOGEL kommentierte
deutsche Übersetzung dieser "Neun Bücher arithmetischer
Technik" [27].
Nicht minder wichtig für die numerische Behnadlung
linearer Gleichungssysteme als das Eliminationsverfahren

ist die Methode, die GAUSS als *indirekte Elimination* bezeichnet und seinem Schüler GERLING in einem Briefe vom 26. Dezember 1823 auseinandersetzt und mit den Worten empfiehlt [28]:

> "Schwerlich werden Sie je wieder direkt eliminieren, jedenfalls, wenn Sie mehr als zwei Unbekannte haben. Das indirekte Verfahren läßt sich halb im Schlafe ausführen, oder man kann während desselben an andere Dinge denken."

Eine beachtenswerte und auch heute noch didaktisch bestens geeignete Darstellung dieses Verfahrens verdanken wir DEDEKIND in der bereits erwähnten Festschrift aus dem Jahre 1801 [14]. Es läßt sich, nachdem dieser Begriff erst einmal durch SOUTHWELL [29] geschaffen worden ist, als ein *Relaxationsverfahren* begreifen, bei dem die Unbekannten nacheinander und dezimalstellenweise "abgearbeitet" werden, und wird dementsprechend in manchen Darstellungen, z.B. bei ZURMÜHL [30] als GAUSS-SOUTHWELL-
-Relaxation bezeichnet.

Aus der DEDEKINDschen Darstellung geht klar hervor, daß GAUSS auch die Idee der Pivotisierung, auch wenn sie nicht explizit erwähnt wird, durchaus geläufig war. Da das Verfahren der großen Klasse der Iterationsverfahren zuzurechnen ist, spielt selbstverständlich auch die Konvergenzfrage eine wichtige Rolle. GAUSS selbst begnügt sich im allgemeinen mit der Feststellung, ob die sukzessive sich ergebenden Näherungswerte in den Grenzen der angestrebten Genauigkeit "zum Stehen kommen". HOUSEHOLDER [31] vermutet aber, daß GAUSS bereits eine Fehlerabschätzung kannte, die später (1884, 1892) von dem russischen Mathematiker NEKRASOV [32] aufgestellt worden ist. Dazu ist noch zu bemerken, daß sowohl in dem von GAUSS in seinem Brief an GERLING vorgeführten Beispiel als auch in

dem Beispiel, das DEDEKIND anführt, die Koeffizientenmatrix symmetrisch ist, daß sie, wenn auch schwach, diagonal-
dominant ist und daß in der Zerlegung

$$(19) \qquad A = D-A \, , \quad D = \mathrm{diag}(a_{11},\dots,a_{nn})$$

beide Bestandteile D und A positiv sind.

Das Verfahren der indirekten Elimination kann als eine
Vorstufe des Iterationsverfahrens angesehen werden, das
L. SEIDEL 1874 [33] unter dem Titel

> – Über ein Verfahren, die Gleichungen, auf
> welche die Methode der kleinsten Quadrate führt,
> sowie lineare Gleichungen überhaupt, durch
> successive Annäherung aufzulösen –

veröffentlicht hat, und zwar des *Einzelschrittverfahrens*.
Wie sich zeigen läßt, führt der erste Schritt dieses Verfahrens, wenn man ihn mit einem passenden Startvektor
durchführt, zu demselben Ergebnis wie n Relaxationsschrit-
te, die nacheinander alle Koordinaten in geeigneter
Reihenfolge erfassen.

Bekanntlich läßt sich nun das GAUSS-SEIDELsche Iterationsverfahren, wie es seitdem heißt, auch auf Systeme
nichtlinearer Gleichungen ausdehnen, die in der *Fixpunkt-
form*

$$(20) \qquad \begin{aligned} \underline{\varphi}(\underline{x}) &= \underline{x}, \quad \underline{x} = (x_1,\dots,x_n)^T \\[1ex] \underline{\varphi} &= (\varphi_1,\dots,\varphi_n)^T : R^n \to R^n \end{aligned}$$

gegeben sind, und zwar sowohl in der Form des Gesamt-
schrittverfahrens

$$(21.1) \qquad x_j^{(\nu+1)} = \varphi_j(x_1^{(\nu)}, \ldots, x_n^{(\nu)})$$

als auch in der Form des im allgemeinen wirksameren Einzel-schrittverfahrens

$$(21.2) \qquad x_j^{(\nu+1)} = \varphi_j(x_1^{(\nu+1)}, \ldots, x_{j-1}^{(\nu+1)}, x_j^{(\nu)}, \ldots, x_n^{(\nu)}).$$

Konvergenzaussagen haben dann allerdings im allgemeinen nur noch lokalen Charakter, weil sie sich auf das variable Verhalten der als FRÉCHET-Ableitung des Operators $\underline{\varphi}$ fungierenden JACOBI-Matrix

$$(22) \qquad \varphi' = \left(\frac{\partial \varphi_j}{\partial x_k}\right) \qquad j,k = 1(1)n$$

stützen.

    Ich bin damit bei dem letzten Punkt angelangt, den ich berücksichtigen möchte: bei der Behandlung nichtlinearer Gleichungssysteme. Sind diese als *Nullstellengleichung* in der Form

$$(23) \qquad \underline{f}(\underline{x}) = \underline{o}, x \in R^n, \ \underline{f}: R^n \to R^m$$

gegeben, wobei $m \neq n$ sein kann, so ist, falls eine Lösung $\underline{x}^*$ existiert, mit $\underline{f}(\underline{x}^*) = \underline{o}$ zugleich auch

$$(24.1) \qquad \|\underline{f}(\underline{x}^*)\|^2 = 0$$

und wegen $\|\underline{f}(\underline{x})\|^2 \geq 0$, $x \in R^n$,

$$(24.2) \qquad \|\underline{f}(\underline{x}^*)\|^2 = \min\{\|\underline{f}(\underline{x})\|^2; \ \underline{x} \in R^n\}.$$

Das Ansetzen der notwendigen Minimalbedingung führt dann
mit der JACOBI-Matrix F des Operators $\underline{f}$ zu der Normalgleichung

$$(25) \qquad F^T(\underline{x})\underline{f}(\underline{x}) = \underline{o},$$

die jetzt eine nichtlineare Gleichung für $\underline{x}^*$ ist und, wenn
man sie nach Art des NEWTON-Verfahrens linearisiert, zu
dem Iterationsverfahren führt, das nach der Vorschrift

$$(26) \qquad [F^T(\underline{x}^{(\nu)})F(\underline{x}^{(\nu)})](\underline{x}^{(\nu+1)}-\underline{x}^{(\nu)})+F^T(\underline{x}^{(\nu)})\underline{f}(\underline{x}^{(\nu)})=\underline{o}$$

arbeitet und daher bei jedem Schritt die Lösung eines *linearen* Gleichungssystems für die am jeweils letzten Nä-
herungsvektor $\underline{x}^{(\nu)}$ anzubringende Korrektur $\underline{x}^{(\nu+1)}-\underline{x}^{(\nu)}$
verlangt. Man erkennt die vollkommene Analogie zu der im
linearen Fall auftretenden Normalgleichung (17), nur daß
jetzt an die Stelle der konstanten Matrix A und ihrer
GAUSSschen Transformierten A^TA die variable JACOBI-Matrix F und ihre GAUSSsche Transformierte F^TF getreten
sind. Damit liegt die Berechtigung, dieses Verfahren als
GAUSS-NEWTON-Verfahren zu bezeichnen, auf der Hand.

Für mich persönlich von besonderem Interesse ist,
wenn ich darauf noch zu sprechen kommen darf, ein *ablei-
tungsfrei* arbeitendes Verfahren zur numerischen Lösung von
nichtlinearen Gleichungen (23) im Falle $m=n$, das als eine
Verallgemeinerung der für $n=1$ seit altersher bekannten
regula falsi angesehen, aber auch aus dem einfachen NEWTON-
-Verfahren

$$(27) \qquad \underline{x}^{(\nu+1)} = \underline{x}^{(\nu)}-F^{-1}(\underline{x}^{(\nu)})\underline{f}(\underline{x}^{(\nu)})$$

durch eine bestimmte Finitisierung der JACOBI-Matrix F

hergeleitet werden kann. Sind $\underline{x}^{(\nu)}$, $\nu=0(1)n$, $n+1$ passende Näherungen für $\underline{x}^*$, und $\underline{f}^{(\nu)}$ die zugehörigen Bildvektoren, so sind die Differenzenmatrizen

$$(28) \qquad \Delta X := (\underline{x}^{(1)}-\underline{x}^{(o)},\ldots,\underline{x}^{(n)}-\underline{x}^{(o)})$$

$$\Delta F := (\underline{f}^{(1)}-\underline{f}^{(o)},\ldots,\underline{f}^{(n)}-\underline{f}^{(o)})$$

genau dann gleichzeitig regulär, wenn die JACOBI-Matrix F in einem hinreichend großen Arbeitsbereich des R^n regulär ist. Dann ist das Verfahren durchführbar, das über

$$(29) \qquad \underline{x}^{(n+1)} = \underline{x}^{(o)}-\Delta X(\Delta F)^{-1}\underline{f}^{(o)}$$

einen neuen Näherungsvektor $\underline{x}^{(n+1)}$ liefert. Genau in dieser Gestalt, wenn auch nicht in der hier verwendeten kurzen Bezeichnungsweise ist das Verfahren für den Spezialfall $n=2$ (zwei Gleichungen mit zwei Unbekannten) von GAUSS im Artikel 120 der Theoria motus corporum ... angegeben worden [34]. Jedoch ist diese Stelle in seinem Werke, worauf OSTROWSKI 1960 in seinem Buch "Solution of equations and systems of equations" [35] aufmerksam gemacht hat, offenbar nicht beachtet worden. Für allgemeines n ist dieses Verfahren Anfang der fünfziger Jahre von mir selbst ohne Kenntnis der GAUSSschen Vorläuferschaft entwickelt und ab 1955 in meine Vorlesungen in Dresden eingebaut worden. Unabhängig von mir hat es 1959 WOLFE gefunden und veröffentlicht [36]. In *geometrischer* Einkleidung findet es sich für $n=2$ auch bei WILLERS [37] der sich dabei auf eine Arbeit von SCHEFFERS aus dem Jahre 1915/1916 [38] bezieht, und dieser wiederum greift auf die bereits in den Jahren 1202 und 1228 bei FIBONACCI (LEONARDO von Pisa) zu findenden Anfänge der regula falsi zurück [39]. Man kann es

aber - ebenfalls für $n=2$ - auch schon bei den Chinesen
in den erwähnten "Neun Büchern" andeutungsweise erkennen,
allerdings nicht mit seiner iterativen Forsetzung, die
sich wohl erstmals bei GAUSS findet und für seine Arbeitsweise charakteristisch ist. Sie ist in dem - von OSTROWSKI
nicht mit angeführten - Artikel 121 der Theoria motus
corporum ... enthalten [40].

Gewisse Voraussetzungen, die für die Konvergenz des
Verfahrens notwendig sind, werden von GAUSS im Artikel
122, der unbedingt in die Betrachtungen einbezogen werden
sollte, erkannt und formuliert. Eingehend untersucht und
geklärt worden ist die Konvergenzfrage für allgemeines n
von L. BITTNER (1959 und 1963) [41], [42]. *Überraschender-
weise finden sich in diesem Artikel 122 der Theoria motus
corporum ..., was bisher ebenfalls nicht beachtet worden
zu sein scheint, die Grundgedanken des sogenannten
STEFFENSEN-Verfahrens*, das STEFFENSEN 1933 [43] für den
Fall $n=1$ angegeben hat und das meines Wissens erstmals
1964 von HENRICI [44] für Gleichungen im R^n verallgmeinert
worden ist.

Ich bin damit am Ende meiner Ausführungen angelangt.
Sie sind notwendig fragmentarisch geblieben. Insbesondere habe ich darauf verzichten müssen, GAUSS' umfangreiche
Arbeit über die Theorie der Interpolation, die in seinem
Nachlaß enthalten ist, seine bemerkenswert frühen Arbeiten
über trigonometrische Reihenentwicklungen und seine Methode
der numerischen Quadratur zu berücksichtigen. Was die
letztgenannte angeht, so spielt ihr Grundgedanke: in die
Menge der freien Parameter, die man benutzen kann, um die
Qualität einer Quadraturformel hinsichtlich Genauigkeit
und Stabilität zu verbessern, die Stützstellen einzu-

beziehen, auch in gegenwärtigen Arbeiten zu diesem Problemkreis immer noch eine wesentliche Rolle. Wenn ich demgegenüber vergleichsweise ausführlich auf GAUSS' Beiträge zu den Fragen der numerischen Behandlung linearer und nichtlinearer Gleichungen und von Optimierungsproblemen eingegangen bin, so hat dies selbstverständlich seinen Grund darin, daß mit diesem für viele Anwendungsgebiete höchst bedeutsamen Zweig der Numerischen Mathematik eine Arbeitsgruppe der Sektion Mathematik der Technischen Universität Dresden beschäftigt ist, an deren Entstehung ich nicht ganz unschulding bin und die heute vor allem von den Herren J. W. SCHMIDT und H. SCHWETLICK getragen wird.

Lassen Sie mich am Schluß noch einmal eine Stelle aus dem Briefwechsel zwischen GAUSS und dem Vater BOLYAI zitieren, die mir für die Geisteshaltung des Forschers CARL FRIEDRICH GAUSS besonders charakteristish zu sein scheint und von jedem wissenschaftlich forschenden Menschen zum Vorbild genommen werden kann. GAUSS schreibt am 2. September 1808 an seinen Freund nach Ungarn:

> "Wahrlich, es ist nicht das Wissen, sondern das
> Lernen, nicht das Besitzen sondern das Erwerben,
> nicht das Da-Seyn sondern das Hinkommen, was
> den grössten Genuß gewährt. Wenn ich eine Sache
> ganz ins Klare gebracht und erschöpft habe, so
> wende ich mich davon weg, um wieder ins Dunkle
> zu gehen, so sonderbar ist der nimmersatte
> Mensch, hat er ein Gebäude vollendet, so ist es
> nicht um nun ruhig darin zu wohnen, sondern um
> ein anderes anzufangen."

ANMERKUNGEN UND LITERATURHINWEISE

GW ... bedeutet: CARL FRIEDRICH GAUSS Werke Bd. ...

[1] *GW* X/1, S. 371-374.

[2] nach K.R. BIERMANN, *Fortschr. u. Forsch.* <u>41</u> (1967),
     S. 362f. (Der in GW X/1, S.6, abgedruckte Text weicht
     von dem hier angegebenen etwas ab).

[3] *GW* VII, S. 401ff.

[4] *GW* X/1, S. 24.

[5] Aus dem Briefwechsel zwishen GAUSS und OLBERS zitiert
     nach C.F. GAUSS 1777-1855, *Gedenkband anläßlich des
     100. Todestages am 23. Februar 1955, Leipzig, 1957.*,
     S. 27. Anm. 4.

[6] vgl. Nachbildung des Tagebuchs (*Notizenjournals*)
     von C.F. GAUSS, in *GW* X/1, nach S. 482.
     bzw. *Mathematisches Tagebuch 1796-1814* von Carl
     Friedrich GAUSS, in *Ostwalds Klassiker der exaktes
     Wissenschaften* Bd. 256, Leipzig, 1976.

[7] *GW* X/1, S. 486.

[8] in *GW* X/2, S. 5 des BACHMANNschen Aufsatzes.

[9] in *GW* X/2, S. 3 des STÄCKELschen Aufsatzes.

[10] *Tagebuchnotiz* Nr. 98 (deutsche Übersetzung in der
     Ausgabe von 1976 [6], S. 75). Kommentar von KLEIN
     und SCHLESINGER in *GW* X/1, S. 542f.

[11] *GW* X/1, S. 20.

[12] *GW* X/1, S. 237-242.

[13] A.M. LEGENDRE, *Nouvelles méthodes pour la détermina-
     tion des orbites des comètes,* (1806)

[14] R. DEDEKIND, Gauß in seiner Vorlesung über die Methode
     der kleinsten Quadrate, abgedruckt in: R. DEDEKIND,
     Ges. Werke <u>2</u>, S. 293-306 (1931)

[15] *GW* VI, S. 53-60.

[16] *GW* VII, S. 1-280.

[17] *GW* VI, S. 59.

[18] *GW* X/1, S. 371-374.

[19] *GW* X/1, S. 469-472.

[20] *GW* X/1, S. 473-481.

[21] P. WOLFE, Methods of nonlinear programming, in: ed. J. ABADIE, *Nonlinear Programming*, North Holland Publ., 1967., S. 98-131.

[22] E.M.L. BEALE, On minimizing a convex function subject to linear inequalities, *J. Roy. Stat. Soc.* (B) 17 (1955), S. 173-184.

[23] *GW* VII, S. 307-309.

[24] *GW* VI, S. 64.

[25] *GW* VI, S. 61-64.

[26] A.P. JUSCHKEWITSCH, *Mathematik im Mittelalter,*russ. Orig. 1961., dt. Übers. Leipzig, 1964.

[27] CHIU CHANG SUAN SHU, *Neun Bücher arithmetischer Technik,* (übers. u. erläutert von K. VOGEL) Braunschweig, 1968.

[28] *GW* IX, S. 278-281.

[29] R.V. SOUTHWELL, Relaxation methods in engineering sciences Stress-calculation in frameworks by the method of systematic relaxation of constraints I/III, *Proc. Roy. Soc.* 151 (1935), S. 56-95, 153 (1935), S. 41-76.

[30] R. ZURMÜHL, *Matrizen*, 4. Aufl. Berlin/Heidelberg/New York 1964, S. 347ff.

[31] A.S. HOUSEHOLDER, *The Theory of Matrices in Numerical Analysis,* New York/Toronto/London, 1964., S. 115f.

[32] P.A. NEKRASOV, *Mat. Sb.* 12 (1884), S. 189-204. ders., *Mat. Sb.* 16 (1892), S. 1-18. (zitiert nach [31], S. 231).

[33] L. SEIDEL, in *Münchener Akad. Abhandlungen* 11 (1874), S. 81-108.

[34] *GW VII*, S. 160-162. (Druckfehler bei OSTROWSKI [35]!)

[35] A.M. OSTROWSKI, *Solution of Equations and Systems of Equations*, New York/London, 1960.

[36] P. WOLFE, The secant method for simultaneous non-linear equations, *Comm. ACM* 2 (1959), S. 146f.

[37] F.A. WILLERS, *Methoden der Praktischen Analysis*, 3. Aufl., Berlin, 1957., S. 256ff.

[38] G. SCHEFFERS, Über die Fehlerregel, *Sitzgsber. Berl. Math. Ges.* 1916., S. 29-33.

[39] G. SCHEFFERS [38] verweist hier auf M. CANTOR, *Vorlesungen über Geschichte der Mathematik*, Bd. 2., 2. Aufl., Leipzig, 1900., S. 27ff.

[40] *GW VII*, S. 162ff.

[41] L. BITTNER, Eine Verallgemeinerung des Sekantenverfahrens, *Wiss. Z. T.H. Dresden* 9 (1959), S. 325-329.

[42] L. BITTNER, Mehrpunktverfahren zur Auflösung von Gleichungssystemen, *ZAMM* 43 (1963), S. 111-126.

[43] J.F. STEFFENSEN, Remarks on iteration, *Skand. Aktuarietidskr.* 16 (1933) S. 64-72.

[44] P. HENRICI, *Elements of Numerical Analysis*, New York, London, Sidney, 1964.

H. Heinrich
Technische Universität Dresden
Sektion Mathematik,
8027 Dresden
Zellerscher Weg 12/14.
DDR

A NORM-REDUCING JACOBI-LIKE ALGORITHM FOR THE EIGENVALUES OF NON-NORMAL MATRICES

C.P. HUANG - R.T. GREGORY

ABSTRACT

A norm-reducing Jacobi-like procedure for reducing
a non-normal matrix to a normal matrix is described. In
most instances, since the norm-reduction reduces only
the off-diagonal norm, the matrix which results is
"almost diagonal" as well as "almost normal" and the
diagonal elements are approximations to the eigenvalues
of the original matrix. Experiments indicate that this
algorithm is more accurate than, and about fifty percent
faster than E b e r l e i n's algorithm [3].

1. INTRODUCTION

Let $A = (a_{rs})$ be a matrix of order n and let
$\Lambda = \text{diag}(\lambda_1, \lambda_2, \ldots, \lambda_n)$ be a matrix whose diagonal
elements are the eigenvalues of A. If A is real and
symmetric, the well-known Jacobi algorithm [15, 23]
provides us with a procedure for reducing A to a matrix
arbitrarily close to Λ. It is based on a sequence of

orthogonal similarity transformations, where the pth transformation uses the plane rotation matrix $R(k_p, m_p)$ with pivot pair (k_p, m_p). Hence, if

$$(1.1) \qquad R_q = \prod_{p=1}^{q} R(k_p, m_p),$$

then $R_q^T A R_q$ can be made arbitrarily close to Λ by choosing q sufficiently large and by choosing the pivot pairs appropriately. J a c o b i's algorithm can be extended to (complex) Hermitian matrices by choosing $R(k_p, m_p)$ to be unitary [12].

If A is non-Hermitian, it may or may not be diagonalizable by a similarity transformation. However, by a theorem of S c h u r [25] it is always triangularizable by a unitary similarity transformation. Consequently, G r e e n s t a d t [9], L o t k i n [17, 2], and others [2, 14] proposed generalizations of the Jacobi algorithm based on this fact. They use a finite sequence of unitary similarity transformations in order to reduce A to a matrix arbitrarily close to a triangular matrix

$$(1.2) \qquad T = \begin{bmatrix} \lambda_1 & t_{12} & t_{13} & \cdots & t_{1n} \\ & \lambda_2 & t_{23} & \cdots & t_{2n} \\ & & \lambda_3 & \cdots & t_{3n} \\ & & & \ddots & \vdots \\ & & & & \lambda_n \end{bmatrix}$$

whose diagonal elements are the eigenvalues of A. Computational results using this approach have not been very successful (except for the QR-algorithm [7, 16] which can be said to belong to this class of algorithms).

Another approach to the non-Hermitian case is to

divide such matrices into two classes; those which are
normal (i.e., those matrices N for which $N^H N = N N^H$)
and those which are not normal. Since all normal matrices
are diagonalizable by unitary similarity transformations
G o l d s t i n e   and   H o r w i t z   [8] generalized
the Jacobi algorithm to this case in 1959. In 1962
E b e r l e i n   [3] attacked the non-normal case. Her
approach makes use of the fact that A is normal if and
only if*

$$(1.3) \qquad \|A\|^2 = \|\Lambda\|^2,$$

whereas, if A is not normal.

$$(1.4) \qquad \|A\| > \|\Lambda\|^2.$$

H e n r i c i   [13] calls the quantity $[\|A\|^2 - \|\Lambda\|^2]^{1/2}$
the departure from normality** of A.

   E b e r l e i n's Jacobi-like algorithm [3] uses
norm-reducing similarity transformations to form the
sequence $A = A_0, A_1, A_2, \ldots, A_q$, where the object is to
make the departure from normality of A_q arbitrary
small. In other words, A_q "approaches" a normal matrix.
   To put this more formally, let $A_0, A_1, A_2, \ldots,$ be
a sequence of matrices, each of which is similar to A,
and let $N(A)$ be the set of normal matrices with the
same eigenvalues as A. The sequence $\{A_p\}$ is said to
converge to normality if there exists a sequence $\{N_p\}$
of matrices in $N(A)$ with the property that

---

* The norm $\|\cdot\|$ is the Euclidean matrix norm throughout
  this paper.
** In general, if a norm $\|\cdot\|_\nu$ is used, this quantity is
  called the ν-departure from normality.

$$(1.5) \qquad \lim_{p \to \infty} \| A_p - N_p \| = 0.$$

Then it can be proved (see [20], for example) that $\{A_p\}$ converges to normality if and only if

$$(1.6) \qquad \lim_{p \to \infty} \| A_p \|^2 = \| \Lambda \|^2,$$

or, equivalently, if and only if

$$(1.7) \qquad \lim_{p \to \infty} \left[A_p^H A_p - A_p A_p^H \right] = 0.$$

That convergence in (1.6) is feasible stems from the following theorem of M i r s k y [18].

$$(1.8) \qquad \inf_{Z} \| Z^{-1} A Z \|^2 = \| \Lambda \|^2,$$

where Z is any nonsingular matrix, and the infimum is achieved if and only if A is diagonalizable. Hence, by choosing q large enough and by choosing the similarity transformations appropriately, A_q can be made arbitrarily close to a normal matrix with the same eigenvalues as A.

Since the Euclidean norm is unitarily invariant, that is, since $\| U^H A U \| = \| A \|$ for all A and all unitary U, the only norm-reducing similarity transformations are non-unitary. Consequently, E b e r l e i n [3, 4, 5], and subsequently R u t i s h a u s e r [22] and others [20, 21] use non-unitary transformations for the norm reduction.

In this paper we propose a similar algorithm. It differs from those of Eberlein and the others in that we choose our similarity transformations for transforming $A = A_0$ into A_q somewhat differently. We describe, in Sections 2 and 3, a Jacobi-type procedure for an arbitrary nonsingular A, which involves the generation of a sequence of transformations

$$(1.9) \qquad A_{p+1} = Z^{-1}(k_p, m_p) A_p Z(k_p, m_p), \qquad p=0,1,\ldots,q-1,$$

such that if

$$(1.10) \qquad Z_{q-1} = \prod_{p=0}^{q-1} Z(k_p, m_p),$$

which implies

$$(1.11) \qquad A_q = Z_{q-1}^{-1} A Z_{q-1},$$

then A_q can be made arbitrarily close to a normal matrix (with the same eigenvalues as A) by choosing q sufficiently large and by choosing the pivot pairs (k_p, m_p) appropriately. In other words, $\|A_q\|^2$ can be made arbitrarily close to $\|A\|^2$, or what is more measurable, the absolute values of the elements of the commutator of A_q,

$$(1.12) \qquad C(A_q) = A_q^H A_q - A_q A_q^H,$$

can be made arbitrarily small.

Since our transformations in Section 2 achieve norm reduction by reducing only the off-diagonal norm[*], it is not unusual to expect that in most cases (and this has been our experience) the end result of our algorithm is not only an "almost normal" matrix but an "almost diagonal" matrix as well. Of course, some non-normal matrices

---

[*] See (4.5)

are diagonalizable, but even if A is not diagonalizable we know (see [1], p. 205, for example) that there always exists a matrix $S = S(\varepsilon)$ for which $S^{-1}AS = T$, as in (1.2), and for which

$$(1.13) \qquad \sum_{k<j} |t_{kj}| \le \varepsilon,$$

where $\varepsilon > 0$ is any preassigned positive constant. Thus, if and when A_q is "almost diagonal", the diagonal elements of A_q are approximations to the eigenvalues of A.

In Section 4 we exhibit the numerical results obtained when our algorithm is applied to several real and complex matrices. These results are somewhat better than those of E b e r l e i n [3, 4, 5] in that convergence is more rapid.

2. TRANSFORMATION MATRICES

We begin by looking at two kinds of similarity transformations. The first kind is of the form

$$(2.1) \qquad B = D^{-1}(k)AD(k),$$

where $D(k) = \mathrm{diag}(1,\ldots,1,d_k,1,\ldots,1)$ with the kth diagonal element $d_k \ne 0$ chosen so as to reduce $\|B\|^2$. O s b o r n e [19] uses such similarity tranformations to pre-condition matrices.

Suppose we let

$$(2.2) \qquad
\begin{aligned}
\mu_k &= \left[\sum_{\substack{j=1 \\ j \ne k}}^{n} |a_{kj}|^2 \right]^{1/2} \\[2ex]
\xi_k &= \left[\sum_{\substack{j=1 \\ j \ne k}}^{n} |a_{jk}|^2 \right]^{1/2},
\end{aligned}$$

where we assume $\mu_k \neq 0$ and $\xi_k \neq 0$. If we choose d_k to be

$$(2.3) \qquad d_k = \left[\frac{\mu_k}{\xi_k}\right]^{1/2},$$

then it can be shown that $\|B\|^2$ is minimal and

$$(2.4) \qquad \|A\|^2 - \|B\|^2 = (\mu_k - \xi_k)^2.$$

Thus, whenever $\mu_k \neq \xi_k$ there is a reduction in the Euclidean norm and this reduction is maximal.
If $\xi_k = 0$ and/or $\mu_k = 0$, let

$$(2.5) \qquad \begin{aligned} \hat{\mu}_k &= \left[\sum_{j=1}^{n} |a_{kj}|^2\right]^{1/2} \\[2ex] \hat{\xi}_k &= \left[\sum_{j=1}^{n} |a_{jk}|^2\right]^{1/2}. \end{aligned}$$

If either $\hat{\mu}_k = 0$, or $\hat{\xi}_k = 0$, then A is singular and has a zero eigenvalue. In this case we can deflate the matrix and continue. If neither $\hat{\mu}_k = 0$ nor $\hat{\xi}_k = 0$, we choose d_k to be

$$(2.6) \qquad d_k = \left[\frac{\hat{\mu}_k}{\hat{\xi}_k}\right]^{1/2}.$$

For this value of d_k

$$(2.7) \qquad \|A\|^2 - \|B\|^2 = (\hat{\mu}_k - \hat{\xi}_k)^2 + |a_{kk}|^2\left[d_k - \frac{1}{d_k}\right]^2$$

and, whenever $\hat{\mu}_k \neq \hat{\xi}_k$, we have a reduction (but not necessarily maximal) in the Euclidean norm.

    The second kind of similarity transformation involves two matrices. One is unitary and the other is not. Given a matrix A we write, for the pivot pair (k,m),

$$F = G^H(k,m)AG(k,m)$$

(2.8)

$$B = L^{-1}(k,m)FL(k,m).$$

The unitary rotation matrix $G(k,m)$ has the form

$$
(2.9) \qquad G(k,m) =
\begin{bmatrix}
1 & & & & \vdots & & & \vdots & & & \\
 & \ddots & & & \vdots & & & \vdots & & & \\
 & & \ddots & & \vdots & & & \vdots & & & \\
 & & & 1 & \vdots & & & \vdots & & & \\
\hline
\cdots & \cdots & \cdots & c & \cdots & \cdots & -se^{i\alpha} & \cdots & & k \\
 & & & 1 & & & \vdots & & & \\
 & & & & \ddots & & \vdots & & & \\
 & & & & & \ddots & \vdots & & & \\
 & & & & & 1 & \vdots & & & \\
\hline
\cdots & se^{-i\alpha} & \cdots & \cdots & c & \cdots & \cdots & & m \\
 & & & & & 1 & & & \\
 & & & & & \ddots & & \\
 & & & & & & \ddots & \\
 & & & & & & & 1
\end{bmatrix}
\begin{matrix} \\ \\ \\ \\ k \\ \\ \\ \\ \\ m \end{matrix}
$$

$$\qquad\qquad\qquad\qquad k \qquad\qquad m$$

which differs from the identity matrix in only four elements

$$
(2.10) \qquad
\begin{cases}
g_{kk} = g_{mm} = c \\[4pt]
g_{km} = -\overline{g}_{mk} = -se^{i\alpha},
\end{cases}
$$

where the constants α, and

$$
(2.11) \quad
\begin{aligned}
c &= \cos\theta \\
s &= \sin\theta
\end{aligned}
$$

are real (θ is the rotation angle) and $\bar{x}$ denotes the complex conjugate of x.

We choose $G(k,m)$ so as to annihilate the element f_{mk} in F. In this case the parameters α and θ must satisfy the equation

$$
(2.12) \quad \tan\theta = \frac{2a_{mk}e^{i\alpha}}{t_{km}\pm[t_{km}^2+4a_{km}a_{mk}]^{1/2}} \, ,
$$

where

$$
(2.13) \quad t_{km} = a_{kk}-a_{mm} \, .
$$

Here α is chosen so that $\tan\theta$ is real and θ is chosen to be the smaller (in absolute value) of the two rotation angles. An equation similar to (2.12) exists if f_{km} is to be annihilated rather than f_{mk}.

The matrix $L(k,m)$ is chosen so as to minimize $\|B\|^2$. It differs form the identity matrix in only one off-diagonal element. If (k,m) is the pivot pair in forming F so that f_{mk} is annihilated, then the non-zero off-diagonal element of $L(k,m)$ is in the (k,m) position. Thus,

$$
(2.14) \quad \ell_{km} = \rho e^{i\beta},
$$

where ρ and β are real.

The transformation $B = L^{-1}(k,m)FL(k,m)$ produces changes in the kth row and m^{th} column of F as follows:

$$b_{kj} = f_{kj} - f_{mj}\rho e^{i\beta}$$
$$b_{jm} = f_{jm} + f_{jk}\rho e^{i\beta} \left.\right\} \quad j \neq k,m$$

$$(2.15) \quad b_{kk} = f_{kk}$$

$$b_{mm} = f_{mm}$$

$$b_{km} = f_{km} + \tau_{km}\rho e^{i\beta} ,$$

where

$$(2.16) \quad \tau = f_{kk} - f_{mm} .$$

REMARK. Notice that "pre-treatment" of A by the unitary transformation involving $G(k,m)$ does not reduce the Euclidean norm but it does annihilate f_{mk}. This makes the non-unitary transformation involving $L(k,m)$, which is a norm-reducing transformation, somewhat simpler in the sense that b_{km} is a function of $\rho e^{i\beta}$ rather than $\rho^2 e^{i2\beta}$ and the diagonal elements of F are unchanged. In other words, (2.21) is a *quadratic* function of ρ and (2.23) gives the optimum value of ρ by a simple calculation.

In order to examine the norm reduction in A we define the quantity

$$(2.17) \quad \Delta = \|A\|^2 - \|B\|^2$$

$$= \|F\|^2 - \|B\|^2 ,$$

since there is no change in norm in forming F. In choosing the parameters ρ and β which determine $L(k,m)$ we wish to minimize $\|B\|^2$, or what is equivalent, we wish to maximize Δ.

By an easy computation we find that

$$(2.18) \qquad \Delta = 2\rho \operatorname{Re}[\, v_{km} e^{-i\beta}\,] - \rho^2 u_{km},$$

where

$$(2.19) \qquad
\begin{aligned}
v_{km} &= \sum_{\substack{j=1 \\ j \neq k,m}}^{n} [\, f_{kj}\overline{f}_{mj} - f_{jm}\overline{f}_{jk}\,] - f_{km}\overline{\tau}_{km} \\[2ex]
u_{km} &= |\tau_{km}|^2 + \sum_{\substack{j=1 \\ j \neq k,m}}^{n} [\, |f_{jk}|^2 + |f_{mj}|^2\,],
\end{aligned}$$

and $\operatorname{Re}[x]$ is the real part of the complex number x.

In order to simplify the expression for Δ in (2.18) we choose β so as to satisfy the equation

$$(2.20) \qquad v_{km} e^{-i\beta} = |v_{km}|.$$

This enables us to write

$$(2.21) \qquad \Delta = 2\rho |v_{km}| - \rho^2 u_{km},$$

and if neither $|v_{km}| = 0$ nor $u_{km} = 0$, we observe that Δ is a quadratic function of ρ. This function is positive for

$$(2.22) \qquad 0 < \rho < \frac{2|v_{km}|}{u_{km}},$$

and achieves a maximum when

$$(2.23) \qquad \rho = \frac{|v_{km}|}{u_{km}}.$$

Consider the following special cases in which Λ is not a quadratic function of ρ.

(i) $|v_{km}| = u_{km} = 0$. In this case $\Lambda \equiv 0$ and no norm reduction takes place. Consequently, we set $\rho = 0$, in which case $L(k,m) = I$, and $B = F$ in (2.8). Thus, the transformation using $L(k,m)$ is avoided.

(ii) $|v_{km}| = 0$ but $u_{km} \neq 0$. In this case $\Lambda < 0$ and instead of norm reduction, we see from (2.17) that $\|B\|^2 > \|A\|^2$. Consequently, we set $\rho = 0$ and again we avoid the transformation using $L(k,m)$.

(iii) $|v_{km}| \neq 0$ but $u_{km} = 0$. This cannot happen since $u_{km} = 0$ implies $v_{km} = 0$, from (2.19).

In order to keep ρ bounded we shall choose ρ as follows:

$$(2.24) \quad \rho = \begin{cases} \dfrac{|v_{km}|}{u_{km}}, & 0 < |v_{km}| \leq u_{km}, \\[2ex] 1, & 0 < u_{km} < |v_{km}|. \\[2ex] 0, & 0 = |v_{km}|, \end{cases}$$

in which case, the norm reduction is

$$(2.25) \quad \Delta = \begin{cases} \dfrac{|v_{km}|^2}{u_{km}}, & 0 < |v_{km}| \leq u_{km} \\[2ex] 2|v_{km}| - u_{km}, & 0 < u_{km} < |v_{km}|, \\[2ex] 0, & 0 = |v_{km}|. \end{cases}$$

3. THE ALGORITHM

Our procedure is to apply the transformations defined in (2.1) and (2.8) to the matrix A with the pivot pairs (k,m) chosen in cyclic order. We use (2.1) when $k = m$ and (2.8) when $k \neq m$. An obvious choice of pivot pairs is the *typewriter ordering* $(1,1),(1,2),\ldots$ $\ldots,(1,n),\ (2,1),\ (2,2),\ldots,(2,n),\ldots,(n,n)$.

In order to keep d_k within reasonable bounds for computational purposes we choose to omit the transformation (2.1) or (2.6) unless $d_k \in [\,\|A\|^{-1},\ \|A\|\,]$. Hence, $\|D(k)\|$ and $\|D^{-1}(k)\|$ are bounded, which implies that the transformation (2.1) or (2.6) cannot increase the norm of A. Likewise, (2.22) and (2.24) ensure that ρ is bounded so $\|L(k,m)\|$ and $\|L^{-1}(k,m)\|$ are bounded. Also, since $G(k,m)$ is unitary, $\|A\| = \|F\|$ in (2.8). Hence, the transformation (2.8) cannot increase the norm of A.

(3,1) THEOREM. *Consider the sequence* $A = A_0, A_1, \ldots, A_q$ *where*

$$A_{p+1} = Z^{-1}(k_p,m_p) A_p Z(k_p,m_p), \quad (p=0,1,\ldots,q-1),$$

with $Z(k_p,m_p)$ *defined by*

$$Z(k_p,m_p) = \begin{cases} D(k_p), & k_p = m_p, \\[2mm] G(k_p,m_p)L(k_p,m_p) & k_p \neq m_p. \end{cases}$$

Choose the pivot pairs (k_p,m_p) *using the typewriter ordering. If* A *is nonsingular, then for sufficiently large* q*, the absolute value of each element of*

$$C(A_q) = A_q^H A_q - A_q A_q^H$$

is arbitrarily close to zero.

PROOF. As a consequence of (2.4), (2.7), (2.17), and (2.25) the norms $\|A_p\|^2$ form a monotone decreasing sequence and, because of (1.8), they tend to a limit not smaller than $\|\Lambda\|^2$. Hence

$$(3.2) \qquad \lim_{p\to\infty}[\,\|A_p\|^2 - \|A_{p+1}\|^2\,] = 0.$$

Let (k_p, m_p) be the pivot pair for the pth transformation. There are two cases.

Case I. In this case $k_p = m_p$ and so

$$(3.3) \qquad A_{p+1} = D^{-1}(k_p)A_p D(k_p).$$

Thus, from (2.4), (2.7), and (3.2),

$$(3.4) \qquad \begin{cases} \lim_{p\to\infty} \mu_{k_p} = \xi_{k_p} \\[2mm] \lim_{p\to\infty} \hat{\mu}_{k_p} = \hat{\xi}_{k_p} \end{cases}.$$

This last result implies the absolute value of each diagonal element of $C(A_p) = (c_{rs}^{(p)})$ goes to zero, that is,

$$(3.5) \qquad \lim_{p\to\infty}|c_{kk}^{(p)}| = 0, \qquad k = 1, 2, \ldots, n.$$

It also implies

$$(3.6) \qquad \lim_{p\to\infty} D(k_p) = I.$$

Case II. In this case $k_p \neq m_p$ and so A_{p+1} is obtained by using

$$(3.7) \quad \begin{cases} F_p = G^H(k_p, m_p) A_p G(k_p, m_p) \\ A_{p+1} = L^{-1}(k_p, m_p) F_p L(k_p, m_p). \end{cases}$$

From (2.19), (2.25), and (3.2),

$$(3.8) \quad \lim_{p \to \infty} |v_{k_p m_p}| = 0.$$

Thus, if we let $\rho^{(p)}$ be the parameter which determines $L(k_p, m_p)$, we have

$$(3.9) \quad \lim_{p \to \infty} \rho^{(p)} = 0$$

from (2.24), and this implies

$$(3.10) \quad \lim_{p \to \infty} L(k_p, m_p) = I.$$

Since both $D(k_p)$ and $L(k_p, m_p)$ approach the identity matrix from (3.6) and (3.10) it follows that

$$(3.11) \quad \lim_{p \to \infty} A_{p+1} = F_p.$$

Hence, we examine the commutator of F_p in order to study the commutator of A_{p+1}.

If we let $G_p = G(k_p, m_p)$, for simplicity of notation, we can write

$$(3.12) \quad C(F_p) = F_p^H F_p - F_p F_p^H = G_p^H [A_p^H A_p - A_p A_p^H] G_p = G_p^H C(A_p) G_p$$

and so

$$(3.13) \quad \| C(F_p) \| = \| C(A_p) \|.$$

Since $|v_{k_p m_p}|$ approaches zero, from (3.8), and since the element of F_p in the (m_p, k_p) position is zero (by definition) it follows that the element of

$C(F_p)$ in the (k_p, m_p) position approaches zero. This fact plus (3.5), (3.6), (3.10), and (3.12) imply that

$$(3.14) \qquad \lim_{p \to \infty} \left| c_{km}^{(p)} \right| = 0, \qquad k \neq m.$$

Since both $\left| c_{kk}^{(p)} \right|$ and $\left| c_{km}^{(p)} \right|$, for $k \neq m$, approach zero as p increases without bound, we can make the moduli of the elements of $C(A_q)$ arbitrarily small by choosing q sufficiently large. This completes the proof.

4. DISCUSSION AND NUMERICAL EXAMPLES

Jacobi-like algorithms have not enjoyed the success of the QR-algorithm when used on serial computers. However, S a m e h [24] has demonstrated the efficiency of Jacobi-like algorithms on parallel computers and it is highly improbable that the QR-algorithm can match this efficiency. Consequently, any success in construct-ing a better Jacobi-like algorithm will have practical as well as theoretical interest. This has been motivation in proposing the present algorithm. In the numerical examples below, we obtain better results than E b e r l e i n [3, 4, 5] in that convergence is more rapid.

A modified version of the algorithm described in Sections 2 and 3 has been implemented in FORTRAN IV on the IBM 360/65. Double-precision arithmetic is used and the pivot pairs (k, m) which form a sweep* are chosen in the order $(1,1),(1,2),\ldots,(1,n),(2,2),(2,3),\ldots$ $\ldots,(2,n),\ldots,(n-1,n-1),\ (n-1,n),\ (n,n)$. We use

---

* A sweep is a sequence of $\frac{n}{2}(n+1)$ transformations, in which no pivot pair appears more than once.

transformation (2.1) when $k = m$ and transformation (2.8) when $k \neq m$. The Greenstadt transformation involving $G(k,m)$ in (2.8) is so chosen that $f_{mm} \leq f_{kk}$. Another difference between the algorithm in Sections 2 and 3 and our implementation is that prior to each sweep of transformations the matrix is replaced by its transpose (if necessary) in order to make the norm of the super-diagonal elements greater than the norm of the sub-diagonal elements,* since we never select pivots from the sub-diagonal elements.

In most cases the matrix A_q is "almost diagonal" as well as "almost normal" and, as we mentioned at the end of Section 1, the diagonal elements are approximations to the eigenvalues of A.

An example of a matrix which cannot be reduced to almost diagonal form is the following.

$$(4.1) \qquad A = \begin{bmatrix} a & 0 & b \\ c & a & 0 \\ 0 & d & a \end{bmatrix},$$

with $bcd \neq 0$. When our algorithm is applied to this matrix, we converge either to a normal matrix of the form

$$(4.2) \qquad B = \begin{bmatrix} a & 0 & e \\ e & a & 0 \\ 0 & e & a \end{bmatrix},$$

$$* \left[\sum_{i<j} |a_{ij}|^2 \right]^{1/2} > \left[\sum_{i>j} |a_{ij}|^2 \right]^{1/2}.$$

or a normal matrix of the form

$$(4.3) \qquad C = \begin{bmatrix} a & e & 0 \\ 0 & a & e \\ e & 0 & a \end{bmatrix},$$

where e is not necessarily smaller than a. Notice that A, B and C have a similar zero-nonzero structure. Also, B and C have the form $eP+aI$, where P is a permutation matrix,* so the eigenvalues are easily computed.

To verify that A is transformed into either B or C consider a sweep of six transformations with pivot pairs chosen in the order $(1,1),(1,2),(1,3),(2,2),(2,3),(3,3)$. The first transformation on $A = A_0$ uses $D(1)$ and, as a result, the norm of the first row of A_1 equals the norm of the first column of A_1. Moreover, the diagonal elements of A_0 and A_1 are the same and the zeros in row one and column one are undisturbed. The second transformation in the sweep uses $G(1,2)$ and $L(1,2)$ on A_1. The rotation angle θ in (2.12) is $\frac{\pi}{2}$ and so $G(1,2)$ merely interchanges rows one and two and interchanges columns one and two in forming F in (2.8). The parameter ρ in (2.24) which defines $L(1,2)$ is given by

$$(4.4) \qquad \rho = \frac{|v_{12}|}{u_{12}} = \frac{f_{13}\overline{f}_{23}-f_{32}\overline{f}_{31}-f_{12}\overline{\tau}_{12}}{|\tau_{12}|^2+|f_{31}|^2+|f_{23}|^2} = 0$$

and so $L(1,2) = I$. Other Greenstadt rotations in the sweep will also have rotation angles of $\frac{\pi}{2}$ and in each

---

* This fact was called to our attention by P.J. Eberlein.

case $L(k,m) = I$. Thus, norm reduction only takes place
when the pivot is on the main diagonal. In the process
of converging to a normal matrix we alternate between
matrices of form B and form C and the final form
merely depends on the point at which the iteration is
terminated.

In the examples which follow

$$(4.5) \qquad N = \left[\sum_{i \neq j} |a_{ij}|^2 \right]^{1/2}$$

is the off-diagonal norm. Our stopping criteria (applied
at the end of each sweep) are

    (i)   terminate when $N < 10^{-8}$, or

    (ii)   terminate after thirty sweeps.

In examples 2 through 7 we have referenced each
matrix so that the matrices themselves do not have to
be printed here. In each case our algorithm reduced the
off-diagonal norm below the amount indicated. The
diagonal elements are printed, as approximations to the
eigenvalues, along with a bound on N and the number of
sweeps required. Example 1, on the other hand, illustra-
tes (4.1) and (4.3).

EXAMPLE 1. [4,10]

$$(12 \text{ sweeps}) \quad \begin{bmatrix} 1 & 0 & 0.01 \\ 0.1 & 1 & 0 \\ 0 & 1 & 1 \end{bmatrix} \to \begin{bmatrix} 1 & 0.1 & 0 \\ 0 & 1 & 0.1 \\ 0.1 & 0 & 1 \end{bmatrix}.$$

After twelve sweeps no improvement was obtained during
subsequent transformations. The normal matrix on the
right was obtained to double-precision accuracy.

EXAMPLE 2.   [11, p.86]
This matrix is defective with two pairs of equal eigenvalues $3 \pm \sqrt{5}$. Computed eigenvalues are

$$
\begin{array}{rcl}
5.236068064 & + & 0.0000000383\ i \\
5.236067891 & - & 0.0000000383\ i \\
\text{(7 sweeps) } 0.7639320378 & + & 0.0000000166\ i \\
0.7639320072 & - & 0.0000000166\ i
\end{array}
$$

$$N < 4 \times 10^{-13}$$

EXAMPLE 3.   [11, p.120]
The computed eigenvalues are

$$
\begin{array}{rcl}
127.3866707730673 & + & 132.2782032001213\ i \\
-9.459984021891320 & + & 7.280185836923769\ i \\
\text{(5 sweeps) } 7.073313248823678 & - & 9.558389037045480\ i \\
0.000000000000012 & + & 0.000000000000025\ i \\
0.000000000000006 & + & 0.000000000000005\ i
\end{array}
$$

$$N < 2 \times 10^{-18}$$

These results agree with those in [11] to at least
twelve significant digits.

EXAMPLE 4.   [11, p.90]
This matrix is defective with eigenvalues -1 and two
paris of equal eigenvalues $1.5 \pm \sqrt{12.75}\ i$.
Computed eigenvalues are

$$
\begin{array}{rcl}
1.500000158 & + & 3.570714190\ i \\
1.500000155 & - & 3.570714174\ i \\
\text{(9 sweeps) } 1.499999845 & - & 3.570714255\ i \\
1.499999842 & + & 3.570714239\ i \\
-1.000000000 & + & 0.000000000\ i
\end{array}
$$

$$N < 9 \times 10^{-20}$$

EXAMPLE 5.   [4, example II]

The exact eigenvalues are -1, $\pm i$, $\frac{\sqrt{2}}{2}(1 \pm i)$, $\frac{\sqrt{2}}{2}(-1 \pm i)$.

Computed eigenvalues are

$$
\begin{array}{rl}
-0.9999999999999973 & + \ 0.0000000000000001 \ i \\
-0.7071067811865454 & + \ 0.7071067811865455 \ i \\
0.7071067811865452 & - \ 0.7071067811865452 \ i \\
\text{(6 sweeps)} \quad -0.0000000000000002 & + \ 0.9999999999999971 \ i \\
0.0000000000000000 & - \ 0.9999999999999973 \ i \\
0.7071067811865462 & + \ 0.7071067811865464 \ i \\
-0.7071067811865451 & - \ 0.7071067811865450 \ i
\end{array}
$$

$$N < 8 \times 10^{-16}$$

EXAMPLE 6.   [11, p.121]

The computed eigenvalues are accurate to at least eight significant digits. They are

$$
\begin{array}{rcl}
10.79767645 & + & 8.623381528 \ i \\
-4.966870087 & - & 8.087124772 \ i \\
1.032058125 & + & 9.294132798 \ i \\
8.811309285 & + & 1.549382663 \ i \\
\text{(7 sweeps)} \quad 2.389887590 & + & 7.268070718 \ i \\
5.436448381 & - & 3.971425823 \ i \\
-5.279506163 & - & 2.275963035 \ i \\
4.161748692 & + & 3.137513560 \ i \\
-1.935201450 & - & 3.975093812 \ i \\
-2.447550824 & + & 0.4371261761 \ i
\end{array}
$$

$$N < 5 \times 10^{-18}$$

EXAMPLE 7. [11, p.92]

The computed eigenvalues are

$$
\begin{array}{rl}
 & 32.228891501571 \quad + \; 0.00000000000000 \; i \\
 & 20.198988645876 \quad - \; 0.00000000000000 \; i \\
 & 12.311077400868 \quad + \; 0.00000000000000 \; i \\
 & 6.9615330855670 \quad - \; 0.00000000000000 \; i \\
 & 3.5118559485807 \quad - \; 0.00000000000001 \; i \\
(15 \text{ sweeps}) \quad & 1.5539887091320 \quad - \; 0.00000000000004 \; i \\
 & 0.64350531900581 \quad - \; 0.00000000000029 \; i \\
 & 0.28474972050091 \quad + \; 0.00000000002552 \; i \\
 & 0.14364652066060 \quad - \; 0.0000000015242 \; i \\
 & 0.08122765528 0648 \quad + \; 0.0000000019343 \; i \\
 & 0.049507435270457 \quad + \; 0.00000000006667 \; i \\
 & 0.031028057683804 \quad - \; 0.00000000013285 \; i
\end{array}
$$

$$N < 4 \times 10^{-11}.$$

The following table shows the number of sweeps needed to
bring the matrices in examples 2 through 7 to convergence
by both our procedure and by E b e r l e i n's procedure.

Example	Number of sweeps	
	E b e r l e i n's procedure*	Our procedure
2	16	7
3	7	5
4	24	9
5	8	6
6	18	7
7	30	15

* Data used here are from [3,4,5].

REFERENCES

[1]   R. Bellman, *Introduction to Matrix Analysis*; Second
      Edition, New York, McGraw-Hill, 1970.

[2]   R.L. Causey, Computing eigenvalues of non-Hermitian
      matrices by methods of Jacobi type, *Jour. SIAM*,
      6(1958), 172-181.

[3]   P.J. Eberlein, A Jacobi-like method for the automatic
      computation of eigenvalues and eigenvectors of an
      arbirary matrix, *Jour. SIAM*, 10(1962), 74-88.

[4]   P.J. Eberlein - J. Boothroyd, Solution to the eigen-
      problem by a norm-reducing Jacobi-type method,
      Numer. Math., 11(1968), 1-12.

[5]   P.J. Eberlein, Solution to the complex eigenproblem
      by a norm-reducing Jacobi-type method, *Numer. Math.*,
      14(1970), 232-245.

[6]   P.J. Eberlein, On a new and stable use of elementary
      transformations for the modified *LR* and the *QR*
      algorithms, *presented at IFIP Congress '74*.

[7]   J.G.F. Francis, The *QR*-transformation, Part I and
      II, *The Computer Journal*, 4(1961), 265-271, 332-
      -345.

[8]   H.H. Goldstine - L.P. Horwitz, A procedure for the
      diagonalization of normal matrices, *Jour. ACM*,
      6(1959), 176-195.

[9]   J. Greenstadt, A method for finding roots of
      arbitrary matrices, *Math. Tables Aids Comput.*,
      9(1955), 47-52.

[10] J. Greenstadt, Some numerical experiments in
      triangularizing matrices, *Numer. Math.*, 4(1962),
      187-195.

[11] R.T. Gregory - D.L. Karney, *A Collection of Matrices for Testing Computational Algorithms*; Wiley-Interscience, 1969.

[12] P. Henrici, On the speed of convergence of cyclic and quasicyclic Jacobi methods for computing eigenvalues of Hermitian matrices, *Jour. SIAM*, 6(1958), 144-162.

[13] P. Henrici, Bounds for iterates, inverses, spectral variation and fields of values of non-normal matrices, *Numer. Math.*, 4(1962), 24-40.

[14] C.P. Huang, A Jacobi-type method for triangularizing an arbitrary matrix, *SIAM J. Numer. Anal.*, 12(1975), 566-570.

[15] C.G.J. Jacobi, Über ein leichtes Verfahren, die in der Theorie der Säkularstörungen vorkommenden Gleichungen numerisch aufzulösen, *J. Reine Angew. Math.*, 30(1846), 51-94.

[16] V.N. Kublanovskaya, On some algorithms for the solution of the complete eigenvalue problem, *Zh. Vych. Mat.*, 1(1961), 555-570 (English translation: USSR Comp. Math. and Math. Phys., 2(1962), 637-657).

[17] M. Lotkin, Characteristic values of arbitrary matrices, *Qart. Appl. Math.*, 14(1956), 267-275.

[18] L. Mirsky, On the minimization of matrix norms, *Amer. Math. Monthly*, 65(1958), 106-107.

[19] E.E. Osborne, On pre-conditioning of matrices, *Jour. ACM*, 7(1960), 338-345.

[20] M.H.V. Paardekooper, An eigenvalue algorithm based on norm-reducing transformations, *Doctoral Dissertation, Eindhoven Institute of Technology*, Eindhoven, 1969.

[21] A. Ruhe, On the quadratic convergence of a
     generalization of the Jacobi method to arbitrary
     matrices, *BIT*, 8(1968), 210-231.

[22] H. Rutishauser, Une methode pour le calcul des
     valeurs propres des matrices non symetriques,
     Comptes Rendus, 259(1964), 2758.

[23] H. Rutishauser, The Jacobi method for real symmetric
     matrices, in J.H. Wilkinson, C. Reinsch, *Handbook
     for Automatic Computation*, vol. II, *Linear Algebra*,
     Springer-Verlag, 1971.

[24] A.H. Sameh, On Jacobi and Jacobi-like algorithms
     for a parallel computer, *Math. of Comp.*, 25(1971),
     579-590.

[25] I. Schur, Über die characteristischen Wurzeln einer
     linearen Substitution mit einer Anwendung die
     Theorie der Integral Gleichungen, *Math. Ann.*,
     66(1909), 488-510.

[26] V.V. Voevodin, a short summary in [20].

[27] J.H. Wilkinson, Error analysis of floating-point
     computation, *Numer. Math.*, 2(1960), 319-340.

[28] J. Wolf, Methodes de calcul des valeurs propers
     d'une matrice quelconque par utilisation de
     transformations unitaires, *Rev. Francaise Traitment
     Information, Chiffres*, 9(1966), 309-326.

C.P. Huang
R.T. Gregory
Computer Science Department
The University of Tennessee
Knoxville, Tennessee 37916 U.S.A.

DIRECT INVESTIGATION METHODS FOR THE EXTERIOR BOUNDARY VALUE PROBLEMS OF ELLIPTIC EQUATIONS

A.S. ILINSKII

In the present paper direct methods for solving exterior boundary value problems of elliptic equations with variable coefficients are examined.

For exterior boundary value problems of elliptic equations, it is necessary to formulate conditions for the infinity, which provides the uniqueness of the boundary value problem. For the Laplace-equation, regularity of the solution in the infinity is such a usual condition. For more general elliptic-type equations, suitable conditions must be given in each concrete case which are called, the "radiation-conditions". The concrete form of these conditions depends on the type of the equations and also on the form of the surrounding domain. In case of bodies having a finite boundary, the corresponding conditions for the equations with constant coefficients are investigated in the papers [1] and [2].

A more general approach to the solution of the exterior boundary problem can be found in the papers [3] and [4], where general limit principles are formu-

lated, the "principle of the limit adsorption" and the
"principle of limit amplitude".

    This approach makes it possible to formulate the
conditions of radiation for a wide class of elliptic
equations in an arbitrary domain with boundaries located
in the infinity.

    In the general case of boundaries of complicated
configuration, the solution of the exterior boundary
value problem for equations with variable coefficients
can't be given by analytic methods, only numerical methods can guarantee efficiency, although the careless application of usual numerical methods may lead to inef-
fective numerical algorithms. Therefore we turn to the
solution of the problem in a finite domain of the space.
In the present paper examples of efficiently organized
numerical algorithms are considered for the solution of
exterior boundary value problems for a special class of
elliptic equations. For the sake of better demonstration
the following boundary value problem will be considered:
Find the solution of the following equation in the finite
domain T, with a smooth boundary S:

$$(1) \qquad L[u] = \mathrm{div}(p(x,y,z)\,\mathrm{grad}\ u) + q(x,y,z)u = 0,$$

satisfying the boundary condition on the surface S

$$(2) \qquad p\,\frac{\partial u}{\partial n} - \alpha u = \Phi, \qquad p \in S,$$

where $p(x,y,z) > 0$, $q(x,y,z)$ is a continuous complex
function; $\alpha(p)$, $\Phi(p)$ are given continuous functions on
the S surface.

    Let us suppose the existence of a sphere with a
radius R and a centre inside the domain T, such that

the functions $p(x,y,z)$ and $q(x,y,z)$ are constants outside the sphere, and therefore equation (1) will turn into the Helmholtz equation outside the sphere:

$$(3) \qquad \Delta u + k_0^2 u = 0$$

with constant real coefficient k_0. In this case the radiation condition can be formulated in the Sommerfeld--Rellich form:

$$(4) \qquad \lim_{r \to \infty} \int_{\Sigma_r} \left| \frac{\partial u}{\partial n} - i k_0 u \right|^2 dt = 0, \qquad r > R,$$

where Σ_r is a sequence of spheres containing entirely the sphere with radius R, where the coefficients of equation (1) are constants. We suppose the surface to be so that the solution of the investigated boundary value problem exists and the use of the Green-formulae is possible. The conditions (4) are guaranteeing the uniqueness of our boundary value problem but their use is inconvenient in the numerical methods, because of their asymptotic form. From the view point of numerical methods an other form of the radiation condition is more convenient, this form has been introduced by A.G. S v e s h n i k o v [5] as the "partial condition of radiation".

Since the solution of equation (3) is analytic outside the sphere Σ_R, it permits its expansion in a uniformly and absolutely converging series of the following spherical functions:

$$(5) \qquad u(M) = \sum_{nm} T_{nm} \zeta_n^1 (k_0 r) Y_n^m (\theta, \varphi), \qquad r \geq R,$$

where r, θ, φ are the spherical coordinates of the point M outside the sphere Σ_R,

$$\zeta_n^1(x) = \sqrt{\frac{2}{\pi x}}\, H_{n+1/2}^{(1)}(x)$$

is the spherical function of Hankel, $Y_n^m(\Theta,\varphi)$ is a spherical angular function. Obviously, the function $u(M)$ satisfies the radiation condition. The coefficients T_{nm} are unknown and they determine the behaviour of the solution in the infinity. Formula (5) is one of the forms of analytic continuation of the solution outside the sphere of radius R.

Let us consider the sequence of concentric spheres in the domain outside the sphere Σ_R. On each sphere Σ_r $(r > R)$ a function

$$u_{nm}(r) = \int_{\Sigma_r} u(r,\Theta,\varphi) Y_n^m(\Theta,\varphi)\, d\Omega$$

can be defined.

If the function $u_{nm}(r)$ satisfies the system of "partial radiation conditions"

$$(6) \qquad \frac{\partial u_{nm}}{\partial r} - i\gamma_n u_{nm} = 0 \qquad r \geq R$$

for each integer n and m where

$$\gamma_n = \frac{\dfrac{d}{dr}(\zeta_n^1(k_0 r))}{\zeta_n^1(k_0 r)}\, ,$$

then the function u satisfies the radiation conditions of S o m m e r f e l d - R e l l i c h (4). Formulae (6) represent a system of "partial radiation conditions" for each $r \geq R$.

From the formulae

$$\operatorname{Im}\left\{\frac{d}{dr}\,\zeta_n(k_0 r)\cdot\zeta_n^*(k_0 r)\right\} = \frac{1}{k_0^2 r^2}$$

the equality

$$\operatorname{Im}\int_{\Sigma_r}\frac{\partial u}{\partial u}\,u^*\,d\tau = \frac{1}{k_0}\sum_{nm}|T_{nm}|^2$$

follows for each $r \geq R_0$.

In this way the conditions (6) and (2) make it possible to investigate the exterior boundary value problem in a finite domain D of the space, bounded by the surface S and the spheres Σ_R.

Now it is possible to create a direct method for the solution in the finite domain D. The infiniteness and non-selfadjointness of the boundary value problem are its distinguishing features. However the "partial radiation conditions" have a non-local character, and that doesn't permit the use of certain direct methods such as the Crank-Nicholson-method.

The most natural way of solving a boundary value problem is done by the projection method. The fundamental idea of the method is rather simple, it has appeared in the publications of L.V. K a n t o r o v i c h [6].

Let it be possible to choose a curvilinear coordinate ζ, in domain D, such that the intersection of the domain D with the coordinate-surface $\zeta = $ const be simply connected.

In each intersection of the plane $\zeta = $ const and the domain D we choose a basic system of functions $\{\chi_k(\zeta)\}$ parametrically depending on ζ and satisfying the boundary conditions of the problem. Then the approximate solution of the problem (1)-(4) will be generated in the form of a finite sum

$$u_N = \sum_{k=1}^{N} z_k(\zeta)\chi_k$$

with suitably determined functions $z_k(\zeta)$. In order to find the unknown functions $z_n(\zeta)$ conditions of a projectional type are generally used. They can be choosen so, that fundamental integral correlations, should be satisfied for the approximate solution u_N which are valid for the exact solution of the problem.

Now we regard the concrete scheme of the projectional method for the boundary problem (1)-(4).

We transform the domain D into a spherical layer so, that the sphere Σ_R gets transformed into a sphere of radius $\zeta = 2$ in the new system of coordinates and the surface S into a sphere with $\zeta = 1$, concentric to the previous one. Such a transformation can be carried out by several methods. We shall suppose, that this transformation is smooth. In this case equation (1) turns into an other equation with variable coefficients in the new system of coordinates. The section $\zeta = \text{const}$ turns into a sphere in the new system of coordinates, and it is natural to introduce the spherical harmonics $Y_n^m(\Theta,\varphi)$ because of the geometrical properties of the domain. We seek an approximate solution in the form

$$(7) \qquad u_N = \sum z_{nm}(\zeta)Y_n^m(\Theta,\varphi).$$

We require that the relation

$$(8) \qquad \int_{\zeta=\text{const}} L[u_N]\cdot Y_n^m(\Theta,\varphi)d\Omega = 0 \qquad (nm)=1,\ldots,N$$

be valid on each sphere $\zeta = \text{const}$, concentric to the spheric layer.

Conditions (8) result in a system of N ordinary differential equations for the functions $z_n(\zeta)$. We prescribe the following conditions on the boundary surfaces of the spheric layer:

$$(9) \qquad \int_S \left\{ p\frac{\partial u_N}{\partial n} - \alpha u - \Phi \right\} Y_n^m(\Theta,\varphi)\,d\Omega = 0, \qquad \zeta = 1,$$

$$(10) \qquad \int_{\Sigma_R} \frac{\partial u_N}{\partial n} Y_n^m(\Theta,\varphi)\,d\Omega = z_n(R)\beta_n(R), \qquad \zeta = 2,$$

$$z_n(R) = \alpha_n T_n^N.$$

The given problem for the system of differential equations (8) satisfies this condition. For the functions u_N the following integral equality is valid:

$$(11) \qquad \mathrm{Im} \int_D q|u_N|^2\,dv + \int_S \mathrm{Im}\alpha \left| u_N + \Phi\frac{1}{2i\sqrt{\mathrm{Im}\alpha}} \right|^2 dv +$$

$$+ \frac{1}{k_0}\Sigma|T_k^N|^2 = \frac{1}{4\pi}\int_S \frac{|\Phi|^2}{\mathrm{Im}\alpha}\,dv,$$

which also holds for the exact solution of the original problem. Using the energetic inequalities, it can be shown [7], that the approximate boundary value problem can always be solved and the approximate solutions are uniformly bounded by N in the $L_2(D)$ norm. Comparing the approximate and the exact solution and using the concrete properties of the Y_n^m basis we find that

$$\| u - u_N \|_{W_2^{(1)}} \to 0 \qquad N \to \infty.$$

The given method can be generalized for the solution of exterior boundary value problems with periodic coefficients [8], for domains with boundaries in the infinity [9] and for the solution of some other boundary value problems [10].

REFERENCES

[1] T. Kato, Growth properties of solutions of the re-
     duced wave equations, *Comm. on Pure and Appl. Math.*
     12, 3, (1959)

[2] W. Jüger, Zur Theorie der Schwingungsgleichung mit
     variablen Koeffizienten in Aussengebieten, *Math.
     Zeit.* 102, 1, (1967).

[3] A.A. Samarskii- A.N. Tihonov, On the principle of
     radiation, *Z. Exper. Theoret. Phys.* 18, 2, (1948).
     (in Russian).

[4] A.G. Sveshnikov, On the principle of radiation,
     Dokl. Ak. Nauk, SSSR, 73, 5, (1950), 917-920.

[5] A.G. Sveshnikov, Diffraction on a limited body, *Dokl.
     Ak.   Nauk, SSSR, ser. Math. Phys.* 184, 1, (1968),
     63-65. (in Russian)

[6] L.V. Kantorovich, *Isvestija A.N. SSSR*, N. 5, (1933).
     (in Russian).

[7] A.S. Ilinskii - A.G. Sveshnikov, Direct method for
     difraction problems on local non-uniform body, *Žurn.
     Vyčicl. Mat. i Mat. Fiz.* 11, N°4(1971), 960-968.

[8] A.S. Ilinskii, Method of investigation of the prob-
     lems of wave diffraction on periodic structure; *Ž.
     Vyčicl. Mat. i. Mat. Fiz.*, 14, N° 4. (1974), 1063-
     -1067. (in Russian)

[9] A.S. Ilinskii - A.G. Sveshnikov, Studies methods
     for non-regular waveguides, *Ž. Vyčicl. Mat. i Mat.
     Fiz.*, 8, N° 2. (1968), 363-373. (In Russian)

[10] A.S. Ilinskii, Direct method of periodic systems
     calculation, *Žurn. Vyčicl. Mat. i Mat. Fiz.* 13, N° 1.
     (1973), 119-126. (In Russian)

A.S. Ilinskii
Computer Centre
Moscow State University
117234 Moscow
USSR

PROBLEMS OF CLUSTER ANALYSIS FROM THE VIEWPOINT OF NUMERICAL ANALYSIS

F. JUHÁSZ - K. MÁLYUSZ

Mathematical questions of pattern recognition including cluster analysis, are usually considered from the viewpoint of probability theory. The numerical side of many problems are, however, not less important and it is worth while, therefore, to pay attention to those cases which are simple enough to be handled by means of numerical analysis. We are going to study the following problem (we prefer at first not to formulate it in full mathematical strictness). Let a graph G be decomposed into two subgraphs

(i)   of roughly equal size and
(ii)   by cutting as few edges as possible.

We follow the method of F i e d l e r [1, 2] who characterizes the connectivity of a graph in terms of the second eigenvalue of a matrix canonically assigned to it and proves certain properties of partitions into level sets of the coordinates of the second eigenvector (considered as a function on the set of vertices).

Let us notice that without the restriction (i) we get a strictly imposed and well-investigated problem. For the solution we refer to [3] from the extensive literature on network flows. The difference becomes significant in the case of large graphs with many edges when the later problem seems to have lost relevance to pattern recognition.

The problems of cluster analysis belong to one of two different classes: we have to classify (1) either points in the K-dimensional space (the coordinates are parameters describing some inner properties of the objects) (2) or objects which are characterized only by their connections. The problem of graphs obviously belongs to this second class as a simpliest case when connection is characterized by one-bit information. We give a solution to the problem by transforming it to the first form; in fact, we reduce it to a problem in the one-dimensional space. If we regard the matrix assigned to G as a kind of a covariance matrix it will be reasonable to attempt some "factor analysis".

In the present paper we construct a functional on the set of decompositions and define the optimum S_{opt} as the maximum point of the functional. Then a method will be proposed, and finally we will give account of the numerical experiments performed on the computer CDC 3300.

Let $G=(V,E)$ be a graph where $V=\{1,2,\ldots,n\}$ and E are the set of vertices and edges of G respectively. We define a_{ik} $(i,k\in V)$ by

$$a_{ii} = 1 \quad \text{for every} \quad i\in V$$

$$a_{ik} = \begin{cases} 1 & \text{if} \quad (i,k)\in E \\ 0 & \text{if} \quad (i,k)\notin E \end{cases}$$

For any two subsets $V_1, V_2 \subset V$ we denote by

$$d(V_1, V_2) = \frac{1}{|V_1||V_2|} \sum_{i \in V_1} \sum_{k \in V_2} a_{ik}$$

the ratio of all existing, respectively all possible, edges between V_1 and V_2, where $|V|$ is the number of elements of the set V. We shall call $d(V_1, V_2)$ the density of edges between V_1 and V_2 (also in the case $V_1 \cap V_2 \neq \emptyset$).

Let S be a partition of V into two subsets

$$S : V = X_1 \cup X_2, \quad X_1 \cap X_2 = \emptyset$$

and put $d_{ij} = d(X_i, X_j)$, $|X_i| = n_i$ $(i, j = 1, 2)$. Then the number $d_{11} - 2d_{12} + d_{22}$ is larger if more edges are between elements belonging to identical calsses and small in the opposite case. Obviously $-2 \leq d_{11} - 2d_{12} + d_{22} \leq 2$. Multiplied by $\frac{n_1 n_2}{n^2} \leq \frac{1}{4}$ which is larger when n_1 and n_2 are near to each other we characterize S by the number

$$p(S) = \frac{n_1 n_2}{n^2} (d_{11} - 2d_{12} + d_{22}).$$

In a certain sense, $p(S)$ measures how well does S fulfill our expectations of a good partition.

EXAMPLES

1. If S is a full graph then $d(V_1, V_2) = 1$ for any $V_1, V_2 \subset V$ and so $p_G(S) = 0$ for every partition.

2. Let S be a partition of G such that there are no edges between X_1 and X_2. If G is a regular graph of order $\nu - 1$ then $p_G(S) = \frac{\nu}{n}$.

3. Let S be a partition of a bipartite graph G such that there are no edges inside X_1 and X_2. If G is

regular of order $\nu-1$ then $p_G(S) = -\dfrac{\nu-2}{n}$.

Let e be the all-one vektor i.e. $e_i = 1$ $(i \in V)$. Denote by $P = (p_{ik})_{i,k \in V}$ the matrix of orthogonal projection parallel to e. Then $p_{ik} = \delta_{ik} - \dfrac{1}{n}$.

PROPOSITION. *Let μ be the largest eigenvalue of the matrix PAP. Then*

$$p(S) \leq \frac{\mu}{n}$$

for every partition S.

PROOF. Consider the vector

$$x : x_i = \begin{cases} \left(\dfrac{n_2}{n_1 n}\right)^{\frac{1}{2}} & \text{if} \quad i \in X_1 \\[2em] -\left(\dfrac{n_1}{n_2 n}\right)^{\frac{1}{2}} & \text{if} \quad i \in X_2 \end{cases}.$$

Then $(x,x) = 1$, $(x,Ax) = np(S)$ and $\sum_{i \in V} x_i = (e,x) = 0$. The quadratic forms PAP and A coincide on the subspace orthogonal to e. Q.E.D.

The conclusion is that if μ is not sufficiently large then no good partition can be found.

Unfortunately, it is not easy to find a partition for which $p(S)$ is maximal. In that case, however, when G consists of exactly two subgraphs which are loosely connected with each other, we can effectively find good partitions by maximizing another functional $q(S)$ of similar character.

Let us consider weight distributions

$$\rho : \rho_i \geq 0 \qquad\qquad (i \in V)$$

$\sum_{i\in V} \rho_i^2 > 0$ on the set of vertices. The number $\rho_i\rho_k$ can be considered as weight on the edge (i,k) and ρ_i^2 as weight on the vertex i.

$$d(V_1,V_2;\rho) = \frac{1}{\sum_{i\in V_1}\sum_{k\in V_2}\rho_i\rho_k}\sum_{i\in V_1}\sum_{k\in V_2} a_{ik}\rho_i\rho_k$$

can be called the weighted density of edges between V_1 and V_2.

For any partition $S: V = X_1 \cup X_2$, $X_1 \cap X_2 = \emptyset$ we define

$$q(S) = \max_{\sum_{i\in X_1}\rho_i = \sum_{i\in X_2}\rho_i} \left\{ \frac{\sqrt{n_1 n_2}}{n} \; \frac{\sqrt{\sum_{i\in X_1}\rho_i^2}\,\sqrt{\sum_{i\in X_2}\rho_i^2}}{\sum_{i\in X_1}\rho_i^2} \; \frac{\sum_{i\in X_1}\rho_i}{\sqrt{\sum_{i\in X_1}\rho_i^2}\,\sqrt{n_1}} \right.$$

$$\left. \frac{\sum_{i\in X_2}\rho_i}{\sqrt{\sum_{i\in X_2}\rho_i^2}\,\sqrt{n_2}} \; [d(X_1,X_1;\rho)-2d(X_1,X_2;\rho)+d(X_2,X_2;\rho)] \right\}.$$

We observe that the four multipliers on the right-hand-side are $\leq\frac{1}{2},\leq\frac{1}{2},\leq 1$ and ≤ 1 equality holds if (1) the sizes of X_1 and X_2 are equal, (2) the weights are (in a sense) equally distributed between X_1 and X_2, (3) and (4) the weights are uniformly ditributed on the sets X_1 and X_2 respectively. (The condition $\sum_{i\in X_1}\rho_i=\sum_{i\in X_2}\rho_i$ can be interpreted so that the weight sums of possible edges inside the two classes are equal.)

PROPOSITION

(a) *For every S*

$$p(S) \leq q(S) \leq \frac{\mu}{n}$$

where μ is the largest eigenvalue of PAP.

(b) *There is an optimal partition S_{opt} such that*

$$q(S_{opt}) = \frac{\mu}{n}.$$

(c) *All optimal partitions can be defined in the following way: We take an eigenvector v of μ*

$$PAPv = \mu v$$

and classify i on the basis of the sign of v_i: $i \in V_1$ if $v_i > 0$ and $i \in X_2$ if $v_i < 0$. The value $q(S) = \frac{\mu}{n}$ is reached at the weight distribution $\rho_i = |v_i|$.

PROOF. Let x be the vector $x_i = \rho_i$ if $i \in X_1$ and $x_i = -\rho_i$ if $i \in X_2$. Then $(e,x) = 0$ and the value on the right-hand side is $\frac{(x,Ax)}{n(x,x)}$. Q.e.d.

In numerical practice we are free to move the point i if v_i is near to zero e.g. if i is an isolated point.

REMARK. We easily show that $q(S)$ has no local maximum different from $\frac{\mu}{n}$. In fact, for any S_o, and $S_{opt}: q(S_{opt}) = \frac{\mu}{n}$ there exist a chain of neighbouring partitions (differing each other in the classification of a single vertex)

$$S_o, S_1, \ldots, S_m = S_{opt}, \quad m < n$$

such that

$$q(S_{k-1}) \leqq q(S_k) \quad k=1,\ldots,m.$$

By standard means of perturbation theory we can estimate the error in S_{opt} if the graph G is given with error. As an example we consider the case when G is a regular graph splitting into two connected components, $V=X_1 \cup X_2$, $X_1 \cap X_2 = \emptyset$, $d(X_1 X_2)=0$. In that case the next largest eigenvalue λ is strictly smaller than ν. Put $a=\frac{\nu}{n}$, $c=\frac{\nu-\lambda}{\nu}$ and let δ be the error ratio of a_{ik} $(i,k \in V)$, that is, G and G' differ in $m=\delta n^2$ edges.

Let $S'_{opt}:V=X'_1 \cup X'_2$, $X'_1 \cap X'_2 = \emptyset$ be optimal partition of G'. Then the number of elements being differently classified by S_{opt} and S'_{opt} is of order $\frac{\delta}{ac}$.
The exact statement is the following.

PROPOSITION. *Let k_1, and k_2 be the number of elements of the sets $X_1 \cap X'_2$ and $X_2 \cap X'_1$. Then*

$$k_1\frac{n_2}{n_1} + k_2\frac{n_1}{n_2} \leqq \frac{mn}{(c\nu - 4\sqrt{m})^2}$$

if $\lambda < \nu(1-c)$.

We notice that for computations with P we do not need computer memory for the entries of P.

For practical solution of the eigenvalue problem in question, we tested the Jacobi method and the power method. Without a priori information one cannot say anything about the rate of convergence of the power method. It suits well to a problem of pattern recognition since it is effective whenever the "pattern" exists, that is, when there exists a single, well-recognizable best partition (the first eigenvalue is significantly larger than the next one). Practical arguments for the power method are in our case that A is a $(0,1)$ matrix and so one can substitute additions for multiplications. Moreover, a large graph A can be stored in bits by means of masking technique [4].

We compared the effectivity of the different methods in this special situation by means of test problems. The experiment was as follows. We generated normally distributed points on the plane around two centres. The number of points was equal, the dispersion ellipsoid at each centre was a circle having radius r while the distance of the centres remained always 1. We generated a graph in the following manner: $(i,k) \in E$ if and only if the distance between the i-th and k-th points was smaller than $\frac{1}{2}$. (See the diagram for $r = \frac{1}{3}$ and $n = 640$.)

The error bounds for the methods depend on the nature of the problems. In our case it was enough compute the Jacobi method with the error bound $\frac{1}{\sqrt{n}}$. Using the power method we found that the required accuracy was always reached in 50 iterations.

J	r	$\frac{1}{4}$		$\frac{1}{3}$		$\frac{1}{2}$	
n	error bound	number of iterations	execution time/sec	number of iterations	execution time/sec	number of iterations	execution time/sec
20	0,23	174	4	131	3	133	3
40	0,16	614	21	837	26	667	21
80	0,12	3542	229	4551	281	4379	230

p	r	$\frac{1}{4}$		$\frac{1}{3}$		$\frac{1}{2}$	
n	number of iterations	error bound	execution time/sec	error bound	execution time/sec	error bound	execution time/sec
20	50	10^{-3}	8	2.10^{-5}	8	2.10^{-3}	8
40	50	4.10^{-5}	19	2.10^{-4}	20	6.10^{-3}	19
80	50	2.10^{-5}	51	3.10^{-3}	45	3.10^{-2}	44

p	r	$\frac{1}{3}$	
n	number of iterations	error bound	execution time/sec
160	50	6.10^{-3}	160
320	50	3.10^{-3}	450
640	50	2.10^{-3}	1640

Computational results for the J=Jacobi and the p=power method for the same problem.

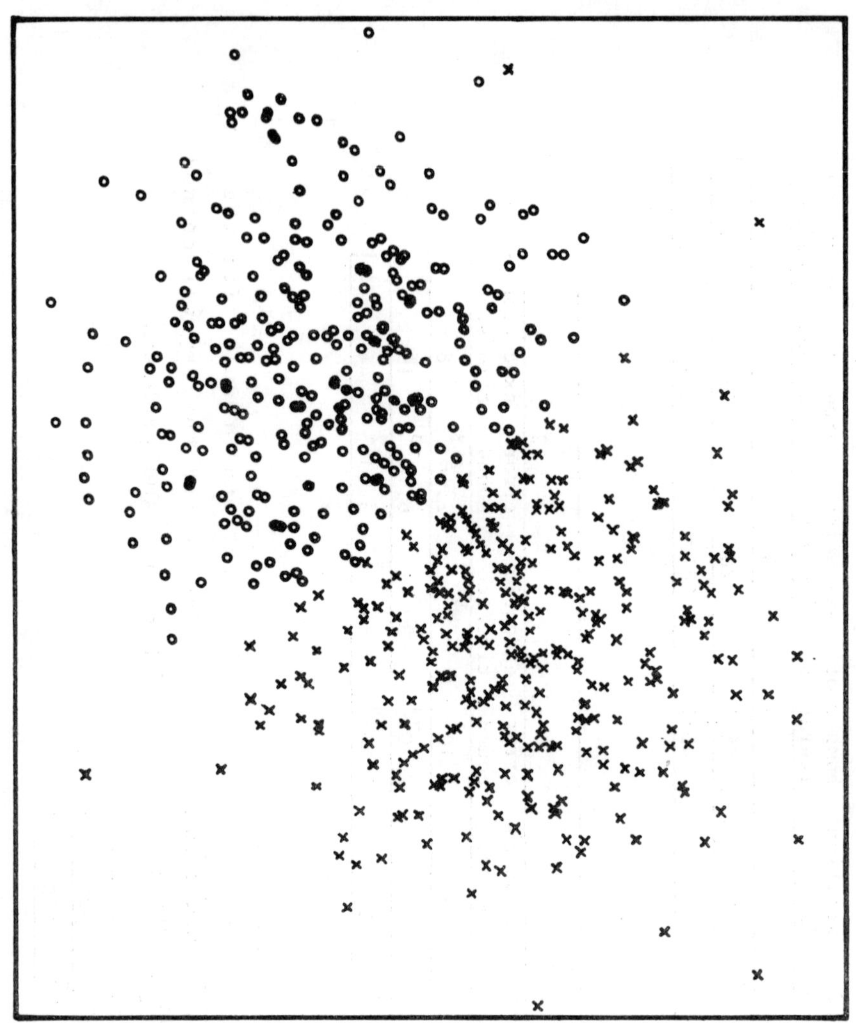

REFERENCES

[1]   M. Fiedler, Algebraic connectivity of graphs,
      Czech. Math. J. 23 (98) 1973, 298-305.

[2]   M. Fiedler, A property of eigenvectors of nonnegative
      symmetric matrices and its application to graph
      theory, *Czech. Math. J.* 25 (100) 1975, 619-633.

[3]   G. M. Adelson-Velskii - E.A. Dinic- A.V. Karzanov,
       twork Flow Algorithms, Nauka, Moscow, 1975.
       (in Russian)

[4]   M. R. Anderberg, *Cluster Analysis for Applications*,
       Academic Press, New York and London, 1973.

[5]   A. A. Dorofeyuk, Algorithms of theaching the
       machine the pattern recognition without teacher
       based on the method of potential functions,
       Automatika i telemechanika, No. 10. 1966, 78-87.
       (in Russian)

F. Juhász
Computer and Automation Institue
of the Hungarian Academy of Sciences
Budapest, Kende u. 13-17.
H-1111

K. Mályusz
Research Institute for Applied Computer Sciences
Budapest, P.O. Box 227.
H-1536

AN EXPLICIT DIFFERENCE SCHEME OF SOLVING THE CAUCHY PROBLEM FOR A DEGENERATE HYPERBOLIC EQUATION

A. KOKOSZKIEWICZ

Let t be a point of a time interval $[0, T]$ and let E_p be the real Euclidean p-space. The coordinates of a point $x \in E_p$ are denoted by $x_1, \ldots, x_p$.

Consider the differential equation

$$\frac{\partial^2 v}{\partial t^2} - \sum_{r,s=1}^{p} a_{rs}(x, t)\frac{\partial^2 v}{\partial x_r \partial x_s} +$$

$$(1) \qquad + \sum_{r=1}^{p} a_r(x, t)\frac{\partial v}{\partial x_r} + a_0(x, t)\frac{\partial v}{\partial t} + a_{-1}(x, t)v =$$

$$= f(x, t), \qquad (x, t) \in G,$$

where $G = E_p \times [0, T]$.

We assume that the condition

$$\sum_{r,s=1}^{p} a_{rs}(x, t)\xi_r \xi_s \geq 0$$

holds for $(x, t) \in G$ and for all real $\xi_1, \ldots, \xi_p$. If

strong inequality holds in this condition then we have
a hyperbolic differential equation, and in case of
equality the differential equation degenerates. The
initial conditions are

$$(2) \qquad v(x, 0) = \varphi_{-1}(x), \qquad \frac{\partial v(x, 0)}{\partial t} = \varphi_0(x), \qquad x \in E_p.$$

We assume that the data of the Cauchy problem (1), (2)
are periodic in the space variables $x_1, \ldots, x_p$ with
periods $\ell_1, \ldots, \ell_p$.

The following result was given by F a t e e v a
[1] :

THEOREM 1. *If the data of the differential problem
(1), (2) are sufficiently smooth and if there exist two
positive numbers* μ, ν *such that the differential
inequality*

$$(3) \qquad \sum_{r,s=1}^{p} (\mu a_{rs} + \frac{\partial a_{rs}}{\partial t}) \xi_r \xi_s - \nu \left[\sum_{r=1}^{p} (a_r + \sum_{s=1}^{p} \frac{\partial a_{rs}}{\partial x_s}) \xi_r \right]^2 \geq 0$$

holds for $(x, t) \in G$ *and for all real* $\xi_1, \ldots, \xi_p$, *then
the differential problem (1), (2) has a unique solution
which is also sufficiently smooth.*

We introduce a mesh for the independent variables.
Let τ be the mesh size in the t direction and let
h_s be the mesh size in the x_s direction for $s =$
$= 1, \ldots, p$. Then we define the sets

$$\Omega_\tau = \{t_n = n\tau, \; n=0, \ldots, N, \quad N\tau < T \leq (N+1)\tau, \quad \tau > 0\},$$

$$E_{ph} = \{x_i = (i_1 h_1, \ldots, i_p h_p), \quad i_s = 0, \pm 1, \ldots, N_s h_s = \ell_s,$$
$$s = 1, \ldots, p\},$$

$$\Omega_h = \{x_i \in E_{ph}, \quad 0 \le i_s \le N_s - 1, \quad s = 1, \ldots, p\}.$$

Furthermore we introduce the notations

$$u_i^n = u(x_i, t_n), \quad u_i = u(x_i), \quad u^n = u(t_n),$$

$$\overset{+s}{u_1} = u(i_1 h_1, \ldots, (i_s \pm 1)h_s, \ldots, i_p h_p), \quad u_{x_s, i}^n = \frac{1}{h_s}(\overset{+s}{u_i}^n - u_i^n),$$

$$u_{\bar{x}_s, i}^n = \frac{1}{h_s}(u_i^n - \overset{-s}{u_i}^n), \quad u_{t, i}^n = \frac{1}{\tau}(u_i^{n+1} - u_i^n),$$

$$u_{\bar{t}, i}^n = \frac{1}{\tau}(u_i^n - u_i^{n-1}),$$

$$u_{\overset{\circ}{x}_s, i}^n = \frac{1}{2}(u_{x_s, i}^n + u_{\bar{x}_s, i}^n), \quad u_{\overset{\circ}{t}, i}^n = \frac{1}{2}(u_{t, i}^n + u_{\bar{t}, i}^n),$$

$$\overset{+N_s}{u_i} = u(i_1 h_1, \ldots, (i_s + N_s)h_s, \ldots, i_p h_p), \quad h_{max} = \max_{1 \le s \le p} h_s,$$

$$h_{min} = \min_{1 \le s \le p} h_s.$$

The solution of the Cauchy problem (1), (2) is also periodic with respect to $x_1, \ldots, x_p$ with periods $\ell_1, \ldots, \ell_p$. Therefore it is enough to find it only for $(x, t) \in \{0 \le x_s < \ell_s, \quad s = 1, \ldots, p\} \times [0, T]$.

Now we define the following space: $L_{2h}(E_{ph})$ is the space of all mesh functions u on E_{ph} such that $\overset{+N_s}{u_i} = u_i$ $(s = 1, \ldots, p)$; we set

$$\|u\|_{L_{2h}}(E_{ph}) = (h_1 \ldots h_p \sum_{x_i \in \Omega_h} u_i^2)^{1/2}.$$

We take a three-level explicit difference scheme for the numerical solution of the differential problem (1), (2) and find a mesh function $u \in L_{2h}(E_{ph})$ satisfying

$$u_{\overline{t}t,i}^n - \sum_{r,s=1}^{p} u_{\overline{x}_r x_s,i}^n + \sum_{r=1}^{p} a_{ri}^n u_{\overset{\circ}{x}_r,i}^n +$$

$$(4) \qquad a_{0i}^n u_{\overset{\circ}{t},i}^n + a_{-1i}^n u_i^n = f_i^n, \qquad x_i \in \Omega_h,$$

$$1 \leq n \leq N-1,$$

$$(5) \qquad u_i^0 = \varphi_{-1i}, \qquad u_i^1 = \psi_i, \qquad x_i \in \Omega_h,$$

where ψ_i can be expressed by φ_{-1i}, φ_{0i}, f_i^0 such that the above sceheme has a truncation error of order 2.

The main result can be stated as follows:

THEOREM 2. *Let* u *be the solution of the difference scheme* (4), (5). *If the data of the differential problem* (1), (2) *are sufficiently smooth, and if there exist three positive numbers* α, β, γ *such that for all real* $\xi_1, \ldots, \xi_p$ *and for sufficiently small* h_{max} *and* τ *the inequalities*

$$\frac{\tau}{(h_{min})^2} < \gamma,$$

$$(6) \qquad \sum_{r,s=1}^{p} (\alpha a_{rsi}^n + a_{rst,i}^n) \xi_r \xi_s -$$

$$-\beta \left[\sum_{r=1}^{p} (a_{ri}^n + \sum_{s=1}^{p} a_{rsx_s,i}^n) \xi_r \right]^2 \geq 0$$

hold for $x_i \in E_{ph}$ *and for* $0 \leq n \leq N-1$*, then the estimate*

$$\|u^k\|_{L_{2h}(E_{ph})} \leq m\left(\sum_{s=-1}^{0} \|\varphi_s\|^2_{L_{2h}(E_{ph})} + \right.$$

$$\left. + \tau \sum_{j=0}^{k} \|f^j\|^2_{L_{2h}(E_{ph})}\right)$$

is true for $0 \leq k \leq N$ *and for sufficiently small* h_{max}*, and* τ*, where* m *is some constant. Here* α*,* β*,* γ *and* m *do not depend on* $h_1, \ldots, h_p$ *and* τ*.*

This theorem shows that the difference scheme (4), (5) is conditionally stable and convergent in the $L_{2h}(E_{ph})$ norm.

REMARK 1. Note that the difference inequality (6) implies the differential inequality (3), namely, if h_{max} and τ tend to zero, then we get (3) from (6).

REMARK 2. In case of $p = 1$ the difference inequality (6) is fulfilled, for example, for the Tricomi equation.

REFERENCE

[1]   G.M. Fateeva, Cauchy problem and boundary value problem for linear and quasilinear degenerate hyperbolic second-order equations, *Dokl. Akad. Nauk SSSR ser Math.* 172, *FASC.* 6(1967), 1278-1281 (in Russian).

A. Kokoszkiewicz, 00-901 Warsaw,
P.K.i N., P. 850,
JJ, University of Warsaw
Poland

A CONTINUOUS METHOD FOR THE NONLINEAR PROGRAMMING PROBLEM WITH EQUALITY CONSTRAINS

MARGIT KOVÁCS

Lyapunov's theorems on asymptotic stability and their generalizations are often applied to prove convergence theorems for the minimization algorithms. In this paper a method will be suggested and the convergence proof of this method will be based upon the asymptotic stability.

The problem is to find a local solution of

$$(1) \qquad \text{minimize} \quad F(x)$$

subject to

$$(2) \qquad Q = \{x \in R^n \mid g(x) = 0\} \ .$$

The functions $F(x) : R^n \to R^1$ and $g(x) : R^n \to R^m$, $m < n$ are defined on the whole space R^n.

To solve the problem (1)-(2), we introduce the following system of differential equations:

$$(3) \qquad \dot{x} = \varphi(x) , \qquad x(o) = x_o ,$$

$$(4) \qquad \varphi(x) = -(E - G^{\#}(x)G(x))F'(x) - G^{\#}(x)g(x) ,$$

where $F'(x)$ — is an $n \times 1$ vector whose jth element is $\partial F(x)/\partial x_j$;

$G(x)$ — is the $m \times n$ Jacobian matrix of the function g, whose ijth element is $\partial g_i(x)/\partial x_j$;

$G^{\#}(x)$ — the pseudoinverse of the matrix $G(x)$;

E — the identity matrix.

Let

$$L(x,y) = F(x) + (y^T, g(x))$$

be the Lagrangian function associated with the problem (1)-(2).

THEOREM. *Let*

(a) $\bar{x}$ be a local solution of the problem (1)-(2),

(b) the functions $F(x)$ and $g(x)$ be three times continuously differentiable in the point $\bar{x}$,

(c) the matrix $G(\bar{x})$ be of full rank,

(d) the second order necessary condition for the minimum be satisfied in the point $\bar{x}$ in form of a strict inequality,

(e) the system (3)-(4) have unique solution for all starting points $x_o \in R^n$ and every solution be defined for all $t \in [0, \infty)$.

Then there exists a neighbourhood N of $\bar{x}$, such that from any starting point $x_o \in N$ the solution of the system (3)-(4) tends to $\bar{x}$ when $t \to \infty$.

PROOF. It will be proved, that the point $\bar{x}$ is asymptotically stable and the region of asymptotic stability is $A(\bar{x})\supset N$. Consider the modified Lagrangian function

$$V_{\mu}(x,\bar{y})=F(x)+(\bar{y}^{T},g(x))+\frac{\mu}{2}\|g(x)\|^{2}-F(\bar{x}), \qquad \mu>0,$$

where $\bar{y}$ is the Lagrangian multiplier belonging to $\bar{x}$.

It is proved in [3], that there is a $\mu_{0}>0$, such that the function $V_{\mu}(x,\bar{y})$ is positive definite for $\bar{x}$ in some neighbourhood of $\bar{x}$, if $\mu\geq\mu_{0}$. Then there exists a positive γ_{0}, such that for all $0<\gamma\leq\gamma_{0}$ the component of the set

$$N_{\mu,\gamma}(\bar{x})=\{x\in R^{n}\mid V_{\mu}(x,\bar{y})\leq\gamma\} ,$$

which contains the point $\bar{x}$, is compact. Choose γ so that the matrix $G(x)$ be of full rank for all $x\in N_{\mu,\gamma}(\bar{x})$. Taking into the consideration, that for all $x\in N_{\mu,\gamma}(\bar{x})$

$$(5) \qquad G(x)\varphi(x) = -g(x),$$

the total time derivative of the scalar function $V_{\mu}(x(t),\bar{y})$ along the solution curves of the differential system (3)--(4) is

$$\dot{V}_{\mu}(x,\bar{y})=-F'^{T}(x)(E-G^{\#}(x)G(x))F'(x)-$$
$$-(F'(x)G^{\#}(x)+\bar{y}^{T}+\mu g^{T}(x),g(x)).$$

Moreover let

$$W(x) = -\dot{V}_{\mu}(x,\bar{y}),$$
$$y(x) = -G^{\#T}(x)F'(x)$$

and

$$\Lambda(x,y(x)) = F'(x) + G^T(x)y(x).$$

For these functions the following equalities hold

$$y(\bar{x}) = \bar{y},$$

$$\Lambda(\bar{x},y(\bar{x})) = 0,$$

and they are twice differentiable in $\bar{x}$. With these notations

$$W(x) = \Lambda^T(x,y(x))\Lambda(x,y(x)) +$$
$$+ (-y^T(x) + \bar{y}^T + \mu g^T(x), g(x)).$$

Then the derivative of $W(x)$ is

$$W'(x) = \Lambda^{T'}(x,y(x))\Lambda(x,y(x)) + \Lambda^T(x,y(x))\Lambda'(x,y(x)) +$$
$$+ G^T(x)(-y(x) + \bar{y} + 2\mu g(x)) - y^{T'}(x)g(x).$$

Consequently

$$W'(\bar{x}) = 0.$$

The second derivative of the function $W(x)$ (omitting the arguments) is:

$$W'' = \Lambda^{T''}\Lambda + 2\Lambda^{T'}\Lambda' + \Lambda^T\Lambda'' + G^{T'}(-y + \bar{y} + 2\mu g) +$$
$$+ G^T(-y' + 2\mu G) - y^{T''}g - y^{T'}G,$$

(here $\Lambda'' \in \mathscr{L}(R^n, \mathscr{L}(R^n, R^n))$ and $y'', G' \in \mathscr{L}(R^n, \mathscr{L}(R^n, R^m)))$. Thus

$$W''(\bar{x}) = 2\Lambda^{T'}(\bar{x},\bar{y})\Lambda'(\bar{x},\bar{y}) - (G^T(\bar{x})y'(\bar{x}) + y^{T'}(\bar{x})G(\bar{x})) +$$
$$+ 2\mu G^T(\bar{x})G(\bar{x}).$$

Since the matrix

$$A = 2\Lambda^{T'}(\bar{x},\bar{y})\Lambda'(\bar{x},\bar{y}) - (G^T(\bar{x})y'(\bar{x}) + y^{T'}(\bar{x})G(\bar{x}))$$

is symmetric, from condition d), provided $G(\bar{x})z=0$ one gets the following relations:

$$z^T A z = 2z^T \Lambda^{T'} \Lambda' z - z^T(G^T y' + y^{T'} G)z =$$

$$= 2z^T \Lambda^{T'} \Lambda' z =$$

$$= 2z^T (L''^T + G^T y', L'' + y^{T'} G)z =$$

$$= 2z^T (L''^T L'')z > 0.$$

One obtains from D e b r e u 's lemma, that $W(x)$ is a strictly locally convex function with $\mu > \mu_1$ and since $W(\bar{x})=0$, it is positive definite in some open neighbourhood of $\bar{x}: N'_\mu(\bar{x})$. If $N_\mu \subset N'_{\mu,\gamma}$, then consider the set $C = N_{\mu,\gamma} \setminus N'_\mu$. If $x \in \varrho \cap C$, then

$$W(x) = F'^{T}(x)(E - G^{\#}(x)G(x))F'(x) > 0.$$

Since $W(x)$ is a continuous function on C, there exists a positive constant δ, such that the function $W(x)$ is positive on the open set

$$N^0(\varrho \cap C) = \{x \in C^0 \mid \|g(x)\| < \delta\}.$$

The set $K = C \setminus N^0(\varrho \cap C)$ is compact, on which $W(x) > 0$, if

$$\mu_2 > \max_K \frac{-F'(x)G^{\#}(x)g(x) - (\bar{y}^T, g(x))}{\|g(x)\|^2},$$

Let $\mu > \max_{i=0,1,2} \mu_i$. Then on the set $N = N_{\mu,\gamma}(\bar{x})$, the

427

function $V_\mu(x,\bar{y})$ along the trajectory of (3)-(4) is strictly decreasing, consequently $N_{\mu,\gamma}(\bar{x})$ is a positively invariant set, on which $V_\mu(x,\bar{y})$ is a Lyapunov function. Using the second theorem of Lyapunov on asymptotical stability one gets that the point $\bar{x}$ is asymptotically stable with a set of attraction $A(\bar{x}) \supset N$.

COROLLARY. *Let the problem (1)-(2) have unique so-lution and the assumption (c) of the Theorem be replaced with the following one:*

(c') let $G(x)$ be of full rank on R^n.
Then the solution of (3)-(4) tends to $\bar{x}$ from any starting point $x_o \in R^n$.

In adddition, some properties of this method may be mentioned:

(α) It follows from equality (5), that $g(x(t)) \to 0$ exponentially for $t \to \infty$.

(β) Since $Q \cap N$ is a positively invariant set, this method is identical with the projected gradient method if $x_o \in Q \cap N$.

(γ) In the general case the numerical computation of pseudoinverse matrices is not very simple. But, under the conditions of the Theorem, matrix $G(x)$ is of full rank in every point of the trajectory, therefore

$$G^{\#}(x) = G^T(x)[G(x)G^T(x)]^{-1} \cdot$$

holds and the system of differential equations (3)-(4) turns into the system

$$\dot{x} = -(E - G^T(x)[G(x)G^T(x)]^{-1}G(x))F'(x) -$$
$$-G^T(x)[G(x)G^T(x)]^{-1}g(x).$$

(δ) *This method is effective in determining such a solution of a system of nonlinear equations, which has some extremal propoerty. For example, the minimal norm solution of the system $g(x)=0$ can be determined by this method with $F(x)=\|x\|^2$.*

REFERENCES

[1]   N.P. Bhatia - G. P. Szegö, *Dynamical Systems: Stability Theory and Applications*, Springer Verlag, 1967.

[2]   A. Albert, *Regression and the Moore-Penrose Pseudoinverse*, Academic Press, 1972.

[3]   K.J. Arrow - L. Hurwicz - H. Uzawa, *Studies in Linear and Non-linear Programming*, Stanford University Press, 1958.

[4]   G. Debreu, Definite and semidefinite quadratic forms, *Econometrica*, 20. (1952), 295-300.

Margit Kovács
Computing Center for Universities
Budapest, Dimitrov tér 8.
H-1093

ITERATIVE LÖSUNG RECHTECKIGER LINEARER GLEICHUNGSSYSTEME UND VERALLGEMEINERTE MATRIXINVERSE

G. MAEß

Die vorliegende Arbeit enthält eine einheitliche Darstellung stationärer und instationärer einstufiger Iterationsverfahren zur Lösung linearer Gleichungssysteme mit rechteckiger Koeffizientenmatrix. Die Beweismethoden lehnen sich an Arbeiten von T a n a b e [17], [18] an (vgl. auch M a r č u k - K u z n e c o v [9]). Zahlreiche bekannte Methoden lassen sich als Spezialfälle in die allgemeine Beschreibung einorden. Zur Ausdehnung der Ergebnisse auf lineare Operatorgleichungen in einem Limesraum und auf mehrstufige Iterationsverfahren vergleiche man B e r g [2].

1. STATIONÄRER FALL

Zur iterativen Lösung des linearen Gleichungssystems

$$(1) \qquad Ax = b \qquad A: R^N \to \mathfrak{R}(A) \subset R^M$$

mit einer $M{\times}N$-Koeffizientenmatrix A betrachten wir die

Gleichung

$$(2) \qquad x = x + Q(b-Ax), \quad Q: R^M \to \Re(Q) \subset R^N$$

wo Q eine $N \times M$-Matrix bezeichnet, die vorzugeben ist und das Iterationsverfahren charakterisiert. Offensichtlich erfüllt jede Lösung von (1) auch (2). Die Umkehrung ist dagegen im allgemeinen nicht richtig.

DEFINITION 1. Genügt ein Vektor $x^* \in R^N$ der Gleichung

$$(2') \qquad Q(b-Ax^*) = 0,$$

so heißt er (bez. Q) *verallgemeinerte Lösung* von (1).

Wenn QA singulär ist, kann (2') mehrere Lösungen besitzen.

DEFINITION 2. Genügt der Vektor x^{**} der Gleichung (2') und gilt außerdem in irgendeiner Norm

$$(3) \qquad \| x^{**} \| = \inf_{x^*} \| x^* \|$$

für alle Lösungen x^* von (2'), so heißt x^{**} (bez. Q) *verallgemeinerte Normallösung* von (1).

Aus (2) ergibt sich ein stationäres einstufiges Iterationsverfahren der Gestalt

$$(4) \qquad x^{n+1} = Tx^n + v$$

mit den zugehörigen Restvektoren

$$(5) \qquad r^{n+1} = Sr^n, \quad r^n := b - Ax^n.$$

Dabei sind T, S und v definiert durch

$$(6) \qquad T := I_N - QA \ , \quad S := I_M - AQ \ , \quad v := Qb \ ,$$

I_N und I_M bezeichnen Einheitsmatrizen entsprechender Ordnung. Durch Multiplikation mit Q bzw. A erhält man aus (6) folgende Vertauschungsregeln

$$(7) \qquad A\,T^n = S^n A \ , \quad T^n Q = Q S^n \ .$$

Ist $x^o \in R^N$ ein vorgegebener Startvektor, so lassen sich n-te Näherung und n-ter Restvektor explizit in der Form

$$(8) \qquad x^n = T^n x^o + B_n b \ , \qquad r^n = S^n r^o$$

angeben, wobei die Matrix B_n die Darstellungen

$$(9) \qquad B_n = \sum_{k=o}^{n-1} T^k Q = \sum_{k=o}^{n-1} Q\,S^k$$

besitzt.

SATZ 1. *Konvergieren die Matrizen* B_n *gegen die Grenzmatrix*

$$(10) \qquad B_\infty := \lim_{n \to \infty} B_n = \sum_{k=o}^{\infty} T^k Q = \sum_{k=o}^{\infty} Q S^k,$$

so existieren die Limites

$$(11) \qquad T^\infty := \lim_{n \to \infty} T^n = I_N - B_\infty A \ , \quad S^\infty = I_M - A B_\infty \ ,$$

$$(12) \qquad x^{\infty} \, . = T^{\infty} x^{0} + B_{\infty} b \ , \qquad r^{\infty} = S^{\infty} r^{0} \ ,$$

und die Matrizen T^{∞} und S^{∞} sind Projektoren mit den Eigenschaften

$$(13) \qquad \begin{aligned} T^{\infty} &= T^{\infty} T = T\, T^{\infty} = (T^{\infty})^{2} \ , \\ S^{\infty} &= S^{\infty} S = S\, S^{\infty} = (S^{\infty})^{2}, \end{aligned}$$

$$(14) \qquad \Re(T^{\infty}) = \Re(QA) \ , \qquad \Re(S^{\infty}) = \Re(AQ) \ ,$$

$$(15) \qquad \Re(T^{\infty}) = \Re(QA) \ , \qquad \Re(S^{\infty}) = \Re(AQ) \ ,$$

Dabei bezeichnen $\Re$ und $\Re$ Wertebereich bzw. Nullraum (Kern) des jeweiligen Operators.

BEWEIS. Aus der Identität

$$(16) \qquad I_{N} - T^{n} = \left(\sum_{k=0}^{n-1} T^{k}\right)(I_{N} - T) = (I_{N} - T)\left(\sum_{k=0}^{n-1} T^{k}\right)$$

und einer entsprechenden für S erhält man mit (6) und (9)

$$(17) \qquad T^{n} = I_{N} - B_{n} A \ , \qquad S^{n} = I_{M} - A B_{n} \ ,$$

woraus sich für $n \to \infty$ (11) und mit (8) auch (12) ergibt. Die erste Gleichung (13) folgt für $n \to \infty$ aus der Identität

$$(18) \qquad T^{2n} = T^{2n-1} T = T\, T^{2n-1} = (T^{n})^{2}.$$

Wegen (6) ist

(19) $\quad Q\ A\ T^n = T^n Q\ A = T^n - T^{n+1}$,

und für $n \to \infty$ folglich

(20) $\quad Q\ A\ T^\infty = T^\infty Q\ A = 0$,

also gilt

$$\mathfrak{K}(T^\infty) \subseteq \mathfrak{N}(QA) \ , \qquad \mathfrak{N}(T^\infty) \supseteq \mathfrak{K}(QA) .$$

Aus (10), (11) und (7) erhält man

(21) $\quad T^\infty = I_N - \sum_{k=o}^{\infty} T^k QA = I_N - \sum_{k=o}^{\infty} QAT^k$,

so daß umgekehrt auch

$$\mathfrak{K}(T^\infty) \supseteq \mathfrak{N}(QA) \ , \qquad \mathfrak{N}(T^\infty) \subseteq \mathfrak{K}(QA)$$

gelten muß. Damit sind die ersten Gleichungen von (14)
und (15) bewiesen. Die entsprechenden Relationen für S^∞
kann man analog herleiten.

SATZ 2.
(a) *Gilt* (10) *und*

(22) $\quad \mathfrak{K}(QA) = \mathfrak{K}(Q) \ , \qquad \mathfrak{N}(AQ) = \mathfrak{N}(Q)$,

so ist B_∞ äußere Matrixinverse (vgl. N a s h e d –
V o t r u b a *[10]) und x^∞ bez. Q verallgemeinerte
Lösung von* (1):

(23) $\quad B_\infty A B_\infty = B_\infty \qquad Q(b - Ax^\infty) = 0$.

(b) *Gilt* (10) *und*

$$(24) \qquad \mathfrak{N}(QA) = \mathfrak{N}(A) \ , \qquad \mathfrak{R}(AQ) = \mathfrak{R}(A) \ ,$$

so ist B_∞ *innere Matrixinverse:*

$$(25) \qquad AB_\infty A = A \ .$$

(c) *Gilt* (10) *und* (24) *und ist* (1) *lösbar,*

$$(26) \qquad b \in \mathfrak{R}(A)$$

so ist x^∞ *Lösung des Ausgangssystems:*

$$(27) \qquad Ax^\infty = b \ .$$

BEWEIS

(a) Wegen (22) existiert für alle $z \in R^M$ ein $x \in R^N$ mit $Qz = QAx$, und nach (20) folgt

$$(28) \qquad T^\infty Qz = T^\infty QAx = 0 \ .$$

Also ist

$$(29) \qquad T^\infty B_\infty = T^\infty \sum_{k=0}^{\infty} Q \, S^k = 0 \ ,$$

woraus sich mit (11) die erste Relation (23) ergibt. Für die zweite benutzen wir, daß r^∞ wegen (12), (14) und (22) in $\mathfrak{R}(S^\infty) = \mathfrak{N}(AQ) = \mathfrak{N}(Q)$ liegt, und folglich $Qr^\infty = Q(b - Ax^\infty) = 0$ ist.

(b) Aus (11) erhält man durch Multiplikation mit A die Gleichung (25), wenn man berücksichtigt, daß AT^∞

wegen $\Re(T^\infty) = \Re(QA) = \Re(A)$ verschwindet.

(c) Ebenso ergibt sich aus (12) mit (11)

$$Ax^\infty = AT^\infty x^O + (I_M - S^\infty)b = b \; ,$$

denn auch $S^\infty b$ verschwindet, weil $b \in \Re(A) = \Re(AQ) = \Re(S^\infty)$ ist.

Abschließend betrachten wir den Fall, daß die Matrizen AQ oder QA Maximalrang haben.

SATZ 3.

(a) *Gilt* (10) *und*

$$(30) \qquad rg \; AQ = M \; ,$$

so genügt B_∞ *den Relationen*

$$(31) \qquad AB_\infty A = A \; , \quad B_\infty AB_\infty = B_\infty \quad (AB_\infty)^T = AB_\infty \; ,$$

ist also eine reflexive Kleinste-Quadrate-Inverse $\overline{A}_{rl}$ (vgl. z.B. R a o - M i t r a [16]), und

$$(32) \qquad x^\infty(0) = B_\infty b$$

ist Kleinste-Quadrate-Lösung von (1).

(b) *Gilt* (10) *und*

$$(33) \qquad rg \; QA = N,$$

so genügt B_∞ *den Relationen*

$$(34) \qquad AB_\infty A = A \; , \; B_\infty AB_\infty = B_\infty \; , \; (B_\infty A)^T = B_\infty A \; ,$$

ist also eine reflexive Kleinste-Norm-Inverse A_{rm}^- , und

$$x^\infty = B_\infty b$$

ist Lösung von (10) mit kleinster euklidischer Norm.

BEWEIS. Wegen (30) ist $\mathfrak{N}(S^\infty) = \mathfrak{R}(AQ) = R^M$, also $S^\infty = 0$, so daß sich aus (11) sofort die drei Relationen (31) ergeben. Analog folgt aus (33) $\mathfrak{N}(T^\infty) = \mathfrak{R}(QA) = R^N$, und damit $T^\infty = 0$ und (34). Daß $A_{rl}^- b$ bzw. $A_{rm}^- b$ Kleinste-Quadrate- bzw. Kleinste-Norm-Lösung des Ausgangssystems sind, ist aus der Theorie der verallgemeinerten Inversen bekannt (vgl. R a o - M i t r a [16], K u h n e r t [4]).

2. INSTATIONÄRER FALL

Das instationäre Iterationsverfahren wird durch die Vorgabe einer Folge von NxM-Matrizen

$$(35) \qquad P_n : R^M \to \mathfrak{R}(P_n) \subset R^N \quad , \qquad n = 1, 2, \ldots,$$

definiert. Die Näherungen berechnen sich aus

$$(36) \qquad x^{n+1} = T_{n+1} x^n + v^{n+1}$$

und die zugehörigen Reste aus

$$(37) \qquad r^{n+1} = S_{n+1} r^n$$

mit

$$(38) \qquad T_n = I_N - P_n A \ , \quad S_n = I_M - A P_n \ , \quad v^n = P_n b \ .$$

DEFINITION 3. Genügt der Vektor $x^* \in R^N$ für alle n
den Gleichungen

$$(39) \quad P_n(b-Ax^*) = 0 \; ,$$

so heißt er (*bez. der Folge* $\{P_n\}$) *verallgemeinerte Lösung*
von (1). Ist x^{**} normkleinster Vektor unter allen x^*, so
heißt er (*bez.* $\{P_n\}$) *verallgemeinerte Normallösung.*

Die Vertauschungsregeln (7) gehen im instationären
Fall über in

$$(40) \quad A T_{n+k} \cdots T_{n+1} T_n = S_{n+k} \cdots S_{n+1} S_n A \; , \quad T_n P_n = P_n S_n$$

$$(n,k \quad \text{nat.}),$$

und n-te Näherung bzw. n-ter Restvektor haben die ex-
plizite Gestalt

$$(41) \quad x^n = \hat{T}_n x^O + \hat{B}_n b, \quad r^n = \hat{S}_n r^O \; .$$

Dabei sind T_n und S_n Produkte der Iterationsmatrizen
T_k bzw. S_k

$$(42) \quad \hat{T}_n = T_n T_{n-1} \cdots T_1, \quad \hat{S}_n = S_n S_{n-1} \cdots S_1,$$

und die Matrizen $\hat{B}_n$ sind durch

$$(43) \quad \hat{B}_n = \sum_{k-1}^{n} Q_{n,k} P_k \; , \quad Q_{n,k} = T_n T_{n-1} \cdots T_{k+1} \; , \quad Q_{n,n} = I_N$$

erklärt.

SATZ 4.

(a) *Konvergieren die Matrizen* $\hat{B}_n$ *gegen die Grenzmatrix*

$$(44) \qquad \hat{B}_\infty \; .= \; \lim_{n\to\infty} \sum_{k=1}^{n} Q_{n,k} P_k \; ,$$

so existieren die Limites

$$(45) \qquad T_\infty = I_N - \hat{B}_\infty A \; , \quad \hat{S}_\infty = I_M - A\hat{B}_\infty$$

$$(46) \qquad x^\infty = \hat{T}_\infty x^0 + \hat{B}_\infty \; , \quad r^\infty = \hat{S}_\infty r^0 .$$

(b) *Gilt* (44) *und*

$$(47) \qquad P_n A \hat{B}_\infty = P_n \; , \quad \hat{B}_\infty A P_n = P_n \; , \quad n \text{ nat.,}$$

so folgt

$$(48) \qquad \mathfrak{R}(AP_n) = \mathfrak{R}(P_n) \; , \; \mathfrak{K}(P_n A) = \mathfrak{K}(P_n) \; , \quad n \quad \text{nat.}$$

die Matrizen $\hat{T}_\infty$ *und* $\hat{S}_\infty$ *sind Projektoren mit den Eigenschaften*

$$(49) \qquad \hat{T}_\infty = \hat{T}_\infty T_n = T_n \hat{T}_\infty = (\hat{T}_\infty)^2 \; , \quad \hat{S}_\infty = \hat{S}_\infty S_n = S_n \hat{S}_\infty = (\hat{S}_\infty)^2 ,$$

$$(50) \qquad \mathfrak{K}(\hat{T}_\infty) = \mathfrak{R}(\hat{B}_\infty A) \; , \quad \mathfrak{K}(\hat{S}_\infty) = \mathfrak{R}(A\hat{B}_\infty) \; ,$$

$$(51) \qquad \mathfrak{R}(\hat{T}_\infty) = \mathfrak{K}(\hat{B}_\infty A) \; , \quad \mathfrak{R}(\hat{S}_\infty) = \mathfrak{K}(A\hat{B}_\infty) \; ,$$

die Matrix $\hat{B}_\infty$ *ist äußere g-Inverse zu A, und* x^∞ *ist (bez.* $\{P_n\}$*) verallgemeinerte Lösung von* (1).

BEWEIS

(a) Analog zu (17) gilt

$$(52) \qquad \hat{T}_n = I_N - \hat{B}_n A \ , \qquad \hat{S}_n = I_M - A\hat{B}_n \ .$$

Um das einzusehen, braucht man nur im Sinne einer vollständigen Induktion $\hat{T}_n$ und $\hat{S}_n$ mit T_{n+1} bzw. S_{n+1} zu multiplizieren und (38), (40) und

$$\hat{B}_{n+1} = T_{n+1} \hat{B}_n + P_{n+1}$$

zu berücksichtigen. Für $n \to \infty$ ergeben sich aus (52) und (41) die Relationen (45) und (46).

(b) Es gilt

$$(53) \qquad \Re(P_n) \supseteq \Re(P_n A) \supseteq \Re(P_n A \hat{B}_\infty) = \Re(P_n) \ ,$$

$$(54) \qquad \Re(P_n) \subseteq \Re(A P_n) \subseteq \Re(\hat{B}_\infty A P_n) = \Re(P_n) \ ,$$

wobei die Gleichheit aus (47) folgt. Damit ist (48) bewiesen. Ebenfalls aus (47) erhält man mit (45)

$$(55) \qquad \hat{T}_\infty P_n = P_n \hat{S}_\infty = 0 \ , \qquad\qquad n \quad \text{nat.}$$

also ist $\hat{T}_\infty T_n = T_n \hat{T}_\infty = \hat{T}_\infty$ und $\hat{S}_\infty S_n = S_n \hat{S}_\infty = \hat{S}_\infty$ für alle n, d.h., $\hat{T}_\infty$ und $\hat{S}_\infty$ sind Projektoren. Wegen (45) sind dann auch $\hat{B}_\infty A$ und $A\hat{B}_\infty$ Projektoren, so daß die Beziehungen (50), (51) trivialerweise erfült sind. Die Gleichung $\hat{B}_\infty A \hat{B}_\infty = \hat{B}_\infty$ erhält man unmittelbar aus (44), wenn man von rechts mit $A\hat{B}_\infty$ multipliziert und (47) berücksichtigt, und schließlich ist wegen (55) für alle n

$$P_n(b - Ax^\infty) = P_n r^\infty = P_n \hat{S}_\infty r^0 = 0 \ ,$$

so daß x^∞ (bez. $\{P_n\}$) verallgemeinerte Lösung ist.

SATZ 5.

(a) *Konvergieren die* $\hat{B}_n$ *gegen* $\hat{B}_\infty$ *und gilt*

$$(56) \qquad \mathfrak{K}(A\hat{B}_\infty) = \mathfrak{K}(A) \quad , \quad \mathfrak{N}(\hat{B}_\infty A) = \mathfrak{N}(A) \quad ,$$

so ist $\hat{B}_\infty$ *innere g-Inverse zu A.*

(b) *Gilt außerdem* $b \in \mathfrak{K}(A)$, *so ist* x^∞ *Lösung von* (1).

BEWEIS

(a) Wegen (56) und (51) ist $\mathfrak{K}(A) = \mathfrak{N}(\hat{S}_\infty)$, also

$$(57) \qquad \hat{S}_\infty A = A\hat{T}_\infty = 0 \ .$$

Folglich ergibt sich $A\hat{B}_\infty A = A$ aus (45) durch Multiplikation mit A.

(b) Ist $b \in \mathfrak{K}(A)$, so existiert eine Lösung x' mit $Ax' = b$, und man erhält aus (46)

$$Ax^\infty = A\hat{T}_\infty x^O + A\hat{B}_\infty Ax' = b \ ,$$

d.h., x^∞ ist Lösung von (1).

Ebenso wie Satz 3 beweist man schließlich.

SATZ 6. *Gelten die Voraussetzungen* (44) *und* (47) *von Satz 4, so ist*

(a) *im Fall*

$$(58) \qquad \mathrm{rg} \ A\hat{B}_\infty = M$$

die Matrix $\hat{B}_\infty$ *eine g-Inverse vom Typ* A_{r1}^- *und* $\overset{\infty}{x}(o) = \hat{B}_\infty b$
Kleinste-Quadrate-Lösung von (1)

(b) *im Fall*

(59) $\quad \text{rg } \hat{B}_\infty A = N$

die Matrix $\hat{B}_\infty$ *eine g-Inverse vom Typ* A^-_{rm} *und* $x^\infty = \hat{B}_\infty b$
Lösung von (1) *mit kleinster euklidischer Norm.*

3. BEISPIELE

1. ADDITIVE AUFSPALTUNG VON A

Analog zu der von V a r g a (1962) im Fall quadratischer Koeffizientenmatrizen vorgeschlagenen Aufspal-
tung

(60) $\quad A = B - C$

kann man auch im rechteckigen Fall verfahren (vgl.
P l e m m o n s [13], [14], B e r m a n - P l e m m o n s
[15], B e r g [12]): Ist

(61) $\quad Q = B^+ \ , \ \mathfrak{K}(A) = \mathfrak{K}(B) \ , \ \mathfrak{N}(A) = \mathfrak{N}(B),$

so konvergiert (4) gegen $A^+ b$, falls für den Spektral-
radius $\rho(B^+ C) < 1$ gilt. Ist

(62) $\quad Q = B^-_1 \ , \ \mathfrak{K}(C) \subseteq \mathfrak{K}(B) \ , \ \mathfrak{K}(B^-_1) \cap \mathfrak{N}(A) = \{0\}$

und $\rho(B^-_1 C) < 1$, so konvergiert (4) gegen $A^-_1 b$, und A^-_1
besitzt die Darstellung

(63) $\quad A^-_1 = (I_N - B^-_1 C)^{-1} B^-_1 \ .$

In ähnlicher Weise kann man auch eine g-Inverse B^-_m zur

Iteration benutzen.

2. PROJEKTION AUF SCHNITTRÄUME VON HYPEREBENEN.

Bei diesem auf K a c z m a r z (1937) zurückgehenden und in letzter Zeit u.a. von T a n a b e [17] und W. P e t e r s [11], [12] untersuchten Projektionsverfahren werden die Matrizen P_n in folgender Weise festgelegt

$$(64) \qquad P_n = A^T D_n \, , \qquad D_n = E_n (E_n^T AA^T E_n)^{-1} E_n^T \, .$$

Dabei bestehen die Matrizen E_n aus gewissen ausgewählten Spalten der Einheitsmatrix I_M. Benutzt man zyklisch eine feste Zahl N_o von Matrizen D_n, so wird das Verfahren stationär mit

$$(65) \qquad Q = \sum_{k=1}^{N_o} Q_{N_o, k} A^T D_k \, ,$$

und es ist $T^\infty = P_{\mathfrak{R}(A)}$ orthogonaler Projektor auf den Nullraum von A, die Grenzmatrix B_∞ existiert, hat die Gestalt

$$(66) \qquad B_\infty = (I_N - T + T^\infty)^{-1} Q$$

und ist eine g-Inverse des Typs A_{rm}^-. Die Iteration (4) konvergiert mit dem Startvektor $x^o = 0$ gegen eine (bez. Q) verallgemeinerte Normallösung $x^\infty(0) = B_\infty b$ der Ausgangsgleichung.

3. SPALTENAPPROXIMATIONSVERFAHREN

Diese Verfahren gehen vermutlich auf d e l a G a r z a (1951) zurück und werden von F a d d e e v und F a d d e e v a (1960) sowie von H o u s e h o l d e r (1964) in ihren bekannten Monographien behandelt. Eine

approximationstheoretische Herleitung findet man bei
K i e s e w e t t e r [3] und A l b r a n d [1].
Verschiedene Varianten werden von K r u t z k e [20]
und M a e ß [5], [6], [7], [8] untersucht.

Die Matrizen P_n haben hier die Gestalt

$$(67) \qquad P_n = D_n A^T, \qquad D_n = E_n (E_n^T A^T A E_n)^{-1} E_n^T ,$$

wo die E_n aus Spalten der Einheitsmatrix I_N bestehen.
Im zyklischen Fall ist

$$(68) \qquad Q = \sum_{k=1}^{N_o} Q_{N_o,k} D_k A^T$$

und S^∞ orthogonaler Projektor auf $\mathfrak{R}(A^T)$. Die Grenzmatrix B_∞ ist eine A_{r1}^--Inverse und explizit in der Form

$$B_\infty = Q(I_M - S + S^\infty)^{-1}$$

angebbar. Der Grenzvektor

$$x^\infty = (I_N - B_\infty A)x^O + B_\infty b$$

ist (bez. A^T) verallgemeinerte Lösung von (1).

Von den instationären Varianten sei besonders das
Gesamtschrittverfahren SPA3 (M a e ß [5], [7]) erwähnt,
das sich auch für Multiprozessor-Bearbeitung eignet.

LITERATUR

[1] H.J. Albrand, Über die Lösung linearer Gleichungs-
    systeme durch Spaltenapproximation, *Wiss. Zeitsch.
    Univ. Rostock Math.-Nat. Reihe* 21, 8 (1972) 755-757.

[2]   L. Berg, General iteration methods for the evaluation
      of linear equations, *Numer. func. anal. opt.* 1,4(1979)

[3]   H. Kiesewetter, *Vorlesungen über Lineare Approximation,*
      DVW, Berlin, 1973.

[4]   F. Kuhnert, *Pseudoinverse Matrizen und die Method
      der Regularisierung,* Teubner-Texte, Leipzig, 1976.

[5]   G. Maeß, Ein Gesamtschrittverfahren zur Lösung li-
      nearer Gleichungssysteme, *Acta Polytechnica, Práce
      ČVUT v Praze* 4 (1973) 65-70.

[6]   G. Maeß, Lösung linearer Gleichungssysteme durch
      Spaltenapproximation, *Wiss. Zeitsch. Univ. Rostock,
      Math.-Nat. Reihe,* 23 (1974) 759-761.

[7]   G. Maeß, Iterative Lösung linearer Gleichungssysteme,
      Nova Acta Leopoldina, Neue F.238,Bd.52,Halle, 1980.

[8]   G. Maeß, Ein stochastischer Projektionsalgorithmus
      zur iterativen Lösung linearer Gleichungssysteme,
      Rostocker Math. Koll., Heft 4 (1977) 47-54.

[9]   G.I. Marčuk - Yu. A.Kuznecov, Iterative methods and
      quadratic functionals, in *Methods of Numerical Mathe-
      matics,* Izdat. "Nauka" Sibirsk. Otdel, Novosibirsk,
      1975., pp. 4-143 (in Russian).

[10] M. Z. Nashed - G. F. Votruba, A unified operator theory
      of generalized inverses, in M.Z. Nashed (ed.) *Genera-
      lized Inverses and Applications,* New York, 1976.
      pp. 1-109.

[11] W. Peters, Lösung linearer Gleichungssysteme durch
      Projektion auf Schnitträume von Hyperebenen und Be-
      rechnung der Verallgemeinerten Inversen, *Beiträge zur
      Num. Math.,* Heft 5, (1976) 129-146.

[12] W. Peters, Eigenschaften einer nach dem Projektions-
      verfahren auf Schnitträume von Hyperebenen berechneten

verallgemeinerten Matrixinversen, *Beiträge zur Num. Math.*, Heft 6, (1977) 127-132.

[13] R. J. Plemmons, Monotonicity and iterative approximations involving rectangular matrices, *Math. Comp.* 26 (1972) 853-858.

[14] R. J. Plemmons, Direct iterative methods for linear systems using weak splittings, *Acta Univ. Carolinae Math. Phys.* 15 (1974) 145-154.

[15] R. J. Plemmons - A. Berman, Cones and iterative methods for best least square solutions of linear systems, *SIAM J. Numer. Anal.* 11 (1974) 145-154.

[16] C. R. Rao - S. K. Mitra, *Generalized Inverses of Matrices and its Applications*, New York, 1971.

[17] K. Tanabe, Projection method for solving a singular system of linear equations and its applications, *Numer. Math.* 17 (1971) 203-214.

[18] K. Tanabe, Characterization of linear stationary iterative processes for solving a singular system of linear equations, *Numer. Math.* 22 (1974) 349-359.

[19] K. Tanabe, Neumann-type expansion of reflexive generalized inverses of a matrix and the hyperpower iterative method, *Linear Alg. its Appl.* 10 (1975) 163-175.

[20] W. Krutzke, Konvergenzverbesserung der Spaltenapproximation bei der Anwendung auf lineare Gleichungssysteme, *Wiss. Zeitsch. Univ. Rostock, Math.-Naturwiss. Reihe* 24 (1975) 1257-1260.

G. Maeß
Sektion Mathematik
Wilhelm Pieck Universität Rostock
Universitätsplatz 1.

COLLOQUIA MATHEMATICA SOCIETATIS JÁNOS BOLYAI
22. NUMERICAL METHODS, KESZTHELY (HUNGARY), 1977.

AN ITERATIVE, OPTIMALIZED INTERPOLATION METHOD FOR RANDOMLY DISTRIBUTED DATA

ANDREA MESKÖ

ABSTRACT

Processing of data given for randomly distributed arguments (x_k, y_k) often requires interpolation in a regular square grid. Methods, proposed for interpolation in the literature, include solution of partial differential equations, weighted averaging, polynomial fitting, bi-cubic splines etc.

The paper suggests an iterative algorithm which fully utilizes the advantages of a stochastic approach but applies a deterministic interpolation method in the course of the procedure. The interpolated value is obtained as a linear combination of the neighbouring data. Weights are determined from the minimization of the average distortion (in the mean square sense). The solution requires the autocovariance of the data. Its first estimate is determined by a deterministic interpolation method and successively better estimates are obtained by iteration.

The structure of the developed program, as well as

preliminary experiments with the method are also reported.

INTRODUCTION

The problem of contouring randomly spaced data has
been analyzed by several authors in the past. The actual
drawing of contours by plotters when data can be computed
analytically, or when data are given on a regular grid is
a well developed field. Therefore we shall concentrate
to the interpolation problem, which can be formulated as
follows.

The height of a surface $z(x,y)$ on a datum plane is
given in N distinct points

$$z(x_i,y_i) = z_i \qquad\qquad i = 1,2,\ldots,N$$

and we want to determine $z(kd,ld)$, i.e. the height
of the surface at the nodes of a regular grid with spacing
d . The variable z is called height merely because the
problem can then be visualized more easily (Fig. 1.) but
in practical applications, z may stand for various other
quantities as well.

It is clear that without further assumptions the
interpolated values could change quite arbitrarily. Various approaches deviate mainly in the assumptions,
concerning the nature of $z(x,y)$.

In a large variety of practical problems, we may
assume that the surface is a sample of a zero mean two-
-dimensional stochastic process whose statistical properties
do not vary with x and y. Sometimes it becomes a good
approximation after having removed a trend. It can be
easily done by fitting two dimensional polynomials to the
irregularly spaced data by standard methods. An assuption,
combined with some other interpolation schemes, known from

the literature, lead us to an iterative algorithm of the
interpolation problem. Before embarking upon the description of the proposed algorithm, a brief summary of
some known methods is given. The application of the algorithm is illustrated by some examples.

A SUMMARY OF DETERMINISTIC AND STOCHASTIC APPROACHES OF THE INTERPOLATION PROBLEM

Numerous methods have been proposed for the solution
of the interpolation problem which can be grouped into
two subclasses as the deterministic and stochastic
approaches. Some of the most often used methods are
summarized in Table 1.

The deterministic methods may perform really well
when the nature of data correponds to the method. E.g.
the harmonic equation describes the equilibrium of a
membrane (M i l n e [4]). Finite difference algorithms
for the solution of the partial differential equation
are available (M i l n e [4]).

The biharmonic equation (L a n d a u - L i f s c h i t z
[3]) describes the position of an elastic plate fixed at
distance z_i at the i-th point.

The curvature at the point (kd, ld) is defined by

$$C_{kl} = z_{k+1,l} + z_{k,l+1} + z_{k-1,l} + z_{k,l-1} - 4z_{k,l}$$

and it can be verified that the solution of the biharmonic
equation minimalizes curvature in the sense that

$$\Sigma\Sigma c_{kl}^2 = \text{minimum.}$$

Methods for the numerical solution of the partial
differential equation are also available and the iterative

schemes can be easily programmed.

In many cases, however, the nature of the data to be contoured does not correspond either to a membrane or to an elastic plate.

Turning to the other methods (item 2 in Table 1), we must remark that the degree of the polynomial fitted locally to data as well as the number of data used in setting up a normal equation, influence the quality of the contour map significantly. But "satisfactory" values of these parameters can be arrived at after long experimentation and even then they are not really data adaptive. The "best" values depend largely on personal judgement. Weighted averaging (item 3 in Table 1) suffers from similar difficulties. The function $C(x,y)$ which yields the weights, though it has an important bearing on the quality of results, is again subjected to personal judgement.

In most cases the deterministic methods share the disadvantage that they are not data adaptive. They do not utilize the data redundancy i.e. their mutual correlation.

In the stochastic approach interpolated values are also obtained as linear combinations of data

$$z(x_j,y_j) = \sum_{i=1}^{N} c_i z_i,$$

the c_i coefficients, however, are now determined from the requirement that

$$\mathscr{E}\{[z(x_j,y_j) - \sum_{i=1}^{N} c_i z_i(x_i,y_i)]^2\} = minimum.$$

($\mathscr{E}$ denotes the expectation operator.) The minimalization leads to the set of linear equations

$$\sum_{k=1}^{M} c_k r(x_1 - x_k, y_1 - y_k) = r(x_j - x_1, y_j - y_1),$$

$$(l = 1, 2, \ldots, M),$$

where $r(x,y)$ denotes the autocovariance of the stochastic process:

$$r(\xi, \eta) = \mathcal{E}\{z(x,y)z(x+\xi, y+\eta)\}.$$

With a matrix notation, the set of simultaneous linear equations reads as

$$[R]\ c = r\ ,$$

where $[R]$ is an M by M matrix, and

$$c^T = [c_1, c_2, \ldots, c_M]$$
$$r^T = [r_1, r_2, \ldots, r_M]$$

(The upper index T denotes transposition.)

The interpolated values are given by

$$z(x_j, y_j) = [z^T, c] = z^T [R]^{-1} r.$$

The inversion of the matrices is generally time consuming when M is large. Therefore we used $M=8$ which proved to be sufficient. Before trying to invert $[R]$ its regularity was checked. In the very few cases when $[R]$ was singular, the interpolated value was determined from adjacent interpolated grid data.

If $r(\xi, \eta)$ were known analitically, the solution

of the equations would be straightforward. The autocovariance $r(\xi,\eta)$, however, is not available and usually
some assumptions should be made concerning its shape.
The following circularly symmetric functions have often
been suggested

$$r(x,y) = e^{-\frac{x^2+y^2}{\lambda}},$$

$$r(x,y) = e^{-\frac{(x^2+y^2)^{1/2}}{\lambda}},$$

$$r(x,y) = \frac{1}{\left(1 + \frac{x^2+y^2}{\lambda}\right)^{3/2}}, \qquad \text{etc.}$$

When using any of these functions, the only parameter
which can be adapted to the data is λ instead of the
whole autocovariance. On the other hand, we obviously
force $z(x,y)$ to be isotropic and this assumption is not
always realistic.

AN ITERATIVE METHOD BASED ON THE STOCHASTIC
APPROACH

In order to improve the stochastic approach, we
worked out an iterative method which is really data
adaptive but it uses deterministic interpolation in some
steps. The flow chart is given in Fig. 2.

The first step is a deterministic interpolation to
the nodes of a regular square grid. Weighted averaging
proved to be a fast and sufficiently accurate method for
performing the first step.

The second step is the estimation of the autocovariance by

$$r_{i,j} = \frac{1}{mn} \sum_{l=1}^{m-i} \sum_{k=1}^{n-j} z_{lk} z_{l+i,k+j} \, ,$$

which is a biased estimate but it has, in general, smaller mean-square error than the unbiased estimate (P a r z e n [5]).

The $r_{i,j}$ data are obtained again in a regular grid and some deterministic interpolation method, e.g. bicubic spline interpolation, can be used to compute $r(\xi,\eta)$ for any value of the arguments ξ and η.

The interpolation of the original data is repeated again but, now with statistically optimalized coefficients which were derived from $r(\xi,\eta)$. Then the improved estimate of the covariance is computed from the interpolated data.

The results of consecutive interpolations are compared with each other and the process terminates if the mean square deviation becomes smaller than a pre-set limit $\in$ or the number of iterations exceeds a pre-set limit L.

Some of the interpolation methods, listed in Table 1 has been compared to the proposed iterative scheme.

Data to irregularly spaced points were derived from an analitically given $z(x,y)$ and the accurate values at the nodes of a regular grid were computed simultaneously. Then irregularly spaced data were input to various interpolation procedures. The performance of the methods is characterized by the relative interpolation noise variance defined by the expression

$$R.N.V. = \sum_{k,l} (z_{kl} - z_{kl}^{int})^2 / \sum_{k,l} z_{kl}^2 \, .$$

Table 2 lists some of the results. The numbers in the first column denote the following methods

(1)   weighted average with $c(x,y)=\exp\left[-\dfrac{x^2+y^2}{(4d)^2}\right]$,

(2)   linear polynomial fitted to 4 points,

(3)   second order polynomial fitted to 8 points,

(4)   stochastic approach with data adaptive λ and normalized autocovariance $r(\xi,\eta)=\exp\left[-\dfrac{\xi^2+\eta^2}{\lambda^2}\right]$

(5)   stochastic approach iterative method with $N=8$ and data adaptive $r(\xi,\eta)$.

The ratios of the number of data points to the number of grid points (to be interpolated) were 0.1 in Experiments 1 and 2, 0.05 in Experiment 3, and 0.025 in Experiment 4.

The experiments are used as mere illustrations. However, it is convincing that the proposed scheme outperformed all the other methods in each cases thus it merits further investigations.

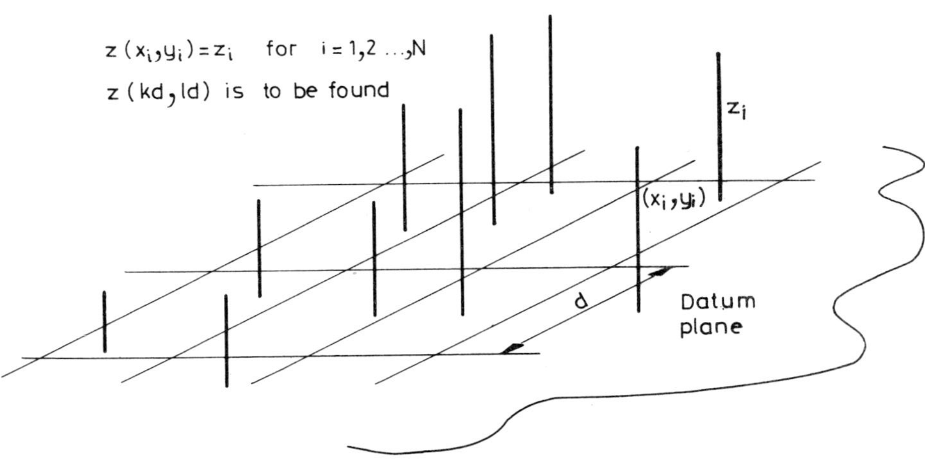

FIGURE 1.

The discrete interpolation problem

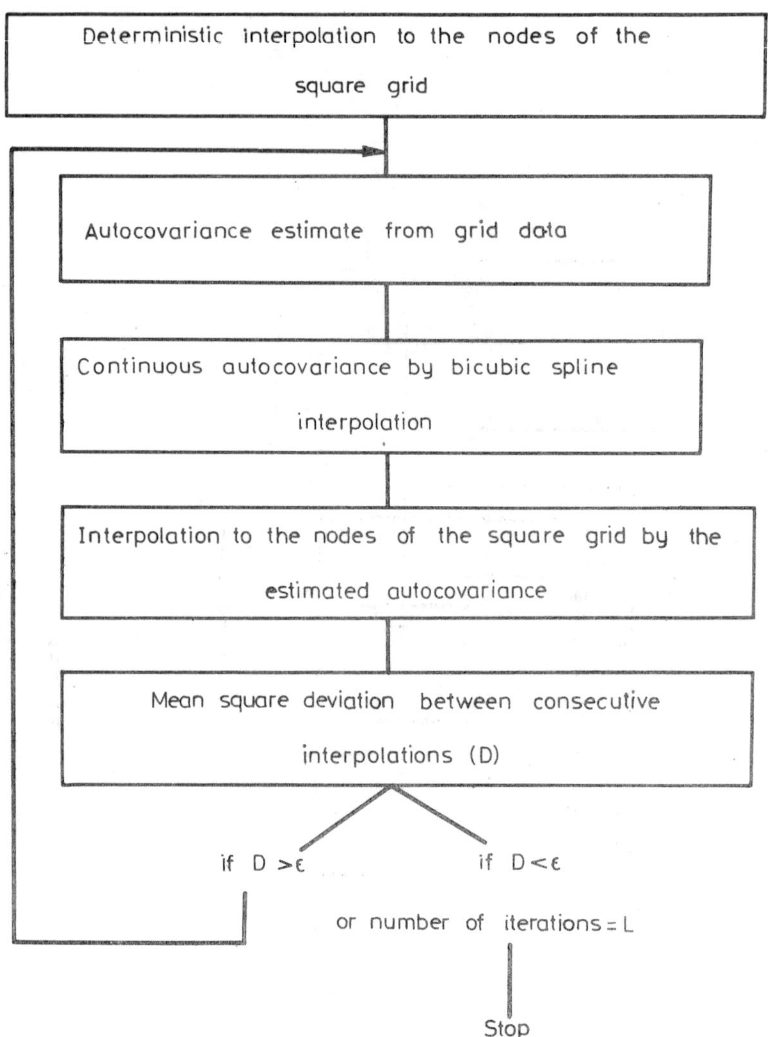

FIGURE 2.

Flow-chart of the proposed iterative scheme for interpolation ($\in$ and L are pre-set limits).

TABLE 1.

METHODS FOR INTERPOLATION OF IRREGULARLY SPACED DATA

1 Numerical solution of partial differential equations

e.g.

Harmonic equation $\dfrac{\partial^2 z}{\partial x^2} + \dfrac{\partial^2 z}{\partial y^2} = 0$

$$z(x_i, y_i) = z_i \qquad (i = 1, 2, \ldots, N)$$

Biharmonic equation $\dfrac{\partial^4 z}{\partial x^4} + 2\dfrac{\partial^4 z}{\partial x^2 \partial y^2} + \dfrac{\partial^4 z}{\partial y^4} = 0$

(solution corresponds to minimum curvature)

$$\sum_{i,k} c_{i,k}^2 = \min$$

2 Local polynomial fitting

(e.g. bicubic spline interpolation)

3 Weighted averaging

$$z(x, y) = \sum_i c_i, z_i$$

$$c_i = c(x - x_i, y - y_i)$$

$c(x, y)$ given continuous function

4 Stochastic approach

($z(x, y)$ is considered a sample of zero mean two dimensional stochastic process)

$$z(x_j, y_j) = \sum_{i=1}^{N} c_i z_i$$

$$\mathscr{E}\left\{\left[z_j - \sum_{i=1}^{N} c_i z_i\right]^2\right\} = \text{minimum}$$

TABLE 2.

R.N.V.[*] IN SOME NUMERICAL EXPERIMENTS

(given in percent)

Experiment No	1	2	3	4
Method No				
1	3.6	2.9	4.8	11.2
2	9.3	8.7	13.9	24.6
3	6.6	6.4	9.9	13.0
4	1.1	0.9	1.2	1.8
5	0.42	0.38	0.48	0.73

[*]*R.N.V.* is defined by

$$R.N.V. = \sum_k \sum_l \left[z(kd,ld) - z^{int}(kd,ld) \right]^2 \bigg/ \sum_k \sum_l \left[z(kd,ld) \right]^2$$

REFERENCES

[1]   J.H. Ahlberg - E.N. Nilson - J.L. Walsh, *The Theory of Splines and Their Applications*, Academic Press, Inc., New York, 1967.

[2]   C. de Boor, Bicubic spline interpolation, *J. Math. and Phys.* 41. (1962), 212-218.

[3]   L. Landau - E. Lifschitz, *Theorie de l'elasticité*, Moscow, 1967.

[4]   W.E. Milne, *Numerical Solutions of Differential Equations*, John Wiley and Sons, New York, 1953.

[5]   E. Parzen, On asimptotically efficient consistent estimates of the spectral density functions of a stationary time series, *J. Roy. Statist. Soc. Ser. B.* 20, (1958), 303-322.

[6]   J.L. Walsh - J.H. Ahlberg - E.N. Nilson, Best approximation properties of the spline fit, *J. of Math. and Mech.* 11. (1962), 225-234.

[7]   E. Wong, *Stochastic Processes in Information and Dynamical Systems*, McGraw-Hill Book Co, Inc., New York, 1971.

[8]   K. Yao, A representation theorem and its applications to spherically invariant random processes, *IEEE Trans. Information Theory* v.IT-19, (1973), 600-608.

Andrea Meskő
Computer and Automation Institute
of the Hungarian Academy of Sciences
Budapest, Kende u. 13-17.
H-1111

ACCELERATED MULTIPLE CALCULATION OF LINEAR RECURRENCES

J. MIKLOŠKO

1. INTRODUCTION

Derivation of a method is given that accelerates the m - multiple repetition of a $(2w+1)$ - term linear recurrence relation. The application of this method for two known algorithms is also described.

2. $(2w+1)$ - TERM LINEAR RECURRENCE RELATION

Let

$$(1) \qquad x_{2w+i} = c_i - \sum_{j=i}^{2w+i-1} a_{ij} x_j, \qquad i = 1, 2, \ldots n$$

be a $(2w+1)$ - term linear recurrence relation, where $\{c_i\}$, $\{a_{ij}\}$ and $\{x_j\}$, $j = 1, 2, \ldots 2w$, are given constants. Assume that $x_{2w+n}^{(k)}$ $k = 1, 2, \ldots m$ is to be calculated from (1) for m different initial conditions $\{x_j^{(k)}\}$ $j = 1, 2, \ldots 2w$; $k = 1, 2, \ldots m$. A simple m-multiple repetition of (1) requires $2wnm$ additions and $2wnm$ multiplications. We are going to

derive algorithms which, utilizing precomputation, lower
the number of arithmetical operations of this problem.

It holds that x_{2w+n} depends on $\{x_j\}$, $j=1,2,\ldots 2w$ linearly, i.e.

$$(2) \qquad x_{2w+n} = A_0 + \sum_{j=1}^{2w} A_j x_j .$$

Thus if $\{A_j\}$, $j=0,1,\ldots 2w$ is computed in the precomputation, the number of operations from the repeated calculation (1) declines.

Consider (1) as a triangular system of n linear equations of n unknowns

$$(3) \qquad Rx = f,$$

where

$$R = \begin{bmatrix}
1 & 0 & \cdots & \cdots & \cdots & \cdots & \cdots & \cdots & 0 \\
a_{2,2w+1} & 1 & & & & & & & \vdots \\
a_{3,2w+1} & a_{3,2w+2} & 1 & & & & & & \vdots \\
\vdots & & \ddots & \ddots & & & & & \vdots \\
a_{2w,2w+1} & a_{2w,2w+2} & \cdots & a_{2w,4w-1} & 1 & & & & \vdots \\
a_{2w+1,2w+1} & a_{2w+1,2w+2} & \cdots & & a_{2w+1,4w} & 1 & & & \vdots \\
0 & \ddots & & \ddots & & \ddots & \ddots & & \vdots \\
\vdots & & \ddots & & & & \ddots & \ddots & 0 \\
0 & \cdots & 0 & a_{nn} & a_{n,n+1} & \cdots & & a_{n,2w+n-1} & 1
\end{bmatrix}$$

$x = (x_{2w+1}, x_{2w+2}, \ldots x_{2w+n})^T$ and $f = (f_1, f_2, \ldots f_n)^T$ with

$$f_i = c_i - \sum_{j=i}^{2w} a_{ij} x_j, \quad i=1,2,\ldots 2w \quad \text{whereby} \quad f_i = c_i \quad \text{for} \quad i > 2w.$$

It holds that $x=R^{-1}f$ and to get x_{2w+n} it suffices to calculate the last row only, i.e. $(y_n, y_{n-1}, \ldots y_1)$ from R^{-1}. It follows from $R^{-1}R=I$ that

$$(4) \qquad y_1=1, \quad y_{2w+k} = -\sum_{j=k}^{2w+k-1} a_{n-j+1,n-k+1}y_j,$$

$$k=-2w+2,-2w+3,\ldots 0,1,\ldots n-2w,$$

where for $j\leq 0$ we define $y_j=0$. For the last component x_{2w+n}, (2) can be applied, where

$$(5) \qquad A_0 = \sum_{i=1}^{n} c_i y_{n-i+1} \quad \text{and} \quad A_j = y_{n+j} = -\sum_{i=1}^{j} a_{ij}y_{n-i+1},$$

$$j=1,2,\ldots 2w,$$

Let the precomputation (4) and (5) and the calculation (2) for m initial conditions be termed algorithm AR. The number of additions in AR is $2w(n+m-1)$, the number of multiplications is $2w(n+m-1)+n-1$.

3. APPLICATION OF ALGORITHM AR

Two examples will show now how it is possible to speed up certain known algorithms by applying algorithm AR.

EXAMPLE 1. To calculate the n-th approximant of the continued fraction i.e.

$$x_{1n} = \frac{b_1 \quad b_2 \quad \cdots \quad b_n}{a_1 + a_2 + \ldots + a_n}$$

frequent use is made of Euler's algorithm, i.e. the calculation of recurrence relations

$$P_{-1}=1, \quad P_0=0, \quad P_j=a_jP_{j-1}+b_jP_{j-2}, \quad j=1,2,\ldots n$$

$$Q_{-1}=0, \quad Q_0=1, \quad Q_j=a_jQ_{j-1}+b_jQ_{j-2}, \quad j=1,2,\ldots n$$

whereby $x_{1n}=P_n/Q_n$.

This algorithm has $2n-3$ additions, $4n-6$ multiplications and 1 division. If algorithm AR is applied to the calculation of these relations, we get for calculation x_{1n}

Algorithm CF1:

1/ $\quad y_1=1, \quad y_2=a_n, \quad y_i=a_{n+2-i}y_{i-1}+b_{n+3-i}y_{i-2}$

$\quad i=3,4,\ldots n+1$

2/ $\quad x_{1n}=b_1y_n/y_{n+1}$.

This algorithm has $n-1$ additions, $2n-2$ multiplications and 1 division. Its only competitor from the point of view of arithmetical operation is backward computation i.e.

Algorithm CF2:

$$x_{n+1,n}=0, \quad x_{jn}=\frac{b_j}{a_j+x_{j+1,n}}, \quad j=n,n-1,\ldots 1$$

having $n-1$ additions and n divisions.

If A is the time of addition, M of multiplication and D of division then it follows from the comparison of the number of arithmetical operations of algorithms CF1 and CF2 that if $M<D/2$ then CF1 is faster than CF2. If in x_{1n} all b_i or a_i are equal to 1, then CF1 outspeeds CF2, if $M<D$.

The analysis of the algorithm CF1 can be found in [1].

EXAMPLE 2. For the computation of the determinant of a pentadiagonal matrix $D_n = det|P|$, where

$$
P = \begin{bmatrix}
b_1 & a_1 & 1 & 0 & \cdot & \cdot & \cdot & \cdot & \cdot & 0 \\
c_2 & b_2 & a_2 & 1 & & & & & & \cdot \\
d_3 & c_3 & b_3 & a_3 & 1 & & & & & \vdots \\
0 & d_4 & c_4 & b_4 & a_4 & 1 & & 0 & & \\
\vdots & & \circ & \cdot & \cdot & \cdot & \cdot & \cdot & \cdot & 1 \\
\vdots & & & d_{n-1} & c_{n-1} & b_{n-1} & a_{n-1} & & & \\
0 & \cdot & \cdot & \cdot & \cdot & 0 & d_n & c_n & b_n &
\end{bmatrix} ,
$$

[3] has suggested

Algorithm D1:

1/ $x_1^{(k)} = x_2^{(k)} = 0$, choose $x_3^{(k)}$, $x_4^{(k)}$, $k=0,1,2$ such that

$$
m_0 = (x_3^{(1)} - x_3^{(0)})(x_4^{(2)} - x_4^{(0)}) - (x_3^{(2)} - x_3^{(0)})(x_4^{(1)} - x_4^{(0)}) \neq 0 ;
$$

2/ with $a_n = 0$ calculate

$$
x_{j+4}^{(k)} = -a_j x_{j+3}^{(k)} - b_j x_{j+2}^{(k)} - c_j x_{j+1}^{(k)} - d_j x_j^{(k)} ,
$$

$$
k=0,1,2; \quad j=1,2,\ldots n ;
$$

3/ $m_n = (x_{n+3}^{(1)} - x_{n+3}^{(0)})(x_{n+4}^{(2)} - x_{n+4}^{(0)}) - (x_{n+3}^{(2)} - x_{n+3}^{(0)})(x_{n+4}^{(1)} - x_{n+4}^{(0)}) ;$

4/ $D_n = m_n / m_0$

Algorithm D1 requires $9n-2$ additions, $12n-8$ multiplications and 1 division. If applying algorithm AR to 2/-4/, we get

Algorithm D2:

1/ $a_n = 0;$

2/ $y_1 = 1,\quad y_2 = 0,\quad y_3 = -b_n,\quad y_4 = -a_{n-2}y_3 - c_n,$

$$
y_i = -a_{n-i+2}y_{i-1} - b_{n-i+3}y_{i-2} - c_{n-i+4}y_{i-3} - d_{n-i+5}y_{i-4} ,
$$

$$
i = 5,6,\ldots n+1, n+2 ;
$$

3/ $z_1 = 0$, $z_2 = 1$, $z_3 = -a_{n-1}$, $z_4 = -a_{n-2} z_3 - b_{n-1}$,

$$z_i = -a_{n-i+2} z_{i-1} - b_{n-i+3} z_{i-2} - c_{n-i+4} z_{i-3} - d_{n-i+5} z_{i-4},$$

$i = 5, 6, \ldots n+1, n+2$;

4/ $x_{n+4}^{(k)} = y_{n+1} x_4^{(k)} + y_{n+2} x_3^{(k)}$, $x_{n+3}^{(k)} = z_{n+1} x_4^{(k)} + z_{n+2} x_3^{(k)}$,

$k = 0, 1, 2,$;

5/ as step 3/ in D1;

6/ as step 4/ in D1.

Algorithm D2 has $6n+1$ additions, $8n-6$ multiplications and 1 division. Calculating D_n by the Gaussian elimination assumes the non-zeroness of all principal minors in P; it has $4n-7$ additions, $7n-11$ multiplications and $n-1$ divisions. If $2A+M<D$ then D2 is almost as fast or respectively faster than the Gaussian elimination.

REMARKS

1. Algorithm AR has not been investigated for numerical stability. Some results concerning the numerical stability of three- and five-term recurrence relations, analogical to the relations in algorithms CF1 and D2 are given in [2] and [3].

2. Some generalisations of algorithm AR and additional examples for its application are in [4].

REFERENCES

[1] J. Mikloško, An algorithm for calculating continued fractions, *Journal Comput. Appl. Math.* (to appear)

[2] J. Mikloško, The numerical computation of three-term recurrence relations and the tridiagonal system of linear equations by the method of shooting, *Ž. Vyčisl. Mat. i Mat. Fyz.*, <u>14</u>, 6, (1974) 1371-1377.

[3] J. Mikloško, A recursive computation of the deter-
     minant of a pentadiagonal matrix, *Journ. Comput.
     Appl. Math.*, 1, 2, (1975) 73-78.

[4] J. Mikloško, A fast algorithm for repeated compu-
     tation of linear recurrence relations, *BIT* 17
     (1977).

J. Mikloško
Institute of Technical Cybernetics
Slovak Academy of Sciences
809 31 Bratislava,
Dúbravska 3,
ČSSR

NUMERICAL SOLUTION OF TWO-POINT-BOUNDARY-VALUE PROBLEMS AND SPLINE FUNCTIONS

FERNANDA ALEIXO OLIVEIRA

ABSTRACT

Approximations to an isolated solution of a second order nonlinear ordinary differential equation with two boundary condition are determined. We begin with quasi-linearization method in order to combine linear approximation techniques with capabilities of the digital computer. These approximations are piecewise polynomial functions of degree three or less and are obtained with collocation method using cardinal spline functions and B-spline functions.

INTRODUCTION

A h l b e r g [1] applies piecewise polynomial functions to linear two-point boundary value problems and suggested that the same process can be applied to nonlinear problems if we first use quasilinearization method.

Here we do so, begining with quasilinearization

techniques as B e l l m a n and K a l a b a [3] do. The nonlinear two-point-boundary value problem (1.1)(1.2) is solved when we solve a sequence of linear two-point-boundary value problems (1.5) (1.6). The corresponding sequence of solutions converges quadratically to the solution of the nonlinear problem as can be seen in [3], [5].

We use collocation method with piecewise polynomial functions in order to solve each linear problem (1.5). (1.6). The theory of collocation procedure for both linear and nonlinear problems is well studied by R u s s e l l and S h a m p i n e [7], (1972)] and by B o o r and S w a r t z [4], (1973)]. We discuss more directly computational aspects, which arise when we choose different base-functions. We applied two different base functions - cardinal spline functions and B-spline functions.

Cardinal spline functions are very efficient, permits interpolation at the partition points but computations involved are not so simple as they are if we use B-spline functions. Solution of the discretized linear problems are simpler if we take B-spline for the base functions. The matrix of the system is tridiagonal and it is relatively simple to set up the equations and solve them.

B o o r and S w a r t z [4] studied problem (1.1) (1.2) applying firstly collocation method and latter solving a sequence of a linear collocation problems associated with a Newton iteration. We did firstly quasi-linearization and after did collocation method to each linear problem. We are comparing the computational advantages of each one the numerical process.

For the special case of the cardinal spline functions

472

as base functions, we finished constructing a sequence
of least square splines which converges with probability
1 to the solution of the linear problem, according to
the result of [2].

Using B-splines we suggest one way of getting some
information about the error function. It seems simple
and useful.

1.

We shall be concerned with the nonlinear ordinary
differential equation

$$(1.1) \qquad u'' = f(x,u,u') \qquad\qquad (a \leqq x \leqq b)$$

subject to the linearly independent boundary conditions

$$(1.2) \qquad \begin{aligned} a_1 u(a) + a_2 u'(a) &= a_0 \\ b_1 u(b) + b_2 u'(b) &= b_0 \end{aligned}$$
$$(a_i, b_i = \text{const.}, \quad i = 0,1,2).$$

There is no convenient or useful technique for
representing the general solution of (1.1) (1.2) in terms
of a finite set of particular solutions as in the linear
case. Consequently, we possess no read means of reducing
the transcendental problem involved in solving (1.1)(1.2)
to an algebraic problem as is the situation in the case
where $f(x,u,u')$ is linear in u and u'.

Our aim is to obtain the solution of (1.1)(1,2),
when it exists, as the limit of a sequence of solutions
of linear differential equations. Each of these can be
solved numerically in a convenient fashion using
collocation methods.

QUASILINEARIZATION TECHNIQUES

Let us assume that we can apply quasilinearization method to (1.1)(1.2). This method is well treated by B e l l m a n and K a l a b a [3] or by G h i z z e t t i [5], where quasilinearization is employed for analitic and computational purposes.

They choose a convenient function $z(x) \in C^2[a,b]$ such that

$$\forall\ x \in [a,b] \qquad z,\ z',\ z''\ \text{ are bounded}$$

and assuming strict convexity for $f(x,u,u')$ they obtain

$$f(x,u,u') = \max_{z(x)}\ \{f(x,z,z') + (u-z)f_u(x,z,z') +$$

$$+ (u'-z')f_{u'}(x,z,z')\}.$$

Now they consider the problem

$$u'' = \max_{z(x)}\ \{f(x,z,z') + (u-z)f_u(x,z,z') + (u'-z')f_{u'}(x,z,z')\}$$

with boundary conditions (1.2).
From that they go to the linear associated problem

$$(1.3) \qquad \omega'' = f(x,z,z') + (\omega-z)f_u(x,z,z') + (\omega'-z')f_u(x,z,z')$$

with the boundary conditions

$$(1.4) \qquad \begin{aligned} a_1\omega(a) + a_2\omega'(a) &= a_o \\ b_1\omega(b) + b_2\omega'(b) &= b_o. \end{aligned}$$

The solution of (1.3)(1.4) exists and it is unique, as easily can be seen. Let us represent it by $\omega(x;z(x))$. With additional conditions we are led to a sequence of linear associated problems as (1.3)(1.4)

$$(1.5) \qquad \omega''_{n+1} = f(x,\omega_n,\omega'_n) + (\omega_{n+1}-\omega_n) f_u(x,\omega_n,\omega'_n) + (\omega'_{n+1}-\omega'_n) f_{u'}(x,\omega_n,\omega'_n)$$

$$(1.6) \qquad \begin{aligned} a_1 \omega_{n+1}(a) + a_2 \omega'_{n+1}(a) &= a_o \\ b_1 \omega_{n+1}(b) + b_2 \omega'_{n+1}(b) &= b_o. \end{aligned}$$

There is no difficult in establishing existence and uniqueness of solution of (1.5) (1.6) which we represent by $\omega_{n+1}(x;\omega_n)$.

Under obvious conditions on the quantities $f_u, f_{u'}$, f_{uu}, $f_{u'u'}$ and $f_{u\,u'}$ it can be prooved that the sequence of solutions $\{\omega_{n+1}(x;\omega_n)\}$ is a monotone non--increasing sequence and that $\{\omega_{n+1}(x;\omega_n)\}$ and $\{\omega'_{n+1}(x;\omega_n)\}$ are quadratically convergent respectivelly to $u(x)$ and $u'(x)$ form (1.1)(1.2).

This is quasilinearization method applied to (1.1) (1.2), as can be seen in [3], page 42 or [5].

So, we begin with a nonlinear problem (1.1)(1.2) and we go to a sequence of associated linear problems (1.5) (1.6). Now, we apply collocation method with piecewise polynomial functions to each one of the linear problems.

According with R u s s e l l and S h a m p i n e [7], collocation with piecewise polynomial functions is a very satisfactory way of solving boundary value problems specially in the linear case.

2.

The differential equation (1.5)(1.6) is taken as an operator equation and collocation conditions are equivalent to a projection of the operator equation into a finite-dimensional subspace. The projection operators are interpolation by polynomials.

We recall that these operators are mappings from C into C.

A classical result of Natanson shows that the projection operators cannot then be uniformly bounded. However, in the special case of interpolation at the Gauss, Legendre or Chebyshev points, the rate of growth of the norms of these operators is known and the collocation method was shown to converge.

Using piecewise polynomial functions the resulting operators from C into C have uniformly bounded norms even if one allows great flexibility in the selection of collocation points.

We use for the norm of a continuous function $r(x)$

$$\| r \|_{\infty} = \max_{a \leq x \leq b} |r(x)|.$$

We shall just write $\| r \|$ unless $\| r \|_{\infty}$ is needed for clarity.

DEFINITION. A function $r(x)$ is in the family $L(\Delta_n, k, m)$ if $r(x)$ is a polynomial of degree k or less on each subinterval of Δ_n and $r(x) \in C^{(m)}[a,b]$. The subfamily $L'(\Delta_n, k, m)$ consists of all functions in $L(\Delta_n, k, m)$ which satisfy the boundary conditions (1.6).

We take the partition Δ_n of $[a,b]$ as

$$\Delta_n : a \equiv x_0 < x_1 < x_2 < \ldots < x_n \equiv b$$

Through out of this work we will take the family $L(\Delta_n, 3, 2)$ and the corresponding subfamily $L'(\Delta_n, 3, 2)$.

To compute an approximat solution $\omega(x)$ of the linear problem (1.5) (1.6) we need a convenient representation for this function. We mention the spline space $L(\Delta_n,3,2)$ which arise using different bases. First, we will consider cardinal splines functions for the base functions of $L(\Delta_n,3,2)$ and latter the B-spline functions.

3.

We introduce cardinal spline functions as a set of $n+3$ linearly independent splines, which are a set of base functions for $L(\Delta_n,3,2)$ in the following way:

Let us consider $A_{\Delta,k}(x)$ $(k=0,1,\ldots,n)$ and $B_{\Delta,k}(x)$ $(k=0,n)$ cubic splines with Knots Δ_n. We take

$$A_{\Delta,k}(x_j) = \delta_{kj} \qquad (j=0,1,\ldots,n),(k=0,1,\ldots,n)$$

$$A'_{\Delta,k}(x_i) = 0 \qquad (i=0,n) \quad , \qquad (k=0,1,\ldots,n)$$

(3.1)

$$B_{\Delta,k}(x_j) = 0 \qquad (j=0,1,\ldots,n),(k=0,n)$$

$$B'_{\Delta,k}(x_i) = \delta_{ki} \qquad (i=0,n) \quad , \qquad (k=0,n) \, ,$$

where δ_{kj} is the Kronecker symbol.

We assume that these cardinal spline functions $A_{\Delta,k}(x)$ satisfy the final conditions:

$$2A''_{\Delta k}(a)+A''_{\Delta k}(x_1)= \frac{6}{h_1} \left[(A_{\Delta k}(x_1)-A_{\Delta k}(a))/h_1-A'_{\Delta k}(a) \right]$$

(3.2)
$$\text{at } x_o=a$$

$$A''_{\Delta k}(x_{n-1})+2A''_{\Delta k}(x_n)= \frac{6}{h_n} \left[A'_{\Delta k}(x_n)-(A_{\Delta k}(x_n)-A_{\Delta k}(x_{n-1}))/h_n \right]$$

$$\text{at } x_n=b$$

The same assumption is made for $B_{\Delta k}(x)$ functions.

We recall that (3.1)(3.2) define cubic spline functions with Knots at Δ_n, according to A h l b e r g [1].

From $L(\Delta_n,3,2)$ we seek a spine interpolate function for the solution of (1.5) (1.6). It will have the representation

$$(3.3) \qquad S_\Delta(\omega;x) = \sum_{j=0}^{n} A_{\Delta j}(x)\omega(x_j) + \omega'(a)B_{\Delta 0}(x) + \omega'(b)B_{\Delta n}(x).$$

We see from that representation that $S_\Delta(\omega;x_i) = \omega(x_i)$; $S_\Delta'(\omega;a) = \omega'(a)$; $S_\Delta'(\omega;b) = \omega'(b)$.

According to A h l b e r g we know that taking for the error function

$$E_\Delta(\omega;x) = \omega(x) - S_\Delta(\omega;x)$$

we have

$$E_\Delta^{(p)}(\omega;x) = O(\|\Delta\|^{2-p}), \qquad p=0,1,2$$

uniformelly for $x \in [a,b]$, as the solution of (1.5) (1.6) is in $C^2[a,b]$. If we rewrite (1.5) (1.6) as an operator equation we have

$$(3.4) \qquad T\omega = r(x)$$

with

$$T\omega = \omega''(x) - f_{u'}(x,z,z')\omega'(x) - f_u(x,z,z')\omega(x)$$

and

$$(3.5) \qquad r(x) = f(x,z,z') - z(x)f_u(x,z,z') - z'(x)f_{u'}(x,z,z').$$

From the definition of the error function we have

$$T \, E_\Delta(\omega;x) = T\omega - T \, S_\Delta(\omega;x)$$

or

$$(3.6) \qquad T \, S_\Delta(\omega;x) - r(x) = - \, G_\Delta(\omega;x)$$

with

$$G_\Delta(\omega;x) = - \, T \, E_\Delta(\omega;x).$$

According with A h l b e r g   we will have $G_\Delta(\omega;x)=O(\|\mathbb{1}\|)$ only if $\omega(x) \in C^2[a,b]$. This result improves if we assume $\omega(x) \in C^4[a,b]$. From (3.3) we rewrite (3.6) getting a system of $n+3$ equations with $n+3$ unknown quantities $\omega(x_i)$ $i=0,1,\ldots,n,\omega'(a),\omega'(b)$:

$$\sum_{j=0}^{n} \omega_j \, T \, A_{\Delta j}(x_i) + \omega'_0 \, T \, B_{\Delta 0}(x_i) + \omega'_n \, T \, B_{\Delta n}(x_i) =$$

$$(3.7) \qquad + G_\Delta(\omega;x_i) \qquad\qquad i=0,1,2,\ldots,n$$

$$a_1\omega_0 + a_2\omega'_0 = a_0$$

$$b_1\omega_n + b_2\omega'_n = b_0 \quad ,$$

where $\omega_j = \omega(x_j)$, $j=0,1,2,\ldots,n.$
With the notation

$$W_\Delta = (\omega'_0,\omega_0,\omega_1,\ldots,\omega_n,\omega'_n)^T$$

$$G_\Delta = (0,G_\Delta(\omega,x_0),\ldots,G_\Delta(\omega;x_n),0)^T$$

$$R_\Delta = (a_0,r(x_0),\ldots,r(x_n),b_0)^T$$

and the matrix H_Δ

$$H_\Delta = \begin{bmatrix}
a_2 & a_1 & 0 & \cdots & 0 & 0 \\
T\,B_{\Delta o}(x_o) & T\,A_{\Delta o}(x_o) & T\,A_{\Delta 1}(x_o) & \cdots & T\,A_{\Delta n}(x_o) & T\,B_{\Delta n}(x_o) \\
\vdots & \vdots & \vdots & & \vdots & \vdots \\
T\,B_{\Delta 0}(x_n) & T\,A_{\Delta 0}(x_n) & T\,A_{\Delta 1}(x_n) & \cdots & T\,A_{\Delta n}(x_n) & T\,B_{\Delta n}(x_n) \\
0 & 0 & 0 & \cdots & b_1 & b_2
\end{bmatrix}$$

we can write (3.7) in matrix notation

$$H_\Delta\, W_\Delta = R_\Delta + G_\Delta \ .$$

The quantities G_Δ are unknown. The method consists in taking $G_\Delta \equiv 0$ and to determine W_Δ^* from $H_\Delta W_\Delta^* = R_\Delta$ as an approximation of W_Δ.

If we know that $\|H^{-1}\|$ exists and it is uniformly bounded with $\|\Delta\| \to 0$ it can be prooved that the solution of the modified equation $H_\Delta W_\Delta^* = R_\Delta$ define spline functions which converge uniformly to $W(x)$ as $\|\Delta\| \to 0$.

HOW TO CONSTRUCT H_Δ

In order to have H_Δ we need to compute $A''_{\Delta k}(x_i)$, $A'_{\Delta k}(x_i)$, $B''_{\Delta k}(x_i)$, $B'_{\Delta k}(x_i)$, $(i=0,1,2,\ldots,n)$. That is obtained from (3.1) and (3.2) using, for instance, the definition equations of an interpolate spline function given by A h l b e r g .

We have to solve $n+3$ systems (each one of $n+1$ linear equations) in order to have $A''_{\Delta k}(x_i)$ $(k=0,1,\ldots,n)$, $B''_{\Delta 0}(x_i)$, $B''_{\Delta n}(x_i)$ $(i=0,1,\ldots,n)$. All these systems have the same tridiagonal matrix, only the second member is different. That tridiagonal matrix is regular, so the solution for each system exists and it is unique. From $A''_{\Delta k}(x_i)$, $B''_{\Delta 0}(x_i)$ and $B''_{\Delta n}(x_i)$ we can easily compute $A'_{\Delta k}(x_i)$, $B'_{\Delta 0}(x_i)$ and $B'_{\Delta n}(x_i)$ according to A h l b e r g.

IMPROVING THE RESULTS

From (3.7) we get a set of points $(x_i, \omega(x_i))$ where $\omega(x_i)$ interpolates the solution of (1.5) (1.6). We can consider now a stochastic approximation procedure in order to approximate $\omega(x)$, solution of (1.5) (1.6). Let us take $(x_i, \omega(x_i))$ as a sampled data where x_i is taken as a random number uniformly distributed in $[a,b]$ and we are looking for an approximation of $\omega(x)$ in the form of a linear combination of some coordinate functions $\varphi_\ell(x)$, $\ell = 1, \ldots, p$. In order to find the coefficients of that combination, we minimize the error function:

$$D(\lambda) = \int_{W*}^{\|\omega\|} \left[\omega(x) - \sum_{\ell=1}^{p} \lambda_\ell \varphi_\ell(x) \right]^2 dx$$

where ω^* is the minimum of $\omega(x)$ and $\|\omega\|$ is the maximum on $[a,b]$. As $\omega(x)$, ω^*, $\|\omega\|$ are unknown, $D(\lambda)$ itself cannot be minimized, therefore it is replaced by the empirical error function:

$$J(\lambda) = \sum_{i=1}^{r} \left[\tilde{\omega}(x_i) - \sum \lambda_\ell \varphi_\ell(x_i) \right]^2$$

where r, $\tilde{\omega}(x_i)$ and x_i are defined in a sequencial process and $\tilde{\omega}(x_i)$ are taken from the cardinal spline process given above.

Let $s_n^r(x)$ be the least squares spline approximation with n equidistributed Knots, such that

$$J(s_n^r) \leq J(s_n)$$

for given r and n.

The sequence $s_{n(r)}^r(x)$, such that $J(s_{n(r)}^r(x)) \leq J(s_{n(r)}(x))$, converges uniformly in $[a,b]$ with probability 1 to $\omega(x)$ provided that the following condition holds (see [2], pg. 9):

$$\lim_{r \to \infty} \frac{n^2(r) \log r}{r} = 0.$$

So, in oder to apply this result to each linear problem (1.5) (1.6), we first use collocation method (3.7) with cardinal splines, getting $\omega(x_i)$, $i=1,2,\ldots,n$ and then we take r of these points $(x_i, \omega(x_i))$, $r<n$ and comput a least square spline $s_{n(r)}^r(x) = \sum_{\ell=1}^{p} \lambda_\ell \varphi_\ell(x)$ which minimizes $J(\lambda)$ for that r and n.

Repeating that, with different $r<n$ we are constructing a sequence $s_{n(r)}^r(x)$ such that $J(s_{n(r)}^r(x)) < J(s_{n(r)}(x))$ and which converges uniformly to $\omega(x)$ on $[a,b]$, with probability 1, if

$$\lim_{r \to \infty} \frac{n^2(r) \log r}{r} = 0.$$

4.

Now we take B-spline functions as a set of base functions of $L(\Delta_n, 3, 2)$. The cubic spline B-functions

$$\{B_{-1}, B_0, \ldots, B_n, B_{n+1}\}$$

may be considered as a set of generating functions of $L(\Delta_n, 3, 2)$. So, $s(x)$ $L(\Delta_n, 3, 2)$, will have the representation

$$(4.1) \qquad s(x) = \sum_{i=-1}^{n+1} c_i \, B_i(x)$$

The B-functions are well studied. S c h o e n b e r g (1966) showed that the B-spline are the unique splines not vanishing and of less compact support, with Knots $x_{-2} < x_{-1} < x_o < \ldots < x_n < x_{n+1} < x_{n+2}$. We need to add four Knots $x_{-2} < x_{-1} < x_o$, $\quad x_n < x_{n+1} < x_{n+2}$. The $B_i(x)$ - functions are defined by

$$(4.2) \qquad B_i(x) = \frac{1}{h^3} \begin{cases} (x-x_{i-2})^3 & \text{if } x \in [x_{i-2}, x_{i-1}] \\[2mm] h^3 + 3h^2(x-x_{i-1}) + 3h(x-x_{i-1})^2 + 3(x-x_{i-1})^3 & \\[1mm] & \text{if } x \in [x_{i-1}, x_i] \\[2mm] h^3 + 3h^2(x_{i+1}-x) + 3h(x_{i+1}-x)^2 - 3(x_{i+1}-x)^3 & \\[1mm] & \text{if } x \in [x_i, x_{i+1}] \\[2mm] (x_{i+2}-x)^3 & \text{if } x \in [x_{i+1}, x_{i+2}] \\[2mm] 0 & \text{elsewhere} \end{cases}$$

The graphicrepresentation is

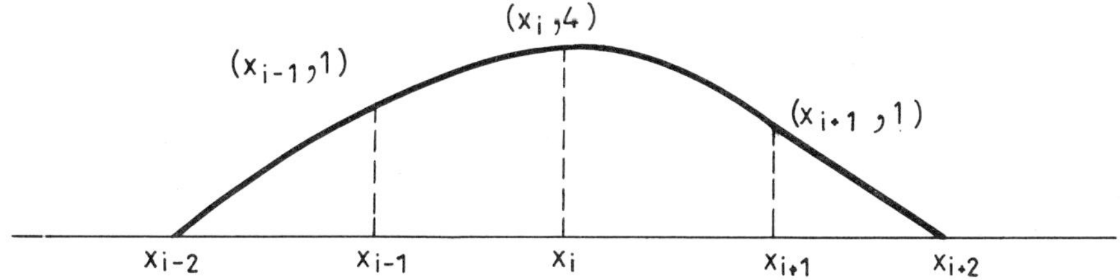

So, $B_i(x)\equiv 0$ for $x \geq x_{i+2}$ and $x \leq x_{i-2}$, $B_i(x)$ is twice continuously differentiable and $B_i(x)$ $L(\Delta_n,3,2)$.

We can state the useful table 1:

	x_{i-2}	x_{i-1}	x_i	x_{i+1}	x_{i+2}
$B_i(x)$	O	1	4	1	O
$B_i'(x)$	O	$\dfrac{3}{h}$	O	$-\dfrac{3}{h}$	O
$B_i''(x)$	O	$\dfrac{6}{h^2}$	$-\dfrac{12}{h^2}$	$\dfrac{6}{h^2}$	O

Again, we take (1.5) (1.6) problem and we seek a numerical approximation of its solution in $L(\Delta_n,3,2)$.

With (1.5) (1.6) as an operator equation $T\omega = r(x)$ with

$$T\omega = \omega''(x) - f_{u'}(x,z,z')\omega'(x) - f_u(x,z,z')\omega(x)$$

$$r(x) = f(x,z,z') - z(x)f_u(x,z,z') - z'(x)f_{u'}(x,z,z')$$

we simplify the notation defining

$$p(x) = -f_{u'}(x,z,z') \quad ; \quad q(x) = -f_u(x,z,z')$$

and we can rewrite

$$T\omega = \omega''(x) + p(x)\omega'(x) + q(x)\,\omega(x).$$

Then, let us assume a representation for the approximate solution $s(x)$ of (1.5) (1.6) in $L(\Delta_n,3,2)$. We will have

$$Ts(x) = \sum_{i=-1}^{n+1} c_i B_i''(x) + p(x)\sum_{i=-1}^{n+1} c_i B_i'(x) + q(x)\sum_{i=-1}^{n+1} c_i B_i(x) .$$

At this moment we have to examine if B_i-functions satisfy the boundary conditions (1.6) of the problem. Usually we have to do some modification in order to get a convenient set of $B_i(x)$-functions satisfying the boundary conditions of the given problem.

This new set $\tilde{B}_i(x)$ of generating functions can be obtained from the set $B_i(x)$ and they span a subfamily $L'(\Delta_n,3,2)$ of $L(\Delta_n,3,2)$ where $L'(\Delta_n,3,2)$ consists of all functions which satisfy the boundary conditions (1.6).

Now, for simplicity we can assume that the boundary conditions (1.6) are just

$$(4.3) \qquad \omega_{n+1}(a) = \omega_{n+1}(b) = 0 \quad \text{and} \quad a=0, \quad b=1.$$

Then, we use $B_{-1}, B_0, B_1, B_{n-1}, B_n, B_{n+1}$ (which are not null at $a=0$, $b=1$) in order to get four additional functions $\tilde{B}_0, \tilde{B}_1, \tilde{B}_{n-1}, \tilde{B}_n$ which satisfy the boundary (4.3). We need some constants a, b, c, d, α, β, γ, δ.

$$\tilde{B}_0(x) = aB_{-1}(x) + bB_0(x) \qquad \tilde{B}_{n-1}(x) = \alpha B_{n-1}(x) + \beta B_n(x)$$

$$\tilde{B}_1(x) = cB_0(x) + dB_1(x) \qquad \tilde{B}_n(x) = \gamma B_n(x) + \delta B_{n+1}(x),$$

where $\tilde{B}_0, \tilde{B}_1, \tilde{B}_{n-1}, \tilde{B}_n$ solve the following interpolation problem

$$\tilde{B}_0(x_0)=0 \qquad \tilde{B}_1(x_0)=0 \qquad \tilde{B}_{n-1}(x_{n+1})=1 \qquad \tilde{B}_n(x_{n-1})=1$$

$$\tilde{B}_0(x_1)=1 \qquad \tilde{B}_1(x_{-1})=1 \qquad \tilde{B}_{n-1}(x)=0 \qquad \tilde{B}_n(x_n)=0.$$

Solving the system we have

$$\tilde{B}_o(x) = B_o(x) - 4B_{-1}(x)$$

$$\tilde{B}_1(x) = B_o(x) - 4B_1(x)$$

$$\tilde{B}_{n-1}(x) = B_n(x) - 4B_{n-1}(x)$$

$$\tilde{B}_n(x) = B_n(x) - 4B_{n+1}(x).$$

We can proove that $\{\tilde{B}_o, \tilde{B}_1, B_2, B_3, \ldots, \tilde{B}_{n-1}, \tilde{B}_n\}$ form a base of $L'(\Delta_n, 3, 2)$.

So, our $s(x)$, approximate solution of (1.5) (4.3) will be

$$s(x) = \sum_{i=o}^{n} c_i \tilde{B}_i(x),$$

where $\tilde{B}_i(x) \equiv B_i(x)$ $\quad 2 \leq i \leq n-2$.

In order to get $\{c_i\}$, $i = 0, 1, \ldots, n$ we do collocation method at the Knot points $x_o = , x_1, \ldots, x_n \equiv 1$, with the operator equation $T\omega = r(x)$

$$Ts(x_j) = s''(x_j) + p(x_j)s'(x_j) + q(x_j)s(x_j) = r(x_j)$$

or

$$(4.4) \quad \sum_{i=o}^{n} c_i \tilde{B}_i''(x_j) + p(x_j)\sum_{i=o}^{n} c_i \tilde{B}_i'(x_j) + q(x_j)\sum_{i=o}^{n} c_i \tilde{B}_i(x_j) =$$

$$= r(x_j) + G(x_j) \qquad j = 0, 1, \ldots, n ,$$

where

$$G(x) = Ts(x) - r(x)$$

$$= -TE(x)$$

(4.5)

$$= -E''(x) - p(x)E'(x) - q(x)$$

$$E(0) = E(1) = 0.$$

Again, as with cardinal spline functions we do'nt know $G(x)$, so instead (4.4) we solve the system (4.5) of $n+1$ equations in $n+1$ unknown $\{c_i\}$ $i=0,1,\ldots,n$.

(4.6)
$$\sum_{i=0}^{n} c_i \tilde{B}_i''(x_j) + p(x_j)\sum_{i=0}^{n} c_i \tilde{B}_i'(x_j) + q(x_j)\Sigma c_i \tilde{B}_i(x_j) = r(x_j)$$

$$j = 0,1,\ldots,n .$$

Let as rewrite (4.6) in the following way:

$$c_0 T\tilde{B}_0(x_0) + c_1 T\tilde{B}_1(x_0) + \ldots + c_n T\tilde{B}_n(x_0) = r(x_0)$$

(4.7)
$$\text{--}$$

$$c_0 T\tilde{B}_0(x_n) + c_1 T\tilde{B}_1(x_n) + \ldots + c_n T\tilde{B}_n(x_n) = r(x_n).$$

According to the definition of the set $\{\tilde{B}_i(x)\}$ $i=0,1,\ldots,n$, and using table 1, we easily see that the matrix of the system (4.6) is tridiagonal. Its organization is very simple, from the table 1. Assuming some conditions on $p(x)$ it can be seen that this matrix is diagonal dominant, so existence and uniqueness of $\{c_i\}$ are guarantied and consequently of $s(x)$ $L'(\Delta_n,3,2)$, solution of $T\omega = r(x)$.

	c_0	c_1	c_2	c_3	$\cdots$	c_{n-2}	c_{n-1}	c_n
x_0	$-\dfrac{36}{h^2}+\dfrac{12p_0}{h}$	$-\dfrac{36}{h^2}-\dfrac{12p_0}{h}$	0	0				
x_1	$\dfrac{6}{h^2}-\dfrac{3p_1}{h}+q_1$	$\dfrac{54}{h^2}-\dfrac{3p_1}{h}-15q_1$	$\dfrac{6}{h^2}+\dfrac{3p_1}{h}+q_1$	0				0
x_2	0	$\dfrac{-24}{h^2}-\dfrac{12p_2}{h}-4q_2$	$\dfrac{-12}{h^2}+4q_2$	$\dfrac{-6}{h^2}+\dfrac{3p_2}{h}+q_2$				0
x_3	0	0	$\dfrac{6}{h^2}-\dfrac{3p_3}{h}+q_3$	$\dfrac{-12}{h^2}+4q_3$				0
				$\dfrac{6}{h^2}-\dfrac{3p_4}{h}+q_4$				0
				0				
$\vdots$								
						$\dfrac{6}{h^2}+\dfrac{3p_{n-3}}{h}+q_{n-3}$		
x_{n-2}	0	0	0			$\dfrac{12}{h^2}+4q_{n-2}$	$\dfrac{-24}{h^2}-12\dfrac{p_{n-2}}{h}-4q_{n-2}$	0
x_{n-3}	0	0	0			$\dfrac{6}{h^2}-\dfrac{3p_{n-1}}{h}+q_{n-1}$	$\dfrac{54}{h^2}-\dfrac{3p_{n-1}}{h}-15q_{n-1}$	$\dfrac{6}{h^2}+\dfrac{3p_{n-1}}{h}+q_{n-1}$
x_n	0	0	0			0	$-\dfrac{36}{h^2}+12\dfrac{p_n}{h}$	$-\dfrac{36}{h^2}-12\dfrac{p_n}{h}$

GETTING SOME INFORMATION FOR THE ERROR-FUNCTION

We see that this $s(x)\equiv \sum_{i=0}^{n} c_i \tilde{B}_i(x)$ gives $G(x_j)=0$, $i=0,1,\ldots,n$, but for other $x \neq x_j$ will not be so. From (4.5) we may get some information for the Error function. We may take points $x_{kj} \in [x_j, x_{j+1}]$ and, as we have $s(x)$, we can compute $G(x_{k_j})$ from $Ts(x)-r(x)$. That is the residue for the operator equation at each x_{k_j}. Going with this and with the equation

$$-E''(x_{k_j})-p(x_{k_j})E'(x_{k_j})-q(x_{k_j})E(x_{k_j})=G(x_{k_j})$$

$$E(0) = E(1) = 0$$

we may approximate $E(x)$ using some known numerical method as the finite difference method, or other. That function $E(x)$ is the Error-function of the computed $s(x)$.

REFERENCES

[1] J.H. Ahlberg - E.N. Nilson - J.L. Walsh, *The Theory of Splines and their Applications*, Academic Press, New York, 1967.

[2] F. Archetti - B. Betro, Recursive stochastique evaluation of the level sets measure optimization problems, *Quaderno dei Gruppi di Ricerce Matematica del C. N. R. Università di Pisa* (1975).

[3] R.E. Bellman - R.E. Kalaba, *Quasilinearization and Nonlinear Boundary - Value Problems*, Elsevier, New York, 1965.

[4] C. de Boor - B. Swartz, Collocation at Gaussian points, *SIAM J. Numer. Anal.* 10 (1973).

[5] A. Ghizzetti, Lezioni sui procedimenti di quasili-
     nearizzazione, *Publicazioni dell'Instituto per le
     Applicazioni del Calcolo,* Roma (1965).

[6] P.M. Prenter, *Splines and Variational Methods,*
     Academic Press, New York, 1975.

[7] R.D. Russel - L.F. Shampine, A collocation method
     for boundary value problems, *Numer. Math.* 19 (1972)
     1-28.

Fernanda Aleixo Oliveira
Department of Mathematics
University of Coimbra
Coimbra, Portugal

THE CHARACTERISTIC VALUES OF THE MATRIX $A+XBX^{-1}$

G.N. de OLIVEIRA

1. MOTIVATION

Let A be an $(n+m)\times(n+m)$ matrix, over a field F, partitioned as follows

$$A = \begin{bmatrix} A_{11} & A_{12} \\ A_{21} & A_{22} \end{bmatrix} ,$$

where A_{11} is $n\times n$ and A_{22} $m\times m$. We shall consider some problems of existence of A when some of the submatrices and the characteristic polynomial $f(\lambda)$ of A are prescribed. These problems have applications for example in control theory [6]. There are seven distinct problems depending on how many and which of the four submatrices we prescribe. Assume we prescribe only A_{11} and the characteristic polynomial. We need some notation.

Let R be an arbitrary square matrix over F and let

$$p_1(\lambda)^{v_1^{(i)}} \;,\ldots,\; p_i(\lambda)^{v_{r_i}^{(i)}} \qquad v_1^{(i)} \leq \ldots \leq v_{r_i}^{(i)}$$

$$i=1,\ldots,t$$

be its elementary divisors.

Let

$$\Theta_{i,k} = \sum_{j=1}^{r_i-k} v_j^{(i)}$$

and

$$\Phi_{R,k}(\lambda) = \prod_{i=1}^{t} p_i(\lambda)^{\Theta_{i,k}}.$$

If $r_i-k < 1$ we take $\Theta_{i,k} = 0$.

It is known that A exists with A_{11} and $f(\lambda)$ prescribed if and only if $\Phi_{A_{11},m}(\lambda)$ divides $f(\lambda)$ [2].

We may wish to prescribe A_{11} and the elementary divisors of A instead of its characteristic polynomial. This is a much more difficult problem whose solution has been discovered by E. M a r q u e s d e S á [5] very recently.

Assume now that we prescribe the blocks A_{11}, A_{22} and the characteristic polynomial $f(\lambda) = \lambda^{n+m} - c\lambda^{n+m-1} +\ldots$. The following result has been proved in [3].

THEOREM 1. *Assume the following conditions are satisfied.*

(1) $\operatorname{tr} A_{11} + \operatorname{tr} A_{22} = c$.

(2) A_{22} *has* m *distinct characteristic values* $\rho_1,\ldots,\rho_m$ *in* F.

(3) The polynomial $f(\lambda)$ has at least $m-1$ roots in F,

$$\xi_2, \ldots, \xi_m, \quad i.e. \quad f(\lambda) = f_1(\lambda)(\lambda-\xi_2)\ldots(\lambda-\xi_m).$$

(4) Let $P_1,\ldots,P_n$ be a partition of the set $\{\rho_2-\xi_2,\ldots,\rho_m-\xi_m\}$ (we require n components but allow empty sets in this partition). We denote by s_i the sum of the elements in P_i, putting $s_i = 0$ whenever P_i is empty. Let $S = \mathrm{diag}(s_1,\ldots,s_n)$. Now our assumption is that there exists a nonsingular matrix U such that with $R = UA_{11}U^{-1}+S$, the polynomial $\Phi_{R,1}(\lambda)$ divides $f(\lambda)$.

Then there exists a matrix A, over F, with prescribed

$$f(\lambda), \quad A_{11} \quad and \quad A_{22}.$$

While conditions (1), (2) and (3) seem simple and reasonably natural, the condition (4) looks complicated and awkward. More important then aesthatical considerations, it is difficult to check whether it is satisfied or not.

Note that the matrix S depends on the way we order the ρ_i's and ξ_i's and on how we partition the set $\{\rho_2-\xi_2,\ldots,\rho_m-\xi_m\}$. Let us assume that S has been fixed. We can pose the following questions:

(a) Under what conditions does there exist a matrix U such that $\Phi_{R,1}(\lambda)$ divides $f(\lambda)$?

(b) Under what conditions does there exist a matrix U such that R is nonderogatory? In this case, $\Phi_{R,1}(\lambda) = 1$ and so divides $f(\lambda)$.

(c) What is the range of the characteristic values
(or characteristic polynomial, elementary
divisors) of $UA_{11}U^{-1}+S$ as U varies?

An answer to this last question would be important
for the clarification of condition (4). In fact if, for
a certain U, R has n distinct characteristic values
in F, the corresponding polynomial $\Phi_{R,1}(\lambda)$ is 1 and
so divides $f(\lambda)$.

We shall examine some of these problems in the
general case in which S is an arbitrary matrix B, not
necessarily diagonal.

2. RESULTS

Let A, B be two k-square matrices over the field
F.

We say that the pair (A,B) is polynomially com-
plete if for every polynomial $f(\lambda) = \lambda^k - c\lambda^{k-1}+\ldots$, over
F, satisfying $c = \operatorname{tr} A + \operatorname{tr} B$, there is a nonsingular
matrix U such that $UAU^{-1}+B$ has characteristic poly-
nomial $f(\lambda)$.

We say that the pair (A,B) is spectrally complete
if for every k-tuple $(v_1,\ldots,v_k)$ of elements of F
satisfying $v_1+\ldots+v_k = \operatorname{tr} A + \operatorname{tr} B$, there is a nonsingular
matrix U such that $UAU^{-1}+B$ has characteristic values
$v_1,\ldots,v_k$.

It is trivial that polynomial completeness implies
spectral completeness. Theorem 4 below gives a reasonably
general condition for spectral completeness. It seems
more difficult to find conditions for polynomial completeness.

If R is a square matrix over F, by $i(R)$ we
denote the number of invariant polynomials of R.

THEOREM 2. *Let A , B be k×k matrices over F. If
there is a nonsingular matrix U such that $UAU^{-1}+B$ is
nonderogatory, then*

$$i(A)+i(B) \leq k+1 \qquad [4].$$

PROOF. Let

$$R_F(A) = \min_{\alpha \in F} \text{rank}(A+\alpha I).$$

Let $\bar{F}$ be an extension of F containing the characteristic values of A. It can be shown that (for
details see [4]),

$$i(A) = k-R_{\bar{F}}(A).$$

It is also easy to see that A is nonderogatory
if and only if $R_{\bar{F}}(A) = k-1$. Assume that for a certain
U, the matrix $R = UAU^{-1}+B$ is nonderogatory. For
$\alpha,\beta \in \bar{F}$, $R+(\alpha+\beta)I = U(A+\alpha I)U^{-1}+(B+\beta I)$ is also nonderogatory.

Therefore

$$\text{rank}(A+\alpha I)+\text{rank}(B+\beta I) \geq k-1.$$

Since this is valid for all α and β in $\bar{F}$ we can
write

$$R_{\bar{F}}(A)+R_{\bar{F}}(B) \geq k-1.$$

Hence

$$i(A)+i(B) \leq k+1.$$

Similarly we can prove

THEOREM 3. *Let A, B be $k \times k$ matrices over F.
Assume that F has more than two elements and that
(A,B) is spectrally complete. Then*

$$i(A)+i(B) \leq k.$$

We omit the proof which can be found in [4].

There are counterexamples showing that the condition of Theorem 3 is not sufficient. However it is not
known whether the condition of Theorem 2 is sufficient.
Probably it is.

Now we present our best result giving a sufficient
condition for spectral completeness [4].

THEOREM 4. *Let A, B be k-square matrices over
F. If one of these matrices is nonderogatory and the
other nonscalar, the pair (A,B) is spectrally complete.*

The proof is very lengthy and cannot be presented
here. The interested reader can find details in [4].
Broadly speaking we can say that the method of proof
consists of constructing U such that $UAU^{-1}+B$ is
triangular with the prescribed characteristic values
down the principal diagonal.

3. APPLICATIONS

Thanks to Theorem 4 we can now replace the condition (4) of Theorem 1 by the simpler condition: *"Assume
that F has more than $n-1$ elements and that one of
the matrices A_{11} and $S = \mathrm{diag}(s_1,\ldots,s_n)$ is nonderogatory".*

In fact suppose that, for example, A_{11} is nonderogatory. If S is scalar, taking $U = I$ we have $\Phi_{R,1}(\lambda) = 1$ and so divides $f(\lambda)$. This is valid even when the number of elements of F is $\leq n$. If S is not scalar, the pair (A_{11}, S) is spectrally complete. Take U so that R has n distinct characteristic values. In this case $\Phi_{R,1}(\lambda) = 1$ and so divides $f(\lambda)$.

For the case of the complex field we can state the following result instead of Theorem 1:

"Assume A_{22} has m distinct characteristic values $\rho_1, \ldots, \rho_m$. Assume that either it is possible to choose $m-1$ roots $\xi_2, \ldots, \xi_m$ of $f(\lambda)$ so that $S = \operatorname{diag}(s_1, \ldots, s_n)$ be nonderogatory or A_{11} is nonderogatory. Assume further that $f(\lambda)$, A_{11} and A_{22} verify the trace condition (1). Then there is a complex matrix with prescribed $f(\lambda)$, A_{11} and A_{22}".

We give now another application.

One of the "seven distinct problems" mentioned at the beginning of this paper is the following: find conditions for the existence of A with the blocks A_{12}, A_{21} prescribed and also the characteristic polynomial $f(\lambda)$. If the field F is algebraically closed, A always exists without any condition on A_{12}, A_{21} and $f(\lambda)$. This is implied by a stronger result due to S. F r i e d l a n d [1]. However if F is not closed, the solution is not known.

It is obvious that when one of the prescribed blocks is the zero block, A exists if and only if $f(\lambda)$ can be written as a product $g(\lambda) h(\lambda)$ with $g(\lambda)$ of degree n and $h(\lambda)$ of degree m. We shall assume that A_{12}, $A_{21} \neq 0$. Then, in certain cases, A does not exist. Indeed it is not difficult to construct a 2×2 counter-example over the real field.

As an application of our studies of the characteristic values of $UAU^{-1}+B$ we give a result (unfortunately valid only for a rather particular case) on this problem. The principal restriction is that n must be equal to m.

Let

$$A = \begin{bmatrix} A_{11} & A_{12} \\ A_{21} & A_{22} \end{bmatrix} ,$$

where the prescribed blocks A_{12}, A_{21} are different from zero and both are of the same size $n \times n$. Assume that the prescribed characteristic polynomial can be written as

$$f(\lambda) = g(\lambda)(\lambda-\xi_1) \ldots (\lambda-\xi_n)$$

with $\xi_i \in F$ and $g(\lambda)$ with coefficients in F. Then there exist blocks A_{11} and A_{22} such that A has $f(\lambda)$ as characteristic polynomial.

PROOF. If U and V are $n \times n$ nonsingular matrices, we can replace A_{12} and A_{21} by $VA_{12}U^{-1}$ and $UA_{21}V^{-1}$ respectively. To prove this consider TAT^{-1} with $T = V \oplus U$. Let $r(> 0)$ be the rank of A_{12}. We can assume that $A_{12} = \text{diag}(1,\ldots,1,0,\ldots,0)$, where the number of 1's is r. We claim that there is no loss of generality if we assume $A_{12}+A_{21}$ nonscalar. If $A_{12}+A_{21}$ is scalar, A_{21} must be of the type $\text{diag}(\mu,\ldots,\mu,\mu+1,\ldots,\mu+1)$, where the number of μ's is r. Firstly suppose that $r < n$. Let V be the matrix obtained from the $n \times n$ identity matrix by interchanging the rows r and $r+1$.

Replace A_{12} and A_{21} by VA_{12} and $A_{21}V^{-1}$ respectively. $VA_{12}+A_{21}V^{-1}$ is not scalar as otherwise $\mu = 0$, $\mu+1 = 0$, a contradiction. Let us finally examine the case $r = n$, i.e., $A_{12} = I_n$. If $A_{12}+A_{21}$ is scalar, then $A_{21} = \Theta I_n$. Let

$$V = \begin{bmatrix} 0 & 1 \\ 1 & 0 \end{bmatrix} \oplus I_{n-2} \ .$$

Clearly $VA_{12}+A_{21}V^{-1}$ cannot be scalar.

Thus in the sequel we suppose that $A_{12}+A_{21}$ is nonscalar.

Let B be a nonderogatory matrix over F with characteristic polynomial $g(\lambda)$ (for example, B may be the companion matrix of $g(\lambda)$). The pair $(A_{12}+A_{21},B)$ is spectrally complete. Choose U so that the characteristic values of $R = A_{12}+A_{21}+UBU^{-1}$ be $\xi_1,\ldots,\xi_n$. Let

$$C = \begin{bmatrix} I & 0 \\ I & I \end{bmatrix} \begin{bmatrix} R & A_{12} \\ 0 & UBU^{-1} \end{bmatrix} \begin{bmatrix} I & 0 \\ -I & I \end{bmatrix}$$

Clearly

$$C = \begin{bmatrix} * & A_{12} \\ A_{21} & * \end{bmatrix}$$

and its characteristic polynomial is $f(\lambda)$.

REFERENCES

[1]   S. Friedland, Matrices with prescribed off diagonal elements, *Israel J. Math.* 11(1972), 184-189.

[2] G.N. de Oliveira, Matrices with prescribed cha-
     racteristic polynomial and a prescribed submatrix,
     III. *Monatsh. Math.* 75(1971), 441-446.

[3] G.N. de Oliveira, Matrices with prescribed cha-
     racteristic polynomial and several prescribed sub-
     matrices, *Linear Mult. Algebra*, 2(1975), 357-364.

[4] G.N. de Oliveira - E. Marques de Sá - J.A. Dias da
     Silva, On the eigenvalues of the matrix $A+XBX^{-1}$,
     Linear Mult. Algebra, 5(1977), 119-128.

[5] E. Marques de Sá, Imbedding conditions for λ-mat-
     rices, *Linear Alg. Appl.*, 24(1979), 33-50.

[6] H.K. Wimmer, Existenzsätze in der Theorie der
     Matrizen und lineare Kontrolltheorie, *Monatsh.
     Math.* 78(1974), 256-263.

G.N. de Oliveira
Instituto de Matemática da
Faculdade de ciências
Universidade de Coimbra
Coimbra, Portugal

COLLOQUIA MATHEMATICA SOCIETATIS JÀNOS BOLYAI
22. NUMERICAL METHODS, KESZTHELY (HUNGARY), 1977.

ON THE PSEUDOINVERSION OF MATRIX PRODUCTS

OLGA POKORNÀ

A necessary and sufficient condition for the validity of the relation $(BA)^+ = A^+B^+$ for pseudoinverse matrices is given. It covers a more general class of matrices than that in [2].

Throughout this paper, A^+ denotes the pseudoinverse matrix of a complex rectangular matrix A in the sense of the Moore – Penrose definition [1]:

A^+ is the pseudoinverse of A if and only if it has the following four properties:

(1) $\qquad AA^+A = A$

(2) $\qquad A^+AA^+ = A^+$

(3) $\qquad (A^+A)^* = A^+A$

(4) $\qquad (AA^+)^* = AA^+ ,$

M^* denoting the conjugate transpose of M.

Let A, B be matrices, the product of which is defined. As is well known, the relation

$$(5) \qquad (BA)^+ = A^+ B^+$$

does not hold in general. For a simple example, let $A = (1,1)^T$, $B = (1,0)$. Then $A^+ = (1/2, 1/2)$, $B^+ = (1,0)^T$; $BA = (1)$, so that $(BA)^+ = (1)$, while $A^+ B^+ = (1/2)$.

The relation (5) holds in special cases, for example when both A and B are of full rank. In this case, AA^* and B^*B are nonsingular matrices, so that $(AA^*)^{-1}$ and $(B^*B)^{-1}$ exist and it holds $A^+ = A^*(AA^*)^{-1}$, $B^+ = (B^*B)^{-1}B^*$, $(BA)^+ = A^*(AA^*)^{-1}(B^*B)^{-1}B^* = A^+ B^+$.

It is clear, that the validity of (5) for particular matrices is useful for numerical computations. We may take use of it for example when we are to compute the pseudoinverse of a matrix which may be expressed as a product of two matrices, both of them being easily pseudoinvertable.

We are going to prove here a theorem giving a necessary and sufficient condition for the validity of (5) under some assumptions on A and B.

Let A be an $m \times n$ matrix, B an $r \times m$ matrix, $\mathrm{rank}(A) = p$ and $\mathrm{rank}(B) = q$.

Let us denote by X the matrix formed by p orthonormal columns - all eigenvectors of the matrix AA^* corresponding to its nonzero eigenvalues $d_1^2, \ldots \ldots, d_p^2$. Thus X is an $m \times p$ matrix, for which

$$X^* X = I_p$$

holds, I_p being the $p \times p$ identity matrix. Further, let S_A be the span of all columns of X.

Similarly, let V be formed by q orthonormal columns - all eigenvectors of B^*B corresponding to its nonzero eigenvalues $f_1^2, \ldots, f_q^2$. Thus V is an $m \times q$ matrix, for which

$$V^*V = I_q$$

holds. Let S_B be the span of all columns of V.

H. S c h w e r d t f e g e r [2] proved a theorem, which may be formulated as follows:

THEOREM H.S. *Let* A, B, p, q, S_A, S_B *have the meaning just shown. Let*

$$q = p \quad \text{and} \quad S_B = S_A.$$

Then (5) holds.

Now, let us consider a more general case: let

$$q \leq p \quad \text{and} \quad S_B \subseteq S_A.$$

As all columns of X form a basis in S_A, it follows from our assumptions on S_B, that each column of the matrix V - as a vector of $S_B \subseteq S_A$ - may be represented in this basis. That means, there exists a $p \times q$ matrix M such that

$$(6) \qquad V = XM$$

holds. We prove the following theorem:

THEOREM. *Let* A, B, p, q, X, V, d_i, S_A, S_B *have the meaning introduced above. Let* $q \leq p$ *and* $S_B \subseteq S_A$ *and let* M *be the matrix given by (6). Then (5) holds*

if and only if the matrices MM^* *and* $D^2 = \mathrm{diag}(d_1^2,\ldots$
$\ldots,d_p^2)$ *commute.*

PROOF. Let $A = XDY^*$ and $B = UFV^*$ be singular
decompositions of A and B; here $D = \mathrm{diag}(d_1,\ldots,d_p)$,
d_i being the singular values of A, X as above, Y an
$n \times p$ matrix, for which $Y^*Y = I_p$ holds, and similarly
$F = \mathrm{diag}(f_1,\ldots,f_q)$, V as above and U an $n \times q$ ·
matrix with $U^*U = I_q$.

Since $I_q = V^*V = M^*X^*XM$, the matrix M fulfils

$$M^*M = I_q.$$

Setting $V = XM$ into the singular decomposition
of B we obtain for B and for BA the following
expressions:

$$B = UFM^*X^*,$$

$$BA = UFM^*DY^*.$$

As $A^+ = YD^{-1}X^*$ and $B^+ = XMF^{-1}U^*$, we have

$$A^+B^+ = YD^{-1}MF^{-1}U^*.$$

It is easy to verify, that the matrix

$$Z = A^+B^+$$

fulfils the relations

$$BA\ Z\ BA = BA,$$

$$Z\ BA\ Z = Z,$$

$$(BA\ Z)^* = BA\ Z.$$

That means, Z has the properties (1), (2) and (4) of
the pseudoinverse of BA. Thus Z is the pseudoinverse
of BA if and only if it also fulfils (3), i.e. if and
only if the matrix $Z\,BA$ is hermitian. The condition
$(Z\,BA)^* = Z\,BA$, i.e. $YDMM^*D^{-1}Y^* = YD^{-1}MM^*DY^*$, holds if
and only if

$$DMM^*D^{-1} = D^{-1}MM^*D,$$

i.e. if and only if

$$(7) \qquad D^2MM^* = MM^*D^2,$$

which completes the proof.

The Theorem H.S. may be regarded as a corollary of
the theorem just proved. Namely if $q = p$ and $S_B = S_A$,
then M is a square unitary matrix, so that $MM^* = I$
commutes with any matrix and the condition of our
theorem is fulfilled. Thus (5) holds.

REMARK. The equality $MM^*D^2 = D^2MM^*$ holds if and
only if the matrix MM^* has — perhaps after some permutation — the block-diagonal form

$$MM^* = \begin{bmatrix} R_1 & & & & \\ & R_2 & & & \\ & & \ddots & & \\ & & & R_t \end{bmatrix},$$

where $t \le p$ is the number of all different singular
values $d_1,\ldots,d_t$ of A and the order of each block
R_i equals to the multiplicity m_i of d_i for $i=1,\ldots$
$\ldots,t$. That means, it holds if and only if the rows of

the matrix M form t groups $H_1,\ldots,H_t$, the group H_i consisting of m_i rows and all rows of H_i being orthogonal to all rows of each H_j for $j \neq i$, $i=1,\ldots$ $\ldots,t$, $j=1,\ldots,t$.

It would be interesting to find a geometrical interpretation of the condition (7), more closely related to the spaces S_A and S_B, but it has not been done till now.

REFERENCES

[1] R. Penrose, A generalized inverse for matrices, *Proc. Cambridge Philos. Soc.*, 51(1955), 406-413.

[2] H. Schwerdtfeger, Remarks on the generalized inverse of a matrix, *Lin. Alg. Appl.*, 1(1968), 325-328.

Olga Pokorná
Mathem.-Phys. Faculty of
Charles University, Prague
Dpt of numerical mathematics
Malostranské nám. 25, Prague 1
Czechoslovakia

COLLOQUIA MATHEMATICA SOCIETATIS JÁNOS BOLYAI
22. NUMERICAL METHODS, KESZTHELY (HUNGARY), 1977.

ABOUT HANDLING LARGE LINEAR SYSTEMS WITH RESPECT TO USING EXTERNAL STORAGE

D. RICHTER

For solving systems of linear equations, many well-known methods exist. All of this methods are described under the assumption that the needed informations are available. Indeed, if the system is relatively small, i.e. if the matrix is sufficiently small in the realization we can make use of this assumption. But for large matrices, all of its elements can't be stored in the mainstorage of a computer and external storage is needed. The used numerical methods should be modified in a natural manner of course and the given problem will be solved.

Beyond the high number of arithmetic operations, against which we can't do anything, the additional problem of many transfers to external storage arises. This paper is concerned with searching for structure of given matrices, such that the necessary transfers are minimal. After finding this structure we deal with reducing the demands of large matrices to demands on so-called "small" matrices.

Each kind of transfers is an additional expense,
hence we have to take into account not the various
technical forms but the quantity of transfers only.

Let us consider the system

(1) $Ax = b$,

where matrix A has dimension n and d bands on
either side of the diagonal, $1 \leq d \leq n$.

We assume in the following considerations that A
is a real, positive definite and symmetric matrix. This
seems to be severe restriction of course, but the obtained results can be carried over to other matrices in
essence.

As the reader knows, for solving system (1) we have
to do the following:

1. Decompose A according to the CHOLESKY factorization,

$$A = LL^T,$$

where L is a real nonsingular lower triangular matrix,

2. Solve the system

$$Ly = b,$$

3. and solve for x

$$L^T x = y.$$

Now the question is, which structure of A guarantees
a minimal number of external transfers if we want to
solve system (1).

Moreover we assume that the input-data are stored
columnwise in an external storage and that the output-
data should also be in an external storage. Beceasue of
symmetry, the lower triangular matrix of A is needed

only. Furthermore let K be the greatest number of real
elements which can be stored in the mainstrorage simultaneously.

According to the CHOLESKY-factorization the triangular matrix L is computed as

$$\ell_{ij} = (a_{ij} - \sum_{k=i-d_i^*}^{j-1} \ell_{ik}\ell_{jk})/\ell_{jj},$$

$$(2) \qquad j = i-d_i^* , \ldots, i-1$$

$$\ell_{ii} = (a_{ii} - \sum_{k=i-d_i^*}^{i-1} \ell_{ik}^2)^{1/2},$$

where $d_i^* = \min\{d-1, i-1\}$, $i=1,\ldots,n$.

These relations show us:

For computing one element ℓ_{ij} in the given data-
structure, elements from at most $d-1$ columns are
needed. That is, when computing one column, $d-1$ columns
are needed that lie to the left of that particular
column.

For the given data-structure we arrive at the
condition

$$(3) \qquad d \leq \sqrt{K} .$$

Under this condition we now obtain for the decomposition
that each element of the lower triangular matrix of A
has to be read exactly once and each computed element of
L also has to be written exactly once.

Of course, thus a number of transfers is reached
which can't be reduced.

To reach this minimal number of transfers when
solving $Ly = b$, too, we conclude analogously

(4) $\qquad d(d+r) \leq K,$

where r is the number of right hand sides of the system (1). Finally the solution of $L^T x = y$ leads to the condition

(5) $\qquad d(r+1) \leq K.$

If the input-data are given row by row then condition (3) remains the same and conditions (4) and (5) are exchanged. Conditions (3), (4) and (5) altogether yield the following conclusion: If the input-data are given column by column or row by row then under the condition

(6) $\qquad d(d+r) \leq K$

we obtain a minimal number of transfers to external storage so that each element of A and each element of L should be transported exactly once.

But generally we can't assume condition (6).

Let n, d, r be given arbitrarily. We want to discuss, which structure of A induces a minimal number of transfers without any conditions on n, d and r.

At first with respect to (2) one can see: Least of all informations (with respect to the already computed elements) are needed for computing certain number of elements if the elements which we have to compute are ordered in a quadratic block or in a triangular block, respectively. We have lower triangular blocks in the main diagonal of course, and we have upper triangular blocks on the left hand side of A. If matrix A is divided into quadratic and triangular blocks (as just mentioned) of dimension m we obtain for L according to (2):

$$L_{ij} = (A_{ij} - \sum_{k=i-d_i}^{j-1} L_{jk}L_{ik}^T)(L_{jj}^{-1})^T,$$

$$(7) \qquad L_{ii}L_{ii}^T = (A_{ii} - \sum_{k=i-d_i}^{i-1} L_{ik}L_{ik}^T),$$

$$i=1,\ldots,\frac{n}{m} \; ; \quad j=i-d_i,\ldots,i-1,$$

where $d_i = \min\{i-1, \frac{d}{m} - 1\}$ and n and d are assumed to be divisible by m. For $i \neq j$ the L_{ij}-s are again quadratic or upper triangular blocks, respectively (triangular only if the block lies on the left hand side of the matrix) and L_{ii} are lower triangular blocks. For computing one block L_{ij} $(i \neq j)$ $2am + \frac{3}{2}m^2 - \frac{m}{2}$ informations are needed. Here a denotes the distance between the left side of A and the beginning of the block. In the case $i = j$, the number of informations is $a \cdot m + \frac{m^2}{2} - \frac{m}{2}$. Therefore, to compute p elements we need

$$(8) \qquad \frac{(2a - \frac{1}{2})p}{m} + \frac{3}{2}P \qquad \left[\frac{(a - \frac{1}{2})P}{m} + \frac{P}{2} , \text{ for } i=j\right]$$

informations. Expression (8) decreases if m increases. If m is greater than $\sqrt{\frac{K}{3}}$, then additional transfers to external storage are necessary to realize the operations in (7), because the site of 3 blocks is not available in the mainstorage in this case. The quantity of additional transfers is

$$(9) \qquad 2(h^2 + 2\sqrt{\frac{K}{3}} h),$$

where h is the difference between m and $\sqrt{\frac{K}{3}}$. Because expression (9) increases more quickly than (8) would decrease near to $\sqrt{\frac{K}{3}}$, an optimal quantity of blocks should be found. Taking into account the modular structure of programming methods, we denote $m = \left[\sqrt{\frac{K}{3}}\right]$ as the optimal quantity.

To summarize:

Partitioning matrices A and L in quadratic and triangular blocks with the dimension $m = \left[\sqrt{\frac{K}{3}}\right]$ guarantees a minimal number of transfers to external storage.

It is seen: Quantity m doesn't depend on the dimension n and the width of band d but only on the available mainstorage capacity. Therefore we have to take into account n and d when not divisible by m. This problem can be solved for instance by enlarging the dimension and the width of band introducing as many zeros as to get divisible quantities at the expense of unessentially more computations and transfers.

In the solution of $Ly = b$ with r right hand sides, we obtain by using the block-structure

$$L_{ii}Y_{ij} = B_{ij} - \sum_{k=i-d_i}^{i-1} L_{ik}Y_{kj}$$

(10)

$$i=1,\ldots,\frac{n}{m}; \quad j=1,\ldots, \left[\frac{r}{m}\right]+1,$$

where $d_i = \min\{\frac{d}{m} - 1, i-1\}$ and in the solution of $L^T x = y$

$$L_{ii}^T X_{ij} = Y_{ij} - \sum_{k=i+1}^{i+d_i} L_{ki}^T Y_{kj}$$

(11)

$$i=1,\ldots,\frac{n}{m}; \quad j=1,\ldots,\left[\frac{r}{m}\right]+1,$$

where $d_i = \min\{\frac{d}{m} - 1, i-1\}$. Here B_{ij}, Y_{ij} and X_{ij} are blocks of dimension (m,m_r) where $m_r = m$ for

$j=1,\ldots,[\frac{r}{m}]$ and $m_r = r-m[\frac{r}{m}]$ for $j = [\frac{r}{m}]+1$.

Analogously to the previous considerations on the decomposition, one concludes: For $m = \left\lceil \sqrt{\frac{K}{3}} \right\rceil$ a minimal expense of transfers to external storage is reached.

By using the developed block-structure, estimations show that for sufficiently large width of band $(d > 4m)$ the expense of transfers is at least m times smaller than at an arbitrary method working with the given data-structure (column by column).

From formulae (7), (10) and (11) one see that for solving system (1) the operations are:

(12)   product of two matrices and subtraction from a third,

(13)   solving a system of equation with triangular matrix L and some right hand sides $(Lx = b)$,

(14)   decomposition of a matrix

when so-called "small" matrices are needed. Based on the given quantities n, d, r and K, the number of executions of these three operations and of transfers to external storage can be determined and with the specific exectuion-times of a computer, predictions can be given for the operation- and transfer-time when solving a given problem.

If we consider the operations with the upper triangular matrices on the left hand side of A as operations of the kind (12), (13) or (14), then we get for the decomposition:

Operation (12) should be done

$$\frac{d^2}{6m^3} (3n-2d) + \frac{d}{2m^2} (n-d) - \frac{d}{6m}$$

times, operation (13) with m right hand sides

$$\frac{d}{2m^2}(2n-d) - \frac{d}{2m}$$

times and operation (14) $\frac{n}{m}$ times. Altogether

$$\frac{d^2}{6m^3}(3n-2d) + \frac{d}{2m^2}(3n-d) - \frac{1}{6m}(12n-5d)$$

quadratic blocks and

$$\frac{d}{m^2}(2n-d) + \frac{1}{m}(4n-3d)$$

triangular blocks should be transported. For solving $Ly = b$ and $L^T x = y$, respectively, one gets

$$\frac{d}{2m^2}(2n-d) - \frac{d}{2m}$$

operations of (12) with an $(m,m)-$ and a (m,r)-matrix and $\frac{n}{m}$ operations of (13) with r right hand sides should be performed. As a result,

$$\frac{d}{2m^2}(2n-d) - \frac{1}{2m}(2n-d)$$

quadratic blocks,

$$\frac{d}{2m^2}(2n-d) + \frac{1}{2m}(4n-d)$$

blocks of dimension (m,r) and $\frac{2n}{m} - \frac{d}{m}$ triangular blocks should be transported.

We have assumed all the time that $m = \left[\sqrt{\frac{K}{3}}\right]$, n and d are divisible by m and r lyes between 1 and m.

Accordingly the realization of the 3 operations (12), (13) and (14) represents a central tool by the following reasons: As we have seen, they determine the

the execution-time. Furthermore it is obvious that their
realization is decisive for the accuracy of the computed
solution of system (1).

REFERENCES

[1] J.H. Wilkinson - C. Reinsch, *Linear Algebra*, Berlin
     Heidelberg, New York, 1971.

D. Richter
Zentrum für Rechentechnik
der Akademie der Wissenschaften der DDR
DDR - 1199 Berlin
Rudower Chaussee 5

ON THE STABILITY OF QUASIDOUBLE STEP SPLINE FUNCTION APPROXIMATIONS FOR SOLUTIONS OF INITIAL VALUE PROBLEMS

S.M. SALLAM

ABSTRACT

In this paper, we present some results on the stability of quasidouble step spline function approximations, $s(x) \in C^{m-2}$, to the initial value problem of ordinary differential equations. It is shown that the method is stable for $m = 4$ and unstable, hence divergent, for $m \geq 5$.

1. INTRODUCTION

Recently, we presented a method in [3] for the construction of a global approximation to the initial value problem in ordinary differential equations. The constructed approximation is a spline $s(x)$ of degree m and continuity class C^{m-2}. It is shown in [3] that the method is not A-stable for $m > 3$. We note that, for $s \in C^{m-1}$ (see [2]), the method is A-stable if $m = 2$ and stable when $m = 3$.

The present paper is an answer to the second part

517

(ii) of the following

CONJECTURE. *The quasidouble step spline function approximations* $s \in C^{m-2}$, $m \geq 2$, *are*

(i)   A-stable, for $m = 2,3$;

(ii)   divergent, for $m \geq 5$.

2. STABILITY CRITERION

As it has been shown in [3, Sec.2] for $n = 1$, the spline function approximation $s(x)$ to the differential equation

$$(2.1) \quad y' = f(x,y), \quad y(0) = y_0,$$

exists and is unique. The answer to (i) is given in [3, Sec.3]. In the interval. $[2\nu h, 2(\nu+1)h]$, $s(x)$ is defined by

$$(2.2) \quad s(x) = \sum_{j=0}^{m-2} \frac{1}{j!}(x-2\nu h)^j s_{2\nu}^{(j)} + \sum_{j=1}^{2}(x-2\nu h)^{m-2+j} a_j^{(\nu)},$$

where

$$\frac{c_{i,\nu}}{i!} = a_{i-m+2}^{(\nu)}, \qquad i = m-1, \; m$$

and the parameters $a_j^{(\nu)}$ are determined according to the relations

$$(2.3) \quad s'((2\nu+r)) = f((2\nu+r)h, \; s(2\nu+r)h), \quad r=1,2.$$

To study the behaviour of the method and hence its stability we need the following [1]

DEFINITION 2.1. A quasidouble step method is said to be stable if all solutions $\{s_{2\nu}\}$ remain bounded, as $\nu \to \infty$, $h \to 0$ while $x_{2\nu} = 2\nu h$ remains fixed when the method is applied to any differential equation of the form

$$(2.4) \qquad y' = \lambda y, \quad y(0) = 1,$$

λ being a constant with negative real part.

Let

$$(2.5) \qquad
\begin{aligned}
\underline{a}_\nu &= (h^{m-1} a_1^{(\nu)}, \; h^m a_2^{(\nu)})^T, \\
\underline{s}_\nu &= (s_{2\nu}, \; \frac{h}{1!} s_{2\nu}^{(1)}, \ldots, \frac{h^{m-2}}{(m-2)!} s_{2\nu}^{(s-2)})^T.
\end{aligned}$$

Applying the method to equation (2.4), thus (2.3), using (2.2) and (2.5), can be written in the matrix form

$$(2.6) \qquad B\underline{a}_\nu = C\underline{s}_\nu \, ,$$

where B is a 2×2 matrix whose elements b_{ij} are given by

$$(2.7) \qquad b_{ij} = (m-2+j-\lambda hi)i^{m-3+j} \, , \qquad i,j = 1,2,$$

and C is a $2\times(m-1)$ matrix whose elements c_{ij} are

$$(2.8) \qquad c_{ij} = \begin{cases} \lambda h, & j=1, \quad i=1,2; \\ (\lambda hi-(j-1))i^{j-2}; & j=2,\ldots,m-1; \quad i=1,2. \end{cases}$$

Also differentiating (2.2) $m-2$ times and substituting $x = 2(\nu+1)h$ yields

$$(2.9) \qquad \underline{s}_{\nu+1} = D\underline{s}_\nu + E\underline{a}_\nu \, ,$$

where D is an $(m-1)\times(m-1)$ upper triangular matrix whose elements d_{ij} are

$$(2.10) \qquad d_{ij} = \begin{cases} \binom{j}{i} 2^{j-1} \, , & i=0,1,\ldots,m-2, \quad j \geq i, \\[2mm] 0 \, , & j < i, \end{cases}$$

and E is an $(m-1)\times 2$ matrix with elements

$$(2.11) \qquad e_{ij} = \binom{m-2+j}{i} 2^{m-2+j-1}, \qquad i=0,1,\ldots,m-2; \; j=1,2.$$

Eliminate $\underline{a}_\nu$ from (2.6) and substitute into (2.9), then

$$(2.12) \qquad \underline{s}_{\nu+1} = A\underline{s}_\nu \, ,$$

where

$$A = D + EB^{-1}C.$$

Let A_0 denote the matrix A when $h = 0$, and let μ_i, $i=1,\ldots,m-1$, be its eigenvalues. By definition (2.1), $|\mu_i| \leq 1$ for stable solutions and $|\mu_i| > 1$ for unstable ones. It can be easily seen that the diagonal elements a_{ii} of A_0 are given by

$$a_{11} = 1,$$

$$a_{ii} = 1 + \binom{m-2}{i-3}\left[2^{m-i+1} - \frac{m+i-3}{i-2}\right]; \qquad i=2,\ldots,m-1.$$

The trace of a matrix is equal to the sum of its

eigenvalues, i.e.,

$$\sum_{i=1}^{m-1} \mu_i = \text{tr}(A_0).$$

But

$$\text{tr}(A_0) = \sum_{i=1}^{m-1} a_{ii} = 3^{m-2} - 3.2^{m-2} + m + 1.$$

Hence

$$(2.13) \qquad \sum_{i=1}^{m-1} \mu_i = 3^{m-2} - 3.2^{m-2} + m + 1.$$

Taking the moduli of both sides of (2.13) we have

$$(2.14) \qquad \sum_{i=1}^{m-1} |\mu_i| \geq \left| \sum_{i=1}^{m-1} \mu_i \right| = \left| m + 1 + 3^{m-2} - 3.2^{m-2} \right| > m$$

for $m \geq 5$. Let $\mu_{\max} = \max_i |\mu_i|$, then (2.14) becomes

$$(m-1)\mu_{\max} > m \, ,$$

and hence $\mu_{\max} > 1$ for $m \geq 5$. Thus we have established the following

THEOREM 2.1. *The quasidouble step spline function approximations* $s \in C^{m-2}$ *are unstable and hence divergent as* $h \to 0$ *for* $m \geq 5$.

In case of $m = 4$, A_0 takes the form

$$A_0 = \begin{bmatrix} 1 & -1/3 & 0 \\ 0 & 0 & 0 \\ 0 & 1 & 1 \end{bmatrix}$$

whose eigenvalues are 0, 1, 1 with corresponding eigenvectors $(-1/3, -1, 1)^T$; $(1, 0, 0)^T$; $(0, 0, 1)^T$, hence

COROLLARY 2.1. *The quasidouble step spline function apprxomations* $s\in C^{m-2}$ *are stable for* $m = 4$.

TEST PROBLEM. The proposed method will be applied to the simple test problem

$$y' = y, \quad y(0) = 1, \qquad x\in[0,5],$$

the exact solution of which is

$$y(x) = e^{x}.$$

We shall apply the method for the cases $m = 3$ and 4, respectively. The entries of the following table give the absolute maximum error over $[0,5]$:

h	0.10	0.05	0.01
Case 1: $m = 3$	1.6×10^{-3}	1.0×10^{-4}	2.6×10^{-7}
Case 2: $m = 4$	3.6×10^{-4}	2.4×10^{-5}	2.7×10^{-9}

ACKNOWLEDGEMENT. The author wishes to express his appreciation to Prof. Tamás Frey for his valuable suggestions and guidance during the preparation of this paper.

REFERENCES

[1]   M.E.A. El Tom, On the numerical stability of spline function approximations to solutions of Volterra integral equations of the second kind, *BIT* 14(1974), 136–143.

[2]   F.R. Loscalzo, An introduction to the application
      of spline functions to initial value problems,
      Theory and Applications of Spline Functions (T.N.E.
      Greville ed.), Academic Press, New York, 1969,
      37-64.

[3]   S. Sallam, Quasidouble step spline function approx-
      imations for solutions of simultaneous first order
      differential equations, (to appear).

S.M. Sallam
Department of Mathematics
Assiut University, Egypt

NEW ITERATIVE METHODS FOR SOLVING DIFFERENCE EQUATIONS
A.A. SAMARSKII

1. Finite difference approximation applied to elliptic equations result in large systems of linear algebraic equations (its order is equal to the number of the mesh points in the grid). In order to solve two or three-dimensional problems, one has to solve systems of equations of order 10^4-10^6. The system of difference euqations is usually ill-conditioned and its matrix is sparse (most of the entries are equal to zero). It is of basic importance in the computing practice that we should have efficient methods (which require minimal computer time) when solving such systems numerically. Efficiency becomes especially important in the case of certain problems of mathematical physics that involves the solution of the same elliptic equation thousand or tens of thousand times, for example, when varying the righ-hand side or the boundary values. In the physics of rarified plasm, for example, the method of large particles for solving the Vlasov equation demands that the potential of the electric field should be determined on each time level

by solving the Poisson equation with a right-hand side
(the density of electric charge per volume) that varies
from level to level.

2. Let us consider some typical problems of elliptic
equations:

$$(1) \qquad Lu = -f_0(x), \quad x = (x_1,\ldots,x_p) \in G,$$

where L is an elliptic operator and G is a p-dimen-
sional domain with boundary Γ. For the sake of simplicity, let us consider the first boundary value problem
(the Dirichlet problem):

$$(2) \qquad u = \mu(x), \qquad x \in \Gamma.$$

Some typical expressions for the elliptic operator
L:

$$(3) \qquad Lu = \Delta u = \sum_{\alpha=1}^{p} \frac{\partial^2 u}{\partial x_\alpha^2} \qquad \text{(the Laplace operator)};$$

$$(4) \qquad \begin{aligned} Lu &= \operatorname{div}(k \operatorname{grad} u) - q(x)u = \\ &= \sum_{\alpha=1}^{p} \frac{\partial}{\partial x_\alpha}\left(k(x)\frac{\partial u}{\partial x_\alpha}\right) - q(x)u, \end{aligned}$$

where $0 < c_1 \le k(x) \le c_2$, $q(x) \ge 0$ and c_1 and c_2
are constants;

$$(5) \qquad Lu = \sum_{\alpha,\beta=1}^{p} \frac{\partial}{\partial x_\alpha}\left(k_{\alpha\beta}(x)\frac{\partial u}{\partial x_\beta}\right).$$

In the latter case the condition of (strong) ellip-
ticity is as follows:

$$c_1 \sum_{\alpha=1}^{p} \xi_\alpha^2 \leq \sum_{\alpha,\beta=1}^{p} k_{\alpha\beta}(x)\xi_\alpha\xi_\beta \leq c_2 \sum_{\alpha=1}^{p} \xi_\alpha^2, \quad c_1 > 0,$$

for every vector $\xi = (\xi_1, \xi_2, \ldots, \xi_p)$.

The operator L is self-adjoint $(L=L^*)$ only if the matrix $(k_{\alpha\beta})$ is symmetric, i.e. $k_{\alpha\beta} = k_{\beta\alpha}$.

3. Let us now have a grid $\overline{\omega}_h = \omega_h + \gamma_h$ on the domain $\overline{G} = G + \Gamma$ and let the differential equation (1) be approximated by the difference equation

$$(6) \qquad \Lambda y = -\varphi(x), \qquad x \in \omega_h,$$

where Λ is the difference operator, $y = y_h(x)$, $x \in \overline{\omega}_h$ is the unknown and $\varphi = \varphi_h(x)$ is the prescribed grid function. On the boundary of the grid, we have the boundary condition

$$(7) \qquad y = \mu(x), \qquad x \in \gamma_h.$$

As result of the discretisation, we substitute a system of algebraic equations for the problem (1)-(2). The algebraic system, the structure of the matrix and, consequently, the set of algebraic problems that have to be solved, all depend on the operator L and also on the manner it has been approximated. This includes the difference operator, the dimension, the shape of domain, the way the grid has been chosen (is it uniform or not) and at last, the way the boundary value condition has been transferred to the difference system. Any of these factors may considerably influence the selection and the characteristics of an appropriate method for solving the arising difference equation.

Further on, the difference problem (6), (7) will be conveniently written in the matrix form

(8) $Av = f,$

where A is a matrix whose order is equal to the number
of the interior mesh points $x \in \omega_h$ or in the equivalent
operator form

(9) $Av = f,$

where A is a linear operator in the space H of the
grid functions which are defined on the mesh points $\omega_h + \gamma_h$
and   vanishing at γ_h. The dimension of H is obviously equal to the order of the matrix A, i.e. the number of interior mesh points.

4. For solving the system of equations (8), both
direct and iterative methods can be used. In recent years
some new methods have been developed which demand revi-
sion of our views about the applicability of certain
iterative methods.

The direct methods for solving difference equations
make use of the structure of the matrix A directly (see
[1]).

1) Factorisation. It is effective in the case of
one-dimensional ($p = 1$) problems when Λ is a three-
-point operator and A is a tridiagonal matrix. For
$p \geq 2$ the matrix factorisation is not economical.

2) The method of decomposition or full reduction is
applicable to the Poisson equation on the rectangle. It
requires $O(N_1 N_2 \log N_2)$ arithmetic operations, where N_α
is the number of mesh points in the grid along the direction of x_α, $\alpha = 1, 2$.

3) Division of variables by fast Fourier transform.
This method, as full reduction, can be employed to the
same problem with the same efficiency. By combining both
methods, the Dirichlet problem (as stated above) for the

Poisson equation in the rectangle can be solved within $Q \approx 5N_1N_2\log_2 N_2$ arithmetic operations (the grid is uniform in x_1 and x_2 with steps h_1 and h_2, respectively).

5. The iterative method of alternating directions (ADI) proposed by P e a c e m a n, R a c h f o r d and D o u g l a s in 1955 and further developed by many authors, played a highly important role in solving elliptic equations. As it turned out, however, it is economical only for a very narrow class of problems. Let $A = A_1 + A_2$, where A_1, $A_2 : H \to H$ are self-adjoint positive-definite operators. Then ADI is written in the following way:

$$\frac{y^{k+\frac{1}{2}} - y^k}{\tau_k^{(1)}} + A_1 y^{k+\frac{1}{2}} + A_2 y^k = f, \qquad k=0,1,\ldots$$

$$\frac{y^{k+1} - y^{k+\frac{1}{2}}}{\tau_k^{(2)}} + A_1 y^{k+\frac{1}{2}} + A_2 y^{k+1} = f,$$

where k is the number of the iteration and $\{\tau_k^{(1)}\}$, $\{\tau_k^{(2)}\}$ are sequences of iteration parameters, selected such that the number of iterations is kept at minimum. Optimum set of parameters ("in the sense of Jordan and Wachspress") has been obtained by myself in the case of commuting A_1 and $A_2 : A_1 A_2 = A_2 A_1$. This is the case of the Poisson equation in a rectangle. For the number of operations the estimate

$$Q(\varepsilon) = 0(N_1 N_2 \log N \log \frac{1}{\varepsilon})$$

holds, where $N = \max(N_1, N_2)$ and $\varepsilon > 0$ is the required precision. In this case, however, direct methods are much more efficient.

If the domain is not rectangular or L has variable coefficients (hence the method of division of variables does not apply), then A_1 and A_2 do not commut $(A_1 A_2 \neq A_2 A_1)$ and the ADI method needs $O(N_1 N_2 N \log \frac{1}{\varepsilon})$ operations. In the three-dimensional case $(p=3)$, the simplest problem is when L is the Laplace operator and G is a parallelepiped. Then $A = A_1 + A_2 + A_3$, where A_1, A_2, A_3 commute with each other. The question of the optimum set of parameters is not yet solved. With the known (cyclic) parameters ADI is less effective than ATM, a method to be discussed below. For equations with variable coeffficients or domains of complicated form, ADI is less effective than SOR.

We remark that the successive over-relaxation method (SOR) is widely applicable and it enjoyes wide popularity. The main difficulty is the choice of the relaxation parameter ω on which the number of iterations depends. Each iterational step is obtained by solving a system with a triangular matrix (as in Seidel's method).

The laternating-triangular method (ATM) was proposed by A.A. S a m a r s k i i [2], [3] and its modified version by E.S. N i k o l a e v and A.B. K u č e r o v in 1976-77 [4] [5]. This algorithm needs the inversion of lower and upper triangular matrices successively. The ATM is universal, i.e. it can be employed to any case, when $A = A^* > 0$. It is effective (the best) in the case of equation $Lu = -f_0$ over arbitrary domain and for operators L with variable coefficients in any dimension p.

In contrast to SOR, ATM has strict mathematical foundation in the context of the general theory of iterative methods. Therefore we are going to set forth the basic results of this general theory. We shall compare the various methods on the model problem, where $Lu = \Delta u$, $p = 2$, $G = \{0 \leq x_1, x_2 \leq 1\}$ is a square, $\Lambda y = -\varphi$ is the five-point "cross" scheme and the grid ω_h is quadratic, i.e. $h_1 = h_2 = h$. With other words, we shall consider the Dirichlet difference problem in the unit square, on a quadratic grid.

6. The last years brought significant results in the theory of iterative methods. We are going to discuss two of these results: 1) choice of Chebyshev parameters so that the iterative process becomes numerically stable, and 2) the universal alternating-triangular method (ATM).

Let us begin with some general facts (see [i]).

Assume that H is a finite-dimensional real linear space with scalar product $(,)$ and A, B, ... are linear operators given in H, A, $B : H \rightarrow H$. The problem is to find the solution of the operator equation

$$(10) \qquad Au = f.$$

Let us write the two-level (one-step) iterative method in the following canonical form:

$$(11) \qquad B \frac{y_{k+1} - y_k}{\tau_{k+1}} + Ay_k = f, \qquad k = 0, 1, \ldots, \quad \forall y_0 \in H,$$

where y_k is the approximation in the k-th iteration, $\tau_k > 0$ are the iteration parameters, $B : H \rightarrow H$ and the operator B is supposed to have the inverse B^{-1}. If $B = E$ is the identity operator then (11) is the explicit method. In the opposite case of $B \neq E$, the method is implicit.

It will be assumed that the following conditions are satisfied:

(12) $\quad A = A^* > 0, \quad B = B^* > 0, \quad \gamma_1 B \leq A \leq \gamma_2 B, \quad \gamma_1 > 0$

that is, A and B are self-adjoint and positive-definit operators; moreover, the smallest (γ_1) and largest (γ_2) eigenvalues of the generalised eigenvalue problem

(13) $\quad Av = \lambda Bv$

are assumed to be known. In the case of an explicit scheme $(B = E)$ γ_1 and γ_2 are the endpoints of the spectrum of the operator A. Otherwise they are endpoints of the spectrum of the operator A in the space H_B.

Let the numbers γ_1 and γ_2 be given. We want to find the parameter values $\tau_1, \tau_2, \ldots, \tau_n$ from the condition that we have minimum number of iterations $n = n(\varepsilon)$ for every given $\varepsilon > 0$, such that

(14) $\quad \| y_n - u \|_D \leq \varepsilon \| y_0 - u \|_D,$

where $D : H \to H$ is some self-adjoint positive operator $D = D^* > 0$ and

$$\| y \|_D = \sqrt{(Dy, y)}.$$

This leads to the classical problem of finding the polynomial of order n with smallest deviation from zero on the interval $[\gamma_1, \gamma_2]$. The solution is the Chebyshev polynomial of degree n.

The (Chebyshev) parameters are (cf. [1])

(15) $\quad \tau_k = \dfrac{\tau_0}{1 + \rho_0 \sigma_k}, \quad k = 1, 2, \ldots, n,$

$$(15) \qquad \tau_0 = \frac{2}{\gamma_1 + \gamma_2} , \qquad \rho_0 = \frac{1-\xi}{1+\xi} , \qquad \xi = \frac{\gamma_1}{\gamma_2} .$$

This solution has been known for a long time and was several times re-discovered.

For the number of iterations we have the estimate

$$n \geq (\ln \frac{2}{\varepsilon})/\ln \frac{1}{\rho_1} , \qquad \rho_1 = \frac{1-\sqrt{\xi}}{1+\sqrt{\xi}} \quad \text{or}$$
$$(16)$$
$$n \geq n_0(\varepsilon) = \frac{1}{2\sqrt{\xi}}\ln \frac{2}{\varepsilon} .$$

Here σ_k are the roots of the Chebyshev polynomial of order n:

$$(17) \qquad \sigma_k = -\cos\beta_k , \qquad \beta_k = \frac{\pi}{2n}\Theta_n(k), \qquad k=1,2,\ldots,n ,$$

where $\Theta_n(k)$ are the numbers $1,3,\ldots,2n-1$.

The estimate (16) with operator $D = AB^{-1}A$ holds independently from the order of the elements in the set $\Theta_n = \{\Theta_n(k)\}$. In case of the "natural" ordering

$$(18) \qquad \Theta_n(k) = 2k-1,$$

$$(19) \qquad \Theta_n(k) = 2n-(2k-1) = 2(n-k)+1, \qquad k=1,2,\ldots,n ,$$

however, the iterative process (11), (15), (17) with parameters (18) or (19) is numerically unstable: it will not converge within the frame of computations which can only use finitely many signs, that is, the method will not work on the computer. Either the intermediate values will accumulate (overflow occurs) or round-off errors will prevail [1], [6].

Proper indexing of the elements of $\Theta_n = \{\Theta_n(k)\}$

makes the process stable. In fact, stable permutations of the set $\Theta_n = \Theta_n^*$ (and the parameters $\{\tau_k^*\}$) have been found: a) for $n = 2^p$, where p is integer, in 1971 in the papers [7] and [1]; b) for arbitrary n, in [8] and [9]. The algorithm proposed in 1972 in the paper [8] is fully described in the book [1]. We mention that the estimate (16) holds for stable sets independently of the choice of the initial approximation.

7. Elaboration of stable Chebyshev iteration (11), (15), (17) is important in solving problems when a difference approximation is applied to boundary value problems of elliptic equations.

We are going to illustrate the rate of convergence of the Chebyshev method for the above model problem. Let us have the explicit scheme $(B = E)$:

$$(20) \qquad \frac{y_{k+1} - y_k}{\tau_{k+1}} + Ay_k = f, \qquad k = 0, 1, \ldots .$$

In this case

$$\gamma_1 = \frac{4}{h^2} \sin^2 \frac{\pi h}{2}, \qquad \gamma_2 = \frac{4}{h^2} \cos^2 \frac{\pi h}{2},$$

$$\xi = \frac{\gamma_1}{\gamma_2} = \mathrm{tg}^2 \frac{\pi h}{2} \approx \frac{\pi^2 h^2}{4}, \qquad \sqrt{\xi} \approx \frac{\pi}{2} h \approx 1{,}57h$$

and for the number of iterations we have

$$(21) \qquad n^{(1)} \geq \frac{3{,}2}{h} \qquad \text{if} \qquad \varepsilon \approx 10^{-4}.$$

In comparison with the simple iteration

$$(22) \qquad \frac{y_{k+1} - y_k}{\tau_0} + Ay_k = f, \qquad k = 0, 1, \ldots, \qquad \tau_0 = \frac{2}{\gamma_1 + \gamma_2},$$

it requires (assuming the same precision $\varepsilon \approx 10^{-4}$)

$$(23) \qquad n^{(2)} \geq \frac{2}{h^2} \qquad \text{iterations}$$

such that one has the following numbers:

	$h = \frac{1}{10}$	$h = \frac{1}{50}$	$h = \frac{1}{100}$
$n^{(1)}$	32	160	320
$n^{(2)}$	200	5000	20000

8. As it is going to be shown, implicit schemes may require less iterations than explicit ones if the "stabilizer" B is chosen accordingly. The simplest example of an implicit scheme is the Seidel iteration. Using the same notation for the operator A and its corresponding matrix $A = (a_{ik})$ we consider A as the sum

$$A = A^- + A^+ + D_0,$$

where D_0 is diagonal, A^- and A^+ are lower and upper triangular matrices (with 0 in the diagonal), respectively. Then according to the Seidel method one puts

$$B = D_0 + A^-, \qquad \tau_k = 1$$

in the canonical form (11). The method, as a rule, converges faster than the simple iteration (though with the same asymptotic order in h for the model problem).

For the SOR method, we have $B = D_0 + \omega A^-$, $\tau_k = \omega$. In case of the model problem, the SOR method needs

$O(\frac{1}{h} \ln \frac{1}{\varepsilon})$ iterations; approximately as many as the explicit Chebyshev method.

As operator B is not self-adjoint $(B \neq B^*)$ for the Seidel or SOR method (B is a lower triangular matrix), hence the general theory, along with the choice of parameters $\{\tau_k^*\}$ does not apply.

In order to apply the general theory (for $B = B^*$) and preserve the algorithm which inverts triangular matrices, we define a new iteration. For this purpose we choose B as a product of a lower and an upper triangular matrix B^- and B^+. Indeed, we can take

$$(24) \qquad B = (E+\omega A_1)(E+\omega A_2)$$

where

$$(25) \qquad A_1 = A^- + \frac{1}{2}D_0, \qquad A_2 = A^+ + \frac{1}{2}D_0$$

such that

$$(26) \qquad \begin{aligned} A_1 + A_2 &= A, \\[4pt] A_1^* &= A_2 \quad \text{if} \quad A^* = A. \end{aligned}$$

In order to compute y_{k+1} we have to solve two equations in succession

$$(E+\omega A_1)\bar{y} = F_k, \qquad F_k = By_k - \tau_{k+1}(Ay_k - f),$$

$$(E+\omega A_2)y_{k+1} = \bar{y},$$

where $B^- = E+\omega A_1$ is a lower triangular and $B^+ = E+\omega A_2$ is an upper triangular matrix. (Hence the name "alternating-triangular method", ATM.)

For the model problem, we have

$$Ay = -\frac{y(i_1+1, i_2)-2y(i_1, i_2)+y(i_1-1, i_2)}{h^2}$$

$$-\frac{y(i_1, i_2+1)-2y(i_1, i_2)+y(i_1, i_2-1)}{h^2},$$

while the operators A_1 and A_2 have the form

$$A_1y = \frac{2y(i_1, i_2)-y(i_1-1, i_2)-y(i_1, i_2-1)}{h^2},$$

$$A_2y = \frac{2y(i_1, i_2)-y(i_1+1, i_2)-y(i_1, i_2+1)}{h^2},$$

where $i_1, i_2 = 1,2,\ldots,N-1$, $hN = 1$, and $y = 0$ for $i_1 = 0,N$ and for $i_2 = 0,N$. The operator A_1 is based on the three-point diagram, , while A_2 comes from the three point diagram . For the determination of $y_{k+1}(i_1, i_2)$, we get the following formulas

$$\overline{\overset{\circ}{y}}(i_1, i_2) = \alpha(\overline{\overset{\circ}{y}}(i_1-1, i_2)+\overline{\overset{\circ}{y}}(i_1, i_2-1))+$$

$$+\beta F_k(i_1, i_2), \qquad 0 < i_1, i_2 < N,$$

$$\overset{\circ}{y}_{k+1}(i_1, i_2) = \alpha(\overset{\circ}{y}_{k+1}(i_1, i_2+1)+\overset{\circ}{y}_{k+1}(i_1+1, i_2))+$$

$$+\beta\overline{\overset{\circ}{y}}(i_1, i_2), \qquad 0 < i_1, i_2 < N,$$

where $\beta = 1/(1+\frac{2\omega}{h^2})$ and $\alpha = \beta\frac{\omega}{h^2}$. The zero over y reminds that $\overset{\circ}{y} = 0$ on the boundary, that is, for $i=0,N$ and $i_2=0,N$.

Suppose now that the right-hand side of $F_k(i_1, i_2)$ is known. We start the computation at the left-hand lower corner, taking $i_1=1$, $i_2=1$. Then $\overline{\overset{\circ}{y}}(1, 1) = \beta F_1(1, 1)$.

After that, moving to the right along the row $i_2=1$, taking $i_1=2,\ldots,N-1$, we get $\overline{\overline{y}}$ along the row. By running through all the rows $i_2=2,3,\ldots,N-1$ we determine $\overline{y}$ in every inner point of the grid ω_h. In order to find $\overset{\circ}{y}_{k+1}$, we select the first right upper mesh point $i_1=N-1$, $i_2=N-1$ such that $y_{k+1}(N-1, N-1) = \beta\overline{\overline{y}}(N-1, N-1)$, then move on along the top row $i_2=N-1$ from left to right as $i_1=N-2, N-3,\ldots,1$. Then changing to the rows $i_2=N-2, N-3,\ldots,2,1$, we obtain $\overset{\circ}{y}_{k+1}(i_1, i_2)$ for the inner mesh points $x\in\omega_h$. The solution y_{k+1} on $\omega_h+\gamma_h$ is

$$y_{k+1} = \begin{cases} \overset{\circ}{y}_{k+1} & \text{for} \quad x\in\omega_h, \\ \\ \mu(x) & \text{for} \quad x\in\gamma_h. \end{cases}$$

The right-hand side of F_k can be computed by

$$F_k = B\overset{\circ}{y}_k + \tau_{k+1}(\Lambda y_k + \varphi).$$

The parameter ω is determined by the formula

$$\omega = \frac{h^2}{\sin \pi h} \approx \frac{h}{\pi} \; .$$

The parameters here are taken from the Chebyshev set $\{\tau_k^*\}$.

Let us now return to the ATM in the general formulation (11), (24), taking

$$(24) \qquad B = (E+\omega A_1)(E+\omega A_2), \quad A_1^* = A_2, \quad A_1+A_2 = A.$$

THEOREM. *Let* $A = A^* > 0$ *and assume that the numbers* $\delta > 0$ *and* $\Delta > 0$ *are known:*

(27) $\qquad A \geq \delta E, \qquad A_1 A_2 \leq \frac{\Delta}{4} A$

i.e.

$$(Ay, y) \geq \delta \|y\|^2, \quad (A_1 A_2 y, y) = \|A_2 y\|^2 \leq \frac{\Delta}{4}(Ay, y)$$

$$\forall y \in H.$$

Then for ATM (11), (24) *with the Chebyshev parameters* $\{\tau_k^*\}$ *and* $\omega = \omega_0 = 2\sqrt{\delta\Delta}$, *the number of iterations is*

(28) $\qquad n \geq n_0(\varepsilon) = (\ln \frac{2}{\varepsilon})/(2\sqrt{2}\sqrt[4]{\eta}, \quad \eta = \frac{\delta}{\Delta} < 1.$

Let us illustrate this theorem for the model problem. We have

(29)
$$\delta = \frac{4}{h^2} \sin^2 \frac{\pi h}{2}, \quad \Delta = \frac{4}{h^2}, \quad \eta = \sin^2 \frac{\pi h}{2}$$

$$n_0(\varepsilon) \approx \ln \frac{2}{\varepsilon}/(3,54\sqrt{h})$$

(for SOR, we have $n_0(\varepsilon) = O(\frac{1}{h})$), and so, for $\varepsilon \approx 10^{-4}$ the estimate

(30) $\qquad n_0(\varepsilon) \approx \dfrac{2,9}{\sqrt{h}}$

holds. For $h = \frac{1}{100}$ ATM requires 29 iterations as compared to 320 for the explicit Chebyshev method, or to 20000 in the case of simple iteration.

Remark that the estimate (29) is valid for any $p \geq 2$ if $L = \Delta$, Λ is a $2p+1$-point second order approximation operator and $\overline{G} = \{0 \leq x_\alpha \leq 1, \quad \alpha=1,2,\ldots,p\}$ is the p-dimensional unit paralelepiped.

Even though ADI gives the estimate $n_0(\varepsilon) = O(\ln \frac{1}{h} \ln \frac{1}{\varepsilon})$ for every $p \geq 2$, it has better asymptotics (in h) for $n_0(\varepsilon)$ then ATM. On the other hand, the number of iterations for $p > 2$ ($p=3,4,\ldots$) is less in the case of ATM if $h \leq \frac{1}{60}$, that is, whenever the number of mesh points is larger than 2.10^5. In the whole volume of computations ATM can be twice more economic than ADI for three-dimensional problems.

In connection with the Dirichlet difference problem over arbitrary domains, ATM was modified by introducing an operator $D = D^* > 0$ into B so that

$$B = (D+\omega A_1)D^{-1}(D+\omega A_2), \qquad \omega > 0$$
(31)
$$A_1^* = A_2, \qquad A_1 + A_2 = A^*.$$

Let us change condition (27) for

(32) $\qquad A \geq \delta D, \quad A_1 D^{-1} A_2 \leq \frac{\Delta}{4} \cdot A_2, \quad \delta > 0, \; \Delta > 0.$

Then the theorems which have been formulated for ATM remain valid for the modified alternating-triangular method (MATM) as well, included the estimate (28).

The operator D has to be chosen on the condition that the number of iterations is kept at minimum and B is efficiently chosen (the number of operations when solving equation $Bv = F$ is at minimum). We may take for $D = (d_{ij})$ a diagonal matrix such that $\eta = \delta/\Delta$ is maximal. The MATM with an appropriately chosen matrix and Chebyshev parameter set $\{\tau_k^*\}$ proved to be highly effective in cases of the Dirichlet difference problem of the Poisson equation over an arbitrary domain and for the equation $\operatorname{div}(k \operatorname{grad} u)=-f_0(x)$ with variable coefficients, over a rectangle [4], [5]. For the Dirichlet problem over an arbitrary domain the number of iterations increases by not more than 5% as compared to the same problem over the square whose side is equal to the diameter of the domain.

Applying the operator D enables us to create efficient iterative methods for problems with rapidly varying coefficients, and for elliptic difference equations of the general type.

As the construction of ATM leads to the form of $A = A^* > 0$ by summing an operator A_1 and its adjoint $A_2 = A_1^*$, $A_1 + A_2 = A$, ATM can be considered universal. Before applying the relations of (28), one has to choose the parameters δ and Δ.

REMARK. The general theory of iterative methods (11) assumes that the endpoints γ_1 and γ_2 of the spectrum of A in H_B are known.

If γ_1 and γ_2 are not known or roughly determined, then iterative methods of the variational type can be applied, like the two-level (one-step) methods of steepest descent and the method of minimal deviations which converge with the same rate as the implicit method of simple iteration with the constant parameter $\tau_k = \tau_0 = 2/(\gamma_1 + \gamma_2)$. The three-level (two-step) method of conjugate gradients converges with the same asymptotic rate $n_0(\varepsilon) = 0(\frac{1}{\sqrt{\xi}} \ln \frac{1}{\varepsilon})$ as the Chebyshev method with exact values of γ_1 and γ_2. Iterative methods of the variational type can be used for determining γ_1 and γ_2 which values, can then be used for the ATM and MATM. We mention here that in [10] a method is proposed for accelerating the methods of steepest descent and minimal deviations.

REFERENCES

[1]   A.A. Samarskii, *Theory of Difference Schemes*, Moscow
      Nauka, 1977. (in Russian).

[2]   A.A. Samarskii, An efficient algorithm for the
      numerical solution of systems of differential and
      algebraic equations, Ž. Vyčisl. Mat. i Mat. Fiz.
      4, N°3 (1964), 580-585, (in Russian).

[3]   A.A. Samarskii, Two-level difference schemes, *Dokl.
      Akad. Nauk SSSR,* 155, N°3 (1969), 524-527, (in
      Russian).

[4]   A.B. Kučerov - E.S. Nikolaev, Alternating-triangular
      iterative method for solving elliptic difference
      equations in a rectangle, Ž. Vyčisl. Mat. i Mat.
      Fiz. 16, N°5 (1976), 1164-1174, (in Russian).

[5]   A.B. Kučerov - E.S. Nikolaev, Alternating-triangular
      iterative method for solving elliptic difference
      equations over arbitrary domains, Ž. Vyčisl. Mat. i
      Mat. Fiz. 17, N°3 (1977), 664-675, (in Russian).

[6]   A.A. Samarskii, *Introduction to the Theory of Dif-
      ference Schemes*, Moscow, Nauka, 1971, (in Russian).

[7]   V.I. Lebedev - S.A. Finogenov, On chosing the sequence
      of iteration parameters in the Chebyshev cyclic
      iteration, Ž. Vyčisl. Mat. i Mat. Fiz. 11, N°2
      (1971), 425-438, (in Russian).

[8]   E.S. Nikolaev - A.A. Samarskii, Selection of itera-
      tional parameters in Richardson's method, Ž. Vyčisl.
      Mat. i. Mat. Fiz. 12, N°4 (1972), 960-973, (in
      Russian).

[9]   V.I. Lebedev - S.A. Finogenov, Solution to the prob-
      lem of ordering the parameters in Chebyschev itera-
      tions, Ž. Vyčisl. Mat. i Mat. Fiz. 13, N°1 (1973),
      18-33, (in Russian).

[10] E.Ś. Nikolaev, Non-linear acceleration of varia-
tional type two-level iterative methods, Ž. Vyčisl.
Mat. i Mat. Fiz. 16, N°6 (1976), 1381-1387, (in
Russian).

A.A. Samarskii
Institute for Applied Mathematics
Moscow A-47
Miusskaja Pl. 4
USSR

COLLOQIA MATHEMATICA SOCIETATIS JÁNOS BOLYAI
22. NUMERICAL METHODS, KESZTHELY (HUNGARY), 1977.

ON THE NUMERICAL SOLUTION OF THE EQUATION OF FILTRATION

E. SCHECHTER

1. INTRODUCTION AND FORMULATION OF THE DIFFERENTIAL PROBLEM

Consider the initial-boundary value problem:

$$(1.1) \qquad u_t = \Delta \varphi(u) \qquad \text{on} \qquad Q = \Omega \times]0, \, T[$$

$$(1.2) \qquad u(x, \, 0) = u_0(x) \qquad x \in \Omega$$

$$(1.3) \qquad u(x, \, t)\big|_S = u_1(x, \, t) \qquad S = \partial\Omega \times [0, \, T]$$

$0 < T < +\infty$. Here $\Omega \subset \mathbf{R}^2$ is a bounded, regular and convex domain. It is known (see e.g. [5] and also [6]) that even for smooth data, this problem has no classical solutions. Throughout this paper we shall suppose that the following assumption (A) holds:

$$(i) \quad u_0 \in C(\overline{\Omega}), \ u_1 \in C(S), \quad u_0, \ u_1 \geq 0.$$

(A)

$$(ii) \quad \varphi \in C^2(\mathbf{R}_+), \ \varphi(u) \quad \text{and} \quad \varphi'(u) > 0 \quad \text{for}$$

545

$$u > 0, \quad \varphi(0) = \varphi'(0) = 0; \varphi''(u) \geq 0,$$
$$u \geq 0.$$

We denote by M a constant such that u_0, $u_1 \leq M$ on their domain of definition.

The following definition extends, to the multidimensional case, that of [6].

DEFINITION. A function $u \in C(\overline{Q})$ is called a *solution* of (1.1)–(1.3) if $u \geq 0$,

$$\frac{\partial \varphi(u)}{\partial x_i} \in L^2(Q) \qquad i = 1, 2,$$

further u satisfies conditions (1.2), (1.3) and

$$\int_Q \left[u \frac{\partial f}{\partial t} - \sum_{i=1}^{2} \frac{\partial f}{\partial x_i} \frac{\partial \varphi(u)}{\partial x_i} \right] dx \, dt +$$

(1.4)

$$+ \int_\Omega f(x, 0) u_0(x) dx = 0,$$

for any $f \in C^1(\overline{Q})$ such that $f|_{S_1} = 0$. Here

$$S_1 = S \cup \{(x, T); \ x \in \overline{\Omega}\}.$$

The uniqueness and existence of a generalized solution of the Cauchy-Dirichlet problem for (1.4) was proved in [6] for the case of one space variable. It seems however, that the generalization of the existence proof of this paper for the case of several space variables is not straightforward. On the other hand, the uniqueness of solutions can be proved in the same manner, under the above conditions, hence we shall always assume that our solution is unique.

The primary aim of this paper is to prove convergence of the simplest explicit difference analogue to
(1.1), and the existence of a solution of (1.4) will be
also a result.

We mention here a very important result of [5],
concerning the existence and uniqueness of the solutions
in a particular case, under more general conditions and
with weaker properties for the solutions.

2. A MAXIMUM PRINCIPLE FOR THE EXPLICIT DIFFERENCE SCHEME

We shall use the following mesh:

$$R_h = \{(x, t) \in \mathbf{R}^3; \quad x_i = k_i h, \quad t = k_0 \tau,$$

$$k_i = 0, \pm 1, \pm 2, \ldots, \quad i = 1, 2, \quad k_0 = 0, 1, \ldots\}$$

with $h, \tau > 0$. We shall also make use of the following notation:

$$\omega_{(kh)} = \{(x, t); \quad k_i h < x_i < (k_i+1)h,$$

$$i = 1, 2, \quad k = (k_1, k_2)\}$$

$$\Omega_h = \bigcup_{\omega \in \Omega} \omega_{(kh)}, \quad \Gamma_h = \partial \Omega_h \quad Q_h = \Omega_h \times \,]0, T[\, .$$

For simplicity, the same notation will stand for the mesh
points of R_h belonging to the respective set. By u_h
we denote the restriction of a function u to a given
set of mesh-points. For convenience, we shall put
$U_{ij}(k)$ or simply $U(k)$ instead of $U(ih, jh, k\tau)$.

The explicit difference analogue to (1.1)-(1.3) is:

$$(2.1) \qquad U_{\bar{t}}(k) = \Delta_h \varphi(U(k-1)) \quad \text{on} \quad Q_h$$

$$(2.2) \qquad U(0) = u_{0h}$$

$$(2.3) \qquad U(k)\big|_{\Gamma_h} = u_2(x,k\tau), \quad x \in \Gamma_h, \quad k=0,1,\dots,K = \left[\frac{T}{\tau}\right].$$

Here

$$U_{\bar{t}}(k) = \frac{U(k)-U(k-1)}{\tau}$$

and Δ_h is the discrete approximation to Δ (the discrete Laplacian). The function $u_2 : S_h \to \mathbf{R}$, is an approximation to u_1. We define it by simple "transportation",

$$u_2(x, t) = u_1(x^*, t) \qquad x \in \Gamma_h,$$

where $x^* \in \partial\Omega$ is the nearest point in the Ox_1 or Ox_2 directions to $x \in S_h = \Gamma_h \times [0, T]$.

Clearly $0 \le u_2 \le M$.

THEOREM 2.1. *Suppose that assumption* (A) *holds and*

$$\lambda = 4 \frac{\tau}{h^2} \varphi'(M) \le 1.$$

Then the solution U *of* (2.1)–(2.3) *satisfies the inequalities:*

$$0 \le U \le M.$$

PROOF. The assertion is evident for $U(0)$. If $k \ge 1$ let us prove first that $U \le M$. Indeed, let us assume that there is a triplet of indices, say (i, j, k), such that $U_{ij}(k) > M$ and $U(\ell) \le M$ for $\ell = 0,\dots,k-1$. Then

$$0 > M - U_{ij}(k) = M - U_{ij}(k-1) + \tau \Delta_h \left[\varphi(M) - \varphi(U_{ij}(k-1)) \right] =$$

$$= \left(1 - 4 \frac{\tau}{h^2} \widetilde{\varphi}'_{ij}(k-1) \right)\left(M - U_{ij}(k-1) \right) +$$

$$+ \frac{\tau}{h^2} \left[\widetilde{\varphi}'_{i+1,\,j}(k-1)\left(M - U_{i+1,\,j}(k-1) \right) + \right.$$

$$+ \widetilde{\varphi}'_{i,\,j+1}(k-1)\left(M - U_{i,\,j+1}(k-1) \right) + \widetilde{\varphi}'_{i-1,\,j}(k-1)\left(M - U(k-1) \right) +$$

$$\left. + \widetilde{\varphi}'_{i,\,j-1}(k-1)\left(M - U_{i,\,j-1}(k-1) \right) \right].$$

which is a contradiction, since the right-hand side is nonnegative. The sign $\sim$ indicates an appropriate "intermediate" value between M and $U(k-1)$.

The same argument can be repeated in order to show that $U \geq 0$. This completes the proof.

3. THE STABILITY OF THE DISCRETE SOLUTION

In the previous section we have already seen that the solution U of (2.1)–(2.3) is stable in the discrete max norm and of course, in any discrete $L^p(Q_h)$ norm if $p \geq 1$. Now we discuss the boundedness of the first order difference quotients in $L(Q_h)$. To begin with here is:

THEOREM 3.1. *Suppose that*

(i)   *Assumption* (A) *holds and* $u_0 \in C^2(\Omega)$,

(ii) $\lambda \leq 1$,

(iii) $\Delta u_0 \geq 0$ *on* Ω *and* u_1 *is nondecreasing in* t *on* $[0,\,T]$.

Then

(j) $U_{\bar{t}}(k) \geq 0$, $\quad k = 1, 2, \ldots, K$, *on* $\overline{\Omega}_h$,

(jj) $\tau h^2 \sum\limits_{Q_h} U_{\bar{t}}(k) \leq Mm(\Omega)$,

where M *is the constant from Theorem 2.1.*

PROOF. By (2.1)

$$(3.2) \qquad U_{\overline{t}}(k) = U_{\overline{t}}(k-1) + \Delta_h [\varphi(U(k-1)) - \varphi(U(k-2))]$$

for $k=2,3,\ldots,K$. Hence, with the notation $\overline{U} \equiv U_{\overline{t}}$,

$$\overline{U}_{ij}(k) = \left(1 - 4\frac{\tau}{h^2} \widetilde{\varphi}_{ij}(k-1)\right)\overline{U}_{ij}(k-1) +$$

$$(3.3) \qquad + \frac{\tau}{h^2}[\widetilde{\varphi}'_{i+1,\,j}(k-1)\overline{U}_{i+1,\,j}(k-1) + \widetilde{\varphi}'_{i-1,\,j}(k-1)\overline{U}_{i-1,\,j}(k-1) +$$

$$+ \widetilde{\varphi}'_{i,\,j+1}(k-1)\overline{U}_{i,\,j+1}(k-1) + \widetilde{\varphi}'_{i,\,j-1}(k-1)\overline{U}_{i,\,j-1}(k-1)],$$

where $\sim$ designates "intermediate" values for $\varphi'(U)$. Since u_0 is convex, it follows that $\Delta_h\varphi(u_0) \geq 0$, if we take the nonnegativity of φ' and φ'' into account. This implies

$$U(1) = \Delta_h\varphi(u_0) \geq 0 \quad \text{on} \quad \Omega_h.$$

Thus according to (3.3) and recalling the monotonicity of u_1, it follows that

$$U_{\overline{t}}(k) \geq 0 \qquad k=1,2,\ldots,K \quad \text{on} \quad \overline{\Omega}_h.$$

To prove (jj) we observe that

$$\tau h^2 \sum_{k=1}^{K} \sum_{\overline{\Omega}_h} U_{\overline{t}} \leq h^2 \left[\sum_{\overline{\Omega}_h} U(K) - \sum_{\overline{\Omega}_h} u_0\right] \leq Mm(\Omega).$$

CONSEQUENCE. Under the assumptions of the theorem, there is a constant C independent of the mesh-sizes, such that

$$(3.4) \qquad \tau h^2 \sum_{k=1}^{K} \sum_{\overline{\Omega}_h} \varphi(U(k))_{\bar{t}} < C.$$

Indeed,

$$\sum_{k=1}^{K} \sum_{\overline{\Omega}_h} \varphi(U(k))_{\bar{t}} = \sum_{k=1}^{K} \sum_{\overline{\Omega}_h} \widetilde{\varphi}'(k) U_{\bar{t}}(k) \leq$$

$$\leq \varphi'(M) \sum_{k=1}^{K} \sum_{\overline{\Omega}_h} U_{\bar{t}}(k) < C.$$

THEOREM 3.2. *Let the conditions of Theorem 3.1 be fulfilled, then there exists a constant C independent of h, such that*

$$(3.5) \qquad h^2 \tau \left[\sum_{k=0}^{K-1} \sum_{\overline{\Omega}_h^*} \left(\varphi^2(U(k)) \right)_{x_1} + \varphi^2(U(k)) \right)_{x_2} \right] < C.$$

With $\overline{\Omega}_h^*$ we have denoted a convex polygonal domain satisfying the following requirements:

(a) $\overline{\Omega}_h^* \subset \Omega$

(b)   The sides of $\partial\Omega_h^*$ are parallel to the coordinate axes and contain nodes of R_h.

Here for example

$$\varphi(U_{ij}(k))_{x_1} = \frac{\varphi(U_{i+1,j}(k)) - \varphi(U_{ij}(k))}{h}$$

PROOF. We shall follow an idea of [2, p.51] (see also [1]). First, let us assume that Ω_h^* is a rectangle which we denote by R_0. Consider a sequence of concentric rectangles

$$R_0 \subset R_1 \subset \ldots \subset R_N \subset \Omega,$$

whose frontiers are $S_0, S_1, \ldots, S_N$, having parallel sides at a constant distance h. The distance between S_0 and S_N, will be denoted by a. We have $Nh = a$. We suppose for simplicity that the sides of the rectangles are parallel to the axes and their centers coincide with the origin. Let the dimensions of R_p be $2a_p$ and $2b_p$, respectively, $p = 0, 1, \ldots, N$.

Now multiplying each equation of (2.1) by $\varphi(U_{ij}(k-1))$ and summing this over R_p gives:

$$(3.6) \quad [U_{\bar{t}}(k), \varphi(U(k-1))]_{R_p} = [\varphi(U(k-1))_{x\bar{x}}, \varphi(U(k-1))]_{R_p} +$$

$$+ [\varphi(k-1))_{y\bar{y}}, \varphi(U(k-1))]_{R_p}, \quad k = 1, \ldots, K.$$

Here $[\ ,\ ]_{\Omega_h}$ stands for the scalar product, on Ω_h, times h^2. We also recall that

$$\Delta_h \varphi(U) = \varphi(U)_{x\bar{x}} + \varphi(U)_{y\bar{y}}.$$

In order to get the desired estimate we transform the right-hand side of (3.6) by the discrete "Green's formula" (see e.g. [2, p. 36] or [4, pp. 280-282]). This becomes

$$-\frac{1}{2}\{[\varphi(U(k-1))_x, \varphi(U(k-1))_x]_{R_p} +$$

$$+ [\varphi(U(k-1))_{\bar{x}}, \varphi(U(k-1))_{\bar{x}}]_{R_p} + [\varphi(U(k-1))_y, \varphi(U(k-1))_y]_{R_p} +$$

$$+ [\varphi(U(k-1))_{\bar{y}}, \varphi(U(k-1))_{\bar{y}}]_{R_p}\} + \gamma_p(k-1),$$

with

$$\gamma_p(k-1) = \frac{1}{2}\{[\varphi(U(k-1)), \varphi(U(k-1))_x]_{x=a_p} +$$

$$+ \left[\varphi(U(k-1)), \ \varphi(U(k-1))_y \right]_{y=b_p} - \left[\varphi(U(k-1)), \ \varphi(U(k-1))_{\bar{x}} \right]_{x=-a_p} -$$

$$- \left[\varphi(U(k-1)), \ \varphi(U(k-1))_{\bar{y}} \right]_{y=-b_p} +$$

$$+ \left[\varphi(U_+(k-1)), \ \varphi(U(k-1))_x \right]_{x=a_p} +$$

$$+ \left[\varphi(U^+(k-1)), \ \varphi(U(k-1))_y \right]_{y=b_p} - \left[\varphi(U_-(k-1)), \ \varphi(U(k-1))_{\bar{x}} \right]_{x=-a_p} -$$

$$- \left[\varphi(U^-(k-1)), \ \varphi(U(k-1))_{\bar{y}} \right]_{y=-b_p} \Big\} , \quad \begin{array}{l} p=0,1,\ldots,N, \\ k=1,2,\ldots,K. \end{array}$$

Here $\quad U_\pm = U_{i\pm1,j} \quad\quad U^\pm = U_{i,j\pm1} \quad$ and, for example,

$$\left[\varphi(U), \ \varphi(U)_x \right]_{x=a_p} = h \sum_{\substack{j \\ x=a_p}} \varphi(U_{ij}) \varphi(U_{ij})_x .$$

We have

$$\left[\varphi(U), \ \varphi(U)_x \right]_{x=a_p} + \left[\varphi(U_+), \ \varphi(U)_x \right]_{x=a_p} =$$

$$= \left[\varphi(U)+\varphi(U_+), \ \varphi(U)_x \right]_{x=a_p} = \sum_{x=a_p} \left(\varphi^2(U_+) - \varphi^2(U) \right) .$$

Similar inequalities hold for the other terms. Therefore

$$\gamma_p = \frac{1}{2} \sum_{S_p} \left(\varphi(U)^2 \big|_{S_{p+1}} - \varphi(U)^2 \right) ,$$

where $\quad \varphi(U_{ij})^2 \big|_{S_{p+1}} = \varphi(U_+) \quad$ if $\quad (x_i, \ y_j) \in S_p .$ Since

$$\left[U_{\bar{t}}, \ \varphi(U) \right] \le \varphi(M) h^2 \sum_{\Omega_h} U_{\bar{t}}$$

there exists, according to Theorem 3.1, a constant $\quad C$ independent of $\quad h \quad$ and $\quad \tau, \quad$ such that

$$\left[U_{\bar{t}}(k), \; \varphi(U(k-1)) \right]_{\Omega_h} < c \qquad \forall k .$$

Hence

$$
\begin{aligned}
(3.7) \qquad & \left[\varphi(U)_x, \; \varphi(U)_x \right]_{R_p} + \left[\varphi(U)_y, \; \varphi(U)_y \right]_{R_p} + \\[4pt]
& + \left[\varphi(U)_{\bar{x}}, \; \varphi(U)_{\bar{x}} \right]_{R_p} + \left[\varphi(U)_{\bar{y}}, \; \varphi(U)_{\bar{y}} \right]_{R_0} \le c + 2\gamma_p, \quad p = 0, 1, \ldots, N .
\end{aligned}
$$

Because of the inequality

$$
\begin{aligned}
& \left[\varphi(U)_x, \; \varphi(U)_x \right]_{R_0} + \left[\varphi(U)_y, \; \varphi(U)_y \right]_{R_0} \le \\[4pt]
& \le \left[\varphi(U)_x, \; \varphi(U)_x \right]_{R_p} + \left[\varphi(U)_y, \; \varphi(U)_y \right]_{R_p}, \qquad p = 1, 2, \ldots, N
\end{aligned}
$$

we have from (3.7):

$$
(3.8) \qquad \left[\varphi(U)_x, \; \varphi(U)_x \right]_{R_0} + \left[\varphi(U)_y, \; \varphi(U)_y \right]_{R_0} \le c + \frac{2}{N} \sum_{p=0}^{N-1} \gamma_p .
$$

On the other hand,

$$
\begin{aligned}
(3.9) \qquad 2 \sum_{p=0}^{N-1} \gamma_p (k-1) &= \sum_{S_{N-1}} \varphi(U)^2 \Big|_{S_N} + \left(\sum_{S_{N-2}} \varphi(U)^2 \Big|_{S_{N-2}} - \right. \\[4pt]
& \left. - \sum_{S_{N-1}} \varphi(U)^2 \right) + \ldots + \left(\sum_{S_{N-n}} \varphi(U)^2 \Big|_{S_{N-n+1}} - \sum_{S_{N-n+1}} \varphi(U)^2 \right) + \\[4pt]
& + \ldots + \left(\sum_{S_0} \varphi(U)^2 \Big|_{S_1} - \sum_{S_1} \varphi(U)^2 \right) - \sum_{S_0} \varphi(U)^2 .
\end{aligned}
$$

All terms in the right hand side are taken on the level $t = (k-1)\tau$, $k = 1, 2, \ldots, K$. This implies

$$
2 \sum_{p=0}^{N-1} \gamma_p (k-1) \le \sum_{S_{N-1}} \varphi(U)^2 \Big|_{S_N} \le \varphi(M)^2 \sum_{S_N} 1 .
$$

Replacing it in (3.8), we get:

$$[\varphi(U)_x, \varphi(U)_x]_{R_0} + [\varphi(U)_y, \varphi(U)_y]_{R_0} \leq C + \varphi^2(M) \sum_{S_N} 1/N.$$

Now, if h is sufficiently small ($Nh = a$, a independent of h), we infer that the left-hand side is bounded by a constant independent of h. Summing up over k and multiplying by τ yields (3.3).

If instead of R_0 we have a polygonal domain Ω_h^*, we repeat the above argument by constructing N similar domains with parallel sides at a distance h and observe that (3.3) remains true. This completes the proof.

The convexity condition on u_0, required in both theorems can be dropped as it is seen from the following theorem:

THEOREM 3.3. *If conditions (i), (ii) of Theorem 3.1 are maintained and u_1 is constant with respect to t, then there exists a constant C independent of h and τ such that:*

$$(3.10) \qquad \tau h^2 \sum_{\overline{Q}_h} |U_{\overline{t}}(k)| < C.$$

PROOF. From (3.3) we deduce that

$$|\overline{U}_{ij}(k)| \leq \left(1 - 4\frac{\tau}{h^2}\widetilde{\varphi}'_{ij}(k-1)\right)|\overline{U}_{ij}(k-1)| +$$

$$+ \frac{\tau}{h^2}\left[\widetilde{\varphi}'_{i+1,\,j}(k-1)|\overline{U}_{i+1,\,j}(k-1)| + \right.$$

$$+ \widetilde{\varphi}'_{i-1,\,j}(k-1)|\overline{U}_{i-1,\,j}(k-1)| +$$

$$\left. + \widetilde{\varphi}'_{i,\,j+1}(k-1)|\overline{U}_{i,\,j+1}(k-1)| + \widetilde{\varphi}'_{i,\,j-1}(k-1)|U_{i,\,j-1}(k-1)|\right].$$

$k=2,3,\ldots,K.$, and recall that all coefficients are non-negative.

Taking into account that $\overline{U}\big|_{\Gamma_h} = 0$ and summing for all i, j we get

$$\sum_{i,j} |\overline{U}_{ij}(k)| \le \sum_{i,j} |\overline{U}_{ij}(k-1)|, \quad k=2,3,\ldots,K.$$

Hence

$$\sum_{\Omega_h} |U_{\overline{t}}(k)| \le \sum_{\Omega_h} |U_{\overline{t}}(1)| \le \sum |\Delta_h u_0|, \quad k=2,3,\ldots,K.$$

Now (3.10) follows at once by summing over k and multiplying by $h^2\tau$. This completes the proof.

It is clear that condition $u_0 \in C^2(\overline{\Omega})$ can be replaced by weaker ones e.g., $u_0 \in W_2^1(\Omega)$. We also mention that Theorem 3.2 may be reformulated in terms of the above theorem.

4. VARIATIONAL FORMULATION OF THE DIFFERENCE
PROBLEM AND AUXILIARY LEMMAS

Suppose now that $U:\overline{Q}_h \rightarrow \mathbf{R}$ satisfies (2.1)-(2.3). Then we get after summation

$$\tau h^2 \sum_{\Omega_h} \sum_{k=1}^{K} [U_{\overline{t}}(k) - \Delta_h(U(k-1))]\psi(x,\, k\tau) = 0,$$

$$\forall \psi \in \mathcal{D}[\Omega \times [0,\, T[\,].$$

Making use of the identity:

$$\tau \sum_{k=1}^{K} U_{\overline{t}}(k)\psi(x,\, k\tau) = -\tau \sum_{k=0}^{K-1} U(k)\psi_t(x,\, k\tau) - U(0)\psi(x,\, 0)$$

$$\text{for } x \in \Omega_h$$

and the discrete Green's formula:

556

$$h^2 \sum_{\Omega_h} \Delta_h \varphi(U(k-1))\psi(x,k\tau) =$$

$$= -h^2 \sum_{\Omega_h} [\varphi_{x_1}(U(k-1))\psi_{x_1}(x, k\tau) + \varphi_{x_2}(U(k-1))\psi_{x_2}(x, k\tau)],$$

we obtain:

$$\tau h^2 \sum_{k=0}^{K-1} \sum_{\Omega_h} [U(k)\psi_t(x, k\tau) - \varphi_{x_1}(U(k))\psi_{x_1}(x, (k+1)\tau) -$$

$$(4.1) \qquad -\varphi_{x_2}(U(k))\psi_{x_2}(x, (k+1)\tau)] + h^2\sum_{\Omega_h} u_0(x)\psi(x, 0) = 0,$$

$$\psi \in \mathcal{D}(\Omega \times [0, T[ ).$$

This is the discrete counterpart of (1.4). Here is now a lemma often used in proofs connected with nonlinear problems (see e.g. [5, p. 12]):

LEMMA 4.1. *Let* $Q = \Omega \times]0, T[$, $\Omega \subset \mathbf{R}^2$ *be bounded and suppose* g_n *and* $g \in L^2(Q)$. *Assume that*

$$\|g_n\|_{L^2(Q)} \leq C, \qquad n = 1, 2, \ldots,$$

C *independent of* n *and* $g_n \to g$ *a.e. on* Q.
  Then $g_n \to g$ *weakly on* $L^2(Q)$.

In the sequel we shall need the following extensions of grid functions [4]:

a) *Constant extension.* This relates to a function defined on the grid-points of $\overline{\Omega}_h$ (or Q_h) the step function $\widetilde{U}$ defined on the domain $\Omega_h \subset \Omega$, by

$$\widetilde{U}(x) = U_{ij} \qquad \forall x \in \omega_{ij},$$

where

$$\omega_{ij} = \{x; \ ih < x_1 < (i+1)h,$$

$$jh < x_2 < (j+1)h\}$$

(and analogously in the cylindrical case).

b) *Linear extension*. This assigns a continuous function to the grid-function defined by:

$$U' : \overline{\Omega}_h \rightarrow \mathbf{R}$$

$$U'(x) = U_{ij} + (U_{ij})_{x_1}(x_1 - ih) + (U_{ij})_{x_2}(x_2 - jh) +$$

$$+ (U_{ij})_{x_2}(x_1 - ih)(x_2 - jh) \quad \text{on} \quad \omega_{ij},$$

and similarly in the cylindrical case.

The following lemmas are all due to L a d y z h e n - s k a y a [4, Ch. VI].

LEMMA 4.2. *Consider a grid-function* $U : \overline{\Omega}_h \rightarrow \mathbf{R}$, *satisfying*

$$(4.2) \qquad h^2 \sum_{\Omega_h} U^2 \leq C,$$

C being a constant independent of h. *Then, if one of the sequences* $\{\widetilde{U}\}$ *or* $\{U'\}$ *is weakly convergent in* $L^2(\Omega)$, *when* $h \rightarrow 0$, *the same is true for the other sequence. The limit is common.*

The lemma remains true if we put Q_h instead of Ω_h.

We introduce the following notation: $\Omega_h^* \supset \Omega_h$ is the domain made up of those $\omega_{(kh)}$ which intersect Ω. Also $\| \cdot \|_{m,\Omega_h}^{(1)}$ $(\| \cdot \|_{m,\Omega_h^*}^{(1)}, \ \| \cdot \|_{m,Q_h}$ etc.) stands for the discrete W_m^1 norm:

$$\| U \|_{m,\,\Omega_h}^{(1)} = \left[\| U \|_{m,\,\Omega_h}^{m} + \| U_x \|_{m,\,\Omega_h}^{m} \right]^{1/m}$$

$$\| U \|_{m,\,\Omega_h} = \left(h^2 \sum_{\Omega_h} | U |^m \right)^{1/m}.$$

LEMMA 4.3. *Suppose that*

$$\| U \|_{1,\,\overline{\Omega}_h^{*}}^{(1)} \leq C,$$

C independent of h. Then, if one of the sequences $\{\widetilde{U}\}$ or $\{U'\}$ is convergent in $L^2(\Omega)$ when $h \to 0$, the other also converges in $L^2(\Omega)$ to the same function.

LEMMA 4.4. *If*

$$\| U \|_{m,\,\overline{\Omega}_h^{*}}^{(1)} < C,$$

then $\{U'\}$ is equibounded in $W_m^1(\Omega)$ and consequently compact in $L^m(\Omega)$.

•   LEMMA 4.5. *Suppose we are given a sequence of sets Ω_h such that*

$$\Omega_h \subset \widetilde{\Omega} \subset \Omega \qquad \text{for} \quad h \to 0.$$

Let $U : \overline{Q}_h \to \mathbf{R}$ also satisfy:

$$(4.3) \qquad h^2 \tau \sum_{k=0}^{K-1} \sum_{\Omega_h} \left(| U | + | U_{x_1} | + | U_{x_2} | + | U_t | \right) \leq C,$$

where C is independent of h and τ.

   Then there exists a subsequence of $U_{ij}(k)$ and a function $u \in W_1^1(Q)$ such that

(i) $U' \to u$ in $L(Q')$ $\forall Q' \subset \widetilde{Q} = \widetilde{\Omega} \times [0, \ T[$

(ii) $U'_{x_i} \to u_{x_i}$ in $L(Q')$ $weakly$ $\forall Q' \subset Q.$

Here $\overline{\widetilde{\Omega}} \subset \Omega$ *and* $Q' = \Omega' \times [0, \ T[, \ \overline{\Omega}' \subset \Omega.$

5. CONVERGENCE AND EXISTENCE

Since the solution U of (2.1) is bounded under the assumptions of Theorem 2.1, it follows that

$$0 \leq \widetilde{U} \leq M, \quad 0 \leq U' \leq M \quad on \quad \overline{Q}_h.$$

Hence both sets of functions are bounded on Q, if we consider them extended by 0 on $\Omega \backslash \Omega_h$. Consequently there is a sequence of steps $h_n \to 0$ such that $\widetilde{U}_{h_n}$ as well as U'_{h_n} converge weakly in $L^2(Q)$, say, to $\chi \in L^2(Q)$ (the limit is common according to Lemma 4.2). In the sequel we shall omit the subscript n that designates subsequences.

LEMMA 5.1. *Suppose that* $U : \Omega_h \to \mathbf{R}$ *is the solution of (2.1)-(2.3) and the sequence* $\widetilde{U}_h$ *is chosen as above. If* u_0 *satisfies condition (A), then there exists a subsequence of* U_h, *also denoted by* U_h, *such that:*

(i) $\widetilde{U}_h \to \chi$ in $L(Q')$, $\forall \Omega' \subset \Omega, \ \overline{\Omega}' \subset \Omega$

(ii) $\varphi(\widetilde{U}_h) \to \varphi(\chi)$ $weakly \ in$ $L^2(Q')$ $\forall \Omega' \subset \Omega$

(iii) $\varphi(\widetilde{U}_h)_{x_i} \to \dfrac{\partial \varphi(\chi)}{\partial x_i}$ $weakly \ in$ $L^2(Q')$ $\forall \Omega' \subset \Omega.$

PROOF. (i) We take Ω_h^* in Lemmas 4.3 and 4.4 such that $\Omega' \subset \overline{\Omega}_h^* \subset \Omega$, which is always possible when h is small (in the Ot direction the grid function can be extended keeping the values of the first and last levels, constant). So by Lemma 4.4 $\widetilde{U}_h \to u$ in $L(\Omega')$ (in fact this is true for a subsequence). But since U is bounded, we must have $u = \chi$.

(ii) The sequence of functions $\varphi(\widetilde{U}_h)$ is bounded (U_h is extended by zero beyond $\partial\Omega_h$), so there exists an element $\Theta \in L^2(\Omega)$ such that

$$\varphi(\widetilde{U}_h) \to \Theta \quad \text{weakly in } L^2(\Omega).$$

Since $\widetilde{U}_h \to \chi$ strongly in $L(\Omega')$, it follows that a subsequence $U_h \to \chi$ a.e. on Ω' and in view of the continuity of φ, we also have $\varphi(\widetilde{U}_h) \to \varphi(\chi)$ a.e. on Ω'. On the other hand,

$$\|\varphi(U_h)\|_{L^2(\Omega)} \le C,$$

with C independent of h. All conditions of Lemma 4.1 are fulfilled and we conclude that $\Theta = \varphi(\chi)$.

(iii) According to Theorem 3.2, the sequences $\varphi(\widetilde{U}_h)_{x_i}$, $i=1,2$ are equibounded. Thus we can choose a subsequence $\varphi(\widetilde{U}_h)_{x_i} \to \mu$ in $L^2(Q')$, $\forall \Omega' \subset \Omega$. But $\varphi(\widetilde{U}_h) \to \varphi(\chi)$ so that $\varphi(\widetilde{U}_h)_{x_i} \to \dfrac{\partial}{\partial x_i}\varphi(\chi)$, $i=1,2$. This completes the proof.

Now we will write (4.1) in the integral form:

$$\int_{Q_h} (\widetilde{U}\psi_t - \widetilde{\varphi}_{x_1}(U)\widetilde{\psi}_{x_1} - \widetilde{\varphi}_{x_2}(U)\widetilde{\psi}_{x_2})\,dx\,dt +$$

$$+ \int_{\Omega_h} \widetilde{u}_0(x)\widetilde{\psi}(x,0)\,dx = 0, \qquad \forall \psi \in \mathcal{D}([0,\,T[\,\times\Omega).$$

When h is chosen small enough so that

$$\operatorname{supp}\psi \quad \text{and} \quad \operatorname{supp}\psi_{x_i} \subset \Omega_h, \quad i=1,2, \quad \forall t \in [0,\,T[\,,$$

this can be written as

$$(5.1) \quad \int_Q \left[\widetilde{U\widetilde{\psi}}_t - \widetilde{\varphi}_{x_1}(U)\widetilde{\psi}_{x_1} - \widetilde{\varphi}_{x_2}(U)\widetilde{\psi}_{x_2} \right] dxdt +$$

$$+ \int_\Omega \widetilde{u}_0(x)\widetilde{\psi}(x,0)\,dx = 0.$$

THEOREM 5.1. *If the conditions of Theorem 3.1 or Theorem 3.3 are assumed, then the above defined function* χ *has the following properties:*

(i) χ *satisfies* (1.4)

(ii) $\chi \in C(\overline{Q})$ *and satisfies* (1.3).

PROOF. (i) Consider the domain Ω' such that $\overline{\Omega}' \subset \Omega$. Suppose that h is small enough so that $\Omega' \subset \Omega_h (\subset \Omega)$ and

$$\text{supp } \psi_{x_1} \cup \text{supp } \psi \cup \text{supp. } \psi_{x_2} \subset \Omega \times [0,\ T[\ ,$$

where ψ is the test function in (5.1). Then according to Lemma 5.1 and taking into account that $\widetilde{\psi}_{x_i}$ tends uniformly to $\frac{\partial \psi}{\partial x_i}$, $\widetilde{\psi}_{\overline{t}}$ tends to $\frac{\partial \psi}{\partial t}$ uniformly, etc., we get after tending to the limit in (5.1):

$$\int_Q \left[\chi \frac{\partial \psi}{\partial t} - \sum_{i=1}^{2} \frac{\partial \psi}{\partial x_i} \frac{\partial \varphi(\chi)}{\partial x_i} \right] dxdt +$$

$$+ \int_\Omega \psi(x,0) u_0(x)\ dx = 0 \qquad \psi \in \mathcal{D}(\Omega \times [0,\ T[\ ),$$

which means that χ satisfies (1.4).

(ii) Consider again the subsequence of discrete functions U_h having the property that $\widetilde{U}_h \to \chi$ in $L^2(Q)$. By Lemma 4.3 it follows that $\widetilde{U}'_h \to \chi$ in $L^2(Q)$. This sequence has a subsequence also denoted by U'_h

possessing the following properties:

(a) $\quad U_h' \in C(\overline{Q}_h)$ and $0 \le U_h'(x,\ t) \le M,$ $\quad (x,\ t) \in \overline{Q}_h$

(b) $\quad \dfrac{\partial U_h'}{\partial x_i}$ $\quad i=1,2,$ $\quad \dfrac{\partial U_h'}{\partial t}$ belong to $L(Q')$.

Property (b) means that the integral of these derivatives are absolutely equicontinuous, which implies, in view of the Newton-Leibniz formula, that U_h' is almost everywhere equicontinuous on any domain Ω'' such that $\overline{\Omega}'' \subset \Omega'$. But our functions are continuous so that U_h' is an equicontinuous family of functions on Ω''. Since this assertion is true for any Ω'' such that $\overline{\Omega}'' \subset \Omega$, it results that $\chi \in C(\Omega)$. Finally in view of the way in which u_2 was defined and because of the uniform convergence of the sequence U_h' on each $\Omega' \subset \Omega$ such that $\overline{\Omega}' \subset \Omega$, it follows that $\chi \in C(\overline{\Omega})$ and that it fulfills (1.3), which completes the proof.

REFERENCES

[1] V.F. Baklanovskaja, A study of the difference method to solve the first boundary-value problem for equations of the non-stationary filtration type, Ž. Vyčisl. Mat. i Mat. Fiz. Supplementary Issue, (1964), pp. 228-241 (in Russian).

[2] R. Courant - K. Friedrichs - H. Lewy, Über die partiellen Differenzengleichungen der mathematischen Physik, Math. Annalen, 100(1928), 32-74.

[3] J.L. Graveleau - P. Jamet, A finite difference approach to some degenerate nonlinear parabolic equations, SIAM J. Appl. Math. 20(2), (1971), 199--223.

[4] O.A. Ladyženskaja, Boundary-value Problems of the Mathematical Physics, Nauka, Moscow, 1974.(in Russian).

[5] J.L. Lions, *Quelques Méthodes de Résolution des Problèmes aux Limites Non Linéaires*, Dunod, Paris, 1969.

[6] O.A. Oleĭnik - A.S. Kalashnikov - Chou Iui-lin, Equations of the non-stationary filtration type, *Izv. AN SSSR Ser. Mat.* 22(5)(1958), 461-469 (in Russian).

E. Schechter
Department of Mathematics
University of Cluj
3400 Cluj, Str. Kogalniceanu 1.
Rumania

FAST FOURIER TRANSFORM AND CONDITIONAL EXPECTATION

F. SCHIPP

1. INTRODUCTION

Let f be a complex-valued function, defined on the set $\{0,1,2,\ldots,M-1\}$ $(M \in \mathbb{N}^* := \{1,2,3,\ldots\})$ and let $\hat{f}$ denote the discrete trigonometric Fourier transform (the discrete trigonometric Fourier coefficients) of f:

$$(1) \qquad \hat{f}(n) := \sum_{k=0}^{M-1} f(k) \exp(-2\pi\, ikn/M), \quad (n=0,1,\ldots,M-1).$$

When computing the values of $\hat{f}$ on the basis of (1), we need $O(M^2)$ operations (additions and multiplications). By applying Fast Fourier Trasform (FFT) algorithms we can achieve the same by using only $O(M \log M)$ operations (see [1], [3]-[11]). As to the history of FFT, see [5]. An analogue of the well-known Cooley-Tukey algorithm can be used for quickly computing the Walsh-Fourier coefficients, too (see the corresponding bibliography of the book [2]).

In this paper we synthesize the various known FFT methods. We show that for a number of orthonormed systems, the Fourier coefficients, similarly to the Fast Fourier Transform, can be computed from a more general algorithm. These orthonormed systems can be represented as product systems of some other systems which have a certain orthogonality property. The diverse known FFT methods such as the trigonometric and Walsh systems are special cases of the method presented here. We get, among others, an efficient procedure for computing the Fourier coefficients with respect to the Vilenkin system [8], [15].

For the description of the above mentioned algorithm, we use the notion of conditional expectation (CE).

2. PRODUCT SYSTEMS

We recall the definition and the most important properties of CE (see [13]).

Let (X, A, P) be a probability measure space and B a sub-σ-algebra of A. For an arbitrary function $f \in L^1 := L^1(X, A, P)$, we denote by $E(f|B)$ the CE of f with respect to B. The CE $E(f|B)$ can be characterized by the following two properties:

$$(2) \quad \begin{aligned} &\text{(i)} \quad E(f|B) \in L^1(X, B, P) \quad \text{and} \\ &\text{(ii)} \quad \int_B f \, dP = \int_B E(f|B) \, dP \quad (\forall B \in B). \end{aligned}$$

Due to the Radon–Nikodym theorem, the function $E(f|B)$ with the above two properties uniquely exists up to a set of zero P-measure, for any integrable function f.

It is known that the CE operator $L^1 \ni f \to E(f|B) \in L^1$ is bounded and linear moreover for any pair of functions $\lambda \in L^1(X, B, P)$ and $f \in L^1$ such that $\lambda f \in L^1$, we have

$$(3) \qquad E(\lambda f|B) = \lambda E(f|B).$$

Further, for arbitrary σ-algebras $C \subseteq B \subseteq A$ the following equalities hold:

$$(4) \qquad E(E(f|C)|B) = E(E(f|B)|C) = E(f|C) \qquad (f \in L^1).$$

On the basis of the interpretation (2) it is obvious that if $B := \{X, \emptyset\}$, the trivial σ-algebra, then $E(f|B) = \int_X f\, dP$, i.e. CE is a generalization of the notion of integral.

In the sequel we fix a monotone nondecreasing sequence

$$A_0 := \{X, \emptyset\} \subseteq \dots \subseteq A_1 \subseteq \dots \subseteq A_n \subseteq \dots \subseteq A$$

of σ-algebras and denote by E_n $(n \in \mathbf{N} := \mathbf{N}^* \cup \{0\})$ the CE operator with respect to A_n.

Let $(m_n, n \in \mathbf{N}^*)$ be a sequence of numbers such that $m_n \in \mathbf{N}$ and $\Phi_n := \{\varphi_n^k : 0 \le k < m_n, k \in \mathbf{N}\}$ $(n=0,1,\dots,N)$ is a function system with the following properties:

$$(5) \qquad \text{(i)} \quad \Phi_n \subseteq L^2(X, A_n, P) \qquad (n=1,2,\dots,N),$$

$$\text{(ii)} \quad E_{n-1}(\varphi_n^k \overline{\varphi_n^r}) = \delta_{kr} \qquad (0 \le k, r < m_n; \ n=1,\dots,N),$$

where δ_{kr} is the Kronecker symbol. If $A_n = A_0$ $(n \in \mathbf{N})$ then (5; ii) means exactly that the systems Φ_n are orthonormed in the usual sense. Owing to (4) evidently

$$\int_X \varphi_n^{k}\overline{\varphi_n^{r}}\, dP = E_0(E_{n-1}(\varphi_n^{k}\overline{\varphi_n^{r}})) = \delta_{kr} \qquad (0 \le k,\ r < m_n),$$

i.e. the systems Φ_n, satisfying (5; ii), are orthonormed. In general, the orthonormedness of the system $\Phi_n \subset L^2 := L^2(X,\ A,\ P)$ does not imply (5; ii).

Condition (5; i) means exactly that the elements of Φ_n are A_n-measurable and quadratically integrable.

In order to define the product system of the systems Φ_k $(k=1,2,\ldots,N)$, let us represent the natural number $0 \le n < M := m_1 m_2 \ldots m_N$ in the form

$$(6) \qquad n = n_1 + n_2 m_1 + n_3(m_1 m_2) + \ldots + n_N(m_1 m_2 \ldots m_{N-1}),$$

where $0 \le n_j < m_j$ and $n_j \in \mathbf{N}$. We define the product system $\Psi := \{\psi_n : 0 \le n < M,\ n \in \mathbf{N}\}$ of the systems Φ_k $(k=1,\ldots,N)$, as follows: for any natural number represented in the form (6) let

$$(7) \qquad \psi_n := \prod_{k=1}^{N} \varphi_k^{n_k}.$$

It can easily be seen that (5) implies that Ψ is an orthonormed system in the usual sense. Indeed, let $0 \le m < n < M$. Then there exists an index $1 \le k_0 \le N$ such that $m_{k_0} \ne n_{k_0}$ and $m_k = n_k$ $(k_0 < k)$, where m_k and n_k denote the digits of the numbers m and n in the representation (6). Due to the properties (3) and (4) of CE, and (5), we have

$$\Delta := E_{k_0-1}(\varphi_{k_0}^{n_{k_0}}\overline{\varphi_{k_0}^{m_{k_0}}} \ldots \overline{\varphi_N^{m_n}}\varphi_N^{n_N}) =$$

$$= E_{k_0-1}(\varphi_{k_0}^{n_{k_0}}\overline{\varphi_{k_0}^{m_{k_0}}} E_{k_0}(|\varphi_{k_0+1}^{n_{k_0}+1}|^2 \ldots E_{N-1}(|\varphi_N^{n_N}|^2) \ldots)) =$$

$$= E_{k_0-1}(\varphi_{k_0}^{n_{k_0}}\overline{\varphi_{k_0}^{m_{k_0}}}) = 0,$$

consequently for $m \neq n$,

$$\int_X \Psi_n \overline{\Psi}_m \, dP = E_0 \left(\varphi_1^{n_1}\overline{\varphi_1^{m_1}}\ldots\varphi_{k_0-1}^{n_{k_0-1}}\overline{\varphi_{k_0-1}^{m_{k_0-1}}}\Delta\right) = 0.$$

Similarly, we get $\int_X |\Psi_n|^2 dP = 1$.

Let $\hat{f}(n)$ $(n=0,1,\ldots,M-1)$ denote the Fourier coefficients of the function $f \in L^2$ with respect to ψ. Taking property (5; i) of the systems Φ_n into account, on the basis of (3), (4) and (7), $\hat{f}$ can be written in the form

$$(8) \quad \hat{f}(n) := \int_X f\overline{\psi}_n \, dP = E_0\left(\overline{\varphi}_1^{n_1} E_1\left(\overline{\varphi}_2^{n_2}\ldots E_{N-1}\left(\overline{\varphi}_N^{n_N} E_N f\right)\right)\right),$$

where the n_k-s are again the digits of n in the representation (6).

We remark that we used only property (5: i) of Φ_n in this expression of $\hat{f}$.

We call the functions $E_{n-1}(\overline{\varphi}_n^{-k} g)$, the A_{n-1}-Fourier coefficients of the function $g \in L^2$ with respect to Φ_n. Evidently, the A_0-Fourier coefficients are identical with the ordinary Fourier coefficients with respect to Φ_1.

Using (8), we can compute the Fourier coefficients of the function $f \in L^2$ with respect to ψ in the following way: first calculate the A_{N-1}-Fourier coefficients of f with respect to Φ_N (there are m_N of them), then the A_{N-2}-Fourier coefficients of each A_{N-1}-messurable function obtained just before. Writing down the A_{N-2}-Fourier coefficients of the (A_{N-2}-measurable, $m_N m_{N-1}$) functions with respect to the system Φ_{N-2}, obviously we get $m_N m_{N-1} m_{N-2}$ functions, each being A_{N-3}-measurable. Continuing this procedure, we obtain the $m_N m_{N-1}\cdots m_1$ Fourier coefficients $\hat{f}(n)$ in the N-th step.

Let us denote by α_n the number of necessary operations (the cost) of calculating the A_{n-1} Fourier coefficients (with respect to the system Φ_n) of the A_n-measurable functions. Thus computing all the Ψ-Fourier coefficients of any A_N-measurable function ($\in L^2$) by means of the previously sketched method, evidently requires

$$(9) \qquad K_N = \alpha_N m_N + \alpha_{N-1} m_N m_{N-1} + \ldots + \alpha_1 m_N m_{N-1} \cdots m_1$$

operations.

3. EXAMPLES

By suitably choosing the systems Φ_n and the σ-algebra sequence (A_n, $n \in \mathbb{N}$), we can obtain every known FFT method. In what follows, we present some of them.

A. *Independent Systems.* Suppose that the systems $\Phi_n \in L^2(X, A, P)$ ($n = 1, \ldots, N$) are independent with each of them being orthonormed in the space L^2, in the usual sense. Let $A_0 := \{X, \emptyset\}$ and let A_n denote the σ-algebra generated by the systems $\Phi_1, \Phi_2, \ldots, \Phi_n$. Since in this case (see e.g. [12])

$$E_{n-1}(\varphi_n^{k-\ell} \varphi_n) = \int_X \varphi_n^{k-\ell} \varphi_n \, dP = \delta_{k\ell},$$

and further (5; i) is satisfied automatically, so these Φ_n ($n = 1, \ldots, N$) satisfy the conditions of the previous section and (8) can be applied to the product system of the Φ_n-s.

Let us examine the following special cases.

a) *Fast-Walsh-FT (FWFT).* Let $X := [0,1)$, A the class of Lebesgue-measurable sets and P the Lebesgue measure. Let the Rademacher system be denoted by

$(r_n, \; n \in \mathbf{N}^*)$, i.e. let

$$r_1(x) := \begin{cases} 1 & (0 \le x \le \tfrac{1}{2}), \\ -1 & (\tfrac{1}{2} \le x < 1), \end{cases} \qquad r_1(x+1) = r_1(x) \quad (x \in \mathbf{R}),$$

$$r_n(x) := r_1(2^{n-1}x) \qquad (n \in \mathbf{N}^*, \; x \in \mathbf{R}).$$

It can easily be seen that the systems $\Phi_n := \{1, r_n\}$ are independent, further, it is obvious that A_n is the σ-algebra generated by the dyadic intervals $J_n :=$ $= \{[k2^{-n}, (k+1)2^{-n}) : k=0,1,\ldots,2^n-1\}$. The product system of the Φ_n-s is the Wals-Paley orthonormed system.

Since the A_n-measurable functions are constant on the intervals from J_n, so they can be characterized by their 2^n values. One can easily see that for any $f \in L^1$,

$$(E_n f)(k2^{-n}) = 2^n \int_{k2^{-n}}^{(k+1)2^{-n}} f \, dP \quad (k=0,1,\ldots,2^n-1), \text{ moreover,}$$

for every A_n-measurable function g,

$$(E_{n-1} g)(k2^{-n}) = \frac{1}{2}\left(g(k2^{-n+1}) + g(k2^{-n+1}+2^{-n})\right)$$

$$(k=0,1,\ldots,2^{n-1}-1)$$

holds. In this case, the number of operations necessary to compute the Φ_n-A_{n-1}-Fourier coefficients of the A_n-measurable functions is obviously proportional to 2^n. Now, since $m_k = 2 \quad (k=1,2,\ldots,N)$, so the cost of the FWFT described above is proportinal to $N2^N$.

Starting from the system $(r_1, \; r_1r_2, \; r_2r_3, \ldots$ $\ldots, r_{N-1}r_N)$ consisting of independent functions, instead of the system $(r_n, \; n=1,2,\ldots,N)$, we get the (original) Walsh system as the product system of the systems $\widetilde{\Phi}_1 =$ $= \{1, r_1\}$, $\widetilde{\Phi}_k = \{1, r_{k-1}r_k\}$ $(k=2,3,\ldots,N)$. Since the systems $\Phi_1,\ldots,\Phi_k$ generate the same σ-algebra as the systems $\widetilde{\Phi}_1,\ldots,\widetilde{\Phi}_k$, so the above results are applicable in this case, too.

    b) FFT (trigonometric). Let $M := m_1 m_2 \cdots m_N$ and suppose the numbers m_j $(j=1,2,\ldots,N)$ are pairwise relative primes. Let $M_j := M/m_j$ $(j=1,2,\ldots,N)$. We can see easily that for any natural numbers $0 \le n,\ x < M$ ther (uniquely) exist numbers $0 \le n_j,\ x_j < m_j$ $(1 \le j \le\ \le N)$ such that

$$(10) \qquad n \equiv \sum_{j=1}^{N} n_j M_j, \qquad x \equiv \sum_{j=1}^{N} x_j M_j \qquad (\text{mod } M).$$

    Let $X := \{0,1,2,\ldots,M-1\}$, A be the powerset of X, P the probability measure defined by $P(\{x\}) := := 1/M$ $(x \in X)$. For the function

$$\varphi_j(x_j) := \exp(2\pi i x_j M_j/m_j) \qquad (0 \le x_j < m_j;\ j=1,\ldots,N)$$

when taking the representation (10) into account, too, we have

$$\varphi_j(x_j) = \exp(2\pi i x_j M_j/m_j) =$$

$$= \exp\left(\frac{2\pi i}{m_j} \sum_{k=1}^{N} x_k M_k\right) = \exp\frac{2\pi i x}{m_j},$$

$$\exp\frac{2\pi i n x}{M} = \varphi_1^{n_1}(x_1)\cdots\varphi_N^{n_N}(x_N).$$

Thus we have expressed the discrete trigonometric system $(\exp 2\pi i n x/M : n,\ x \in X)$ as the product system of the independent systems $\Phi_n := \{\varphi_n^k : 0 \le k \le m_n\}$.

    In this special case, algorithm (8) is identical to the one introduced by I.J. G o o d [10], [11].

c. The Vilenkin System. Let $X_j := \{0,1,..,m_j-1\}$ $(j=1,...,N)$, the probability measure μ_j be defined as $\mu_j(\{x_j\}) := 1/m_j$ $(x_j \in X_j)$, $X := X_1 \times X_2 \times ... \times X_N$ and $\mu := \mu_1 \times ... \times \mu_N$. Evidently the systems

$$\varphi_n^k(x) := \exp(2\pi i x_n k/m_n) \qquad (x=(x_1,...,x_N),$$

$$0 \le k < m_n)$$

are independent and their product system can be identified as the Vilenkin orthonormed system [8], [15]. In this case, by applying the procedure mentioned in the introductory remarks, we obtain an efficient method for computing the Vilenkin-Fourier coefficients.

B. The Cooley-Tukey Algorithm. Let $M := m_1 m_2 ... m_N$, $X := \{0,1,...,M-1\}$, A be the powerset of X and P the probability measure defined by $P(\{x\}) := 1/M$ $(x \in X)$. In this example we do not require the numbers m_i $(i=1,...,N)$ to be pairwise relative primes. Let A_n denote the set of those subsets of X, periodic with respect to $q_n := p_1 p_2 ... p_n$ $(n=1,2,...,N)$, and let $q_0 := 1$, $A_0 := \{X, \emptyset\}$. One can easily see that for any A_n-measurable, complex-valued function, defined on X

$$(E_{n-1}f)(x) = \frac{1}{m_n} \sum_{k=1}^{m_n-1} f(x \dot{+} k q_{n-1}), \qquad (x \in X),$$

where $\dot{+}$ denotes mod M addition.

Let

$$\varphi_n^k(x) = \exp(2\pi i x k/q_n) \qquad (x \in X, \ 0 \le k < m_n;$$

$$n=1,...,N).$$

Then evidently the systems $\Phi_n := \{\varphi_n^k : 0 \le k < m_n\}$

satisfy (5). If the representation of the number $0 \le$
$\le n < M$ is of the form

$$n = n_N + n_{N-1} m_N + \ldots + n_1 (m_2 \ldots m_N) \quad (0 \le n_j < m_j),$$

then

$$\exp 2\pi i n x = \varphi_1^{n_1}(x) \ldots \varphi_N^{n_N}(x) \qquad (x \in X),$$

consequently, we have expressed the discrete trigonometric system as the product system of systems satisfying (5).

In this special case, the procedure described in the previous section is identical to the Cooley-Tukey algorithm [3], [4], [5], [6]. In the case of $m_1 = \ldots = m_N = 2$ we get the algorithm described in [1].

C. A Class of Orthogonal Systems. In what follows, we define a new class of orthogonal systems, and show that the formerly described fast algorithm is applicable to the computation of the Fourier coefficients in this case, too.

In order to construct the system in question, let the interval $I_0 := [0,1)$ be divided into three (not necessarily congruent) subintervals I_1^j $(j=1,2,3)$. Let φ_1^j $(j=1,2,3)$ be (real) functions such that they are constant on the intervals I_1^j and

$$\int_{I_0} \varphi_1^k \varphi_1^\ell \, dx = \delta_{k\ell} |I_0|$$

hold, where $|I_0|$ is the length of I_0. Let us further divide each subinterval I_1^j into three subintervals, getting the subintervals I_2^k $(k=1,2,\ldots,9)$. Consider now the functions φ_2^j $(j=1,2,3)$ that have the properties:

(1) φ_2^j is constant on the subintervals I_2^k,

(2) $\displaystyle\int_{I_1^j} \varphi_2^k \varphi_2^\ell \, dx = \delta_{k\ell} |I_1^j|.$

Continuing this procedure, we can construct an interval system I_n^k $(k=1,2,\ldots,3^n; n=1,2,\ldots,N)$, together with a function system φ_n^j $(j=1,2,3; n=1,2,\ldots,N)$ such that φ_n^j is constant on I_n^k, and

$$\int_{I_{n-1}^k} \varphi_n^j \varphi_n^\ell \, dx = \delta_{j\ell} |I_{n-1}^k|.$$

Let A_n denote the σ-algebra, generated by the intervals I_n^k $(k=1,2,\ldots,3^n)$. On the basis of their definition, the systems $\{\varphi_n^j : j=1,2,3\}$ $(n=1,2,\ldots,N)$ evidently satisfy (5), consequently, we can apply procedure (8) to the computation of the Fourier coefficients of the system in question.

REFERENCES

[1] N.S. Bakhvalov, *Numerical Methods*, Nauka Moscow, 1973, (in Russian).

[2] K.G. Beauchamp, *Walsh Functions and Their Applications*, Academic Press, London-New York-San Francisco, 1975.

[3] P. Bloomfiled, *Fourier Analysis of Time Series: An Introduction*, John Wiley, London-New York-Sydney--Toronto, 1976.

[4] J. Coffy, Description d'algorithmes de transformation de Fourier. Contributions a la theorie des systemes *IRIA* (1970), 171-200.

[5] J.W. Cooley - P.A. Lewis - P.D. Welch, Historical notes on the fast Fourier transform, *Proc. IEEE* 55(1967), 1675-1677.

[6] J.W. Cooley - P.A.W. Lewis, P.D. Welch, The fast Fourier transform and its application to time series analysis. Stat. methods for digital computers (Vol. III., 1977), 377-423.

[7] J.W. Cooley - J.W. Tukey, An algorithm for the machine calculation of complex Fourier series, *Math. Comp.* 19(1965), 297-301.

[8] A.V. Efimov - A.F. Karakulin, On the continuous analogue of periodic multiplicative orthonormed systems, *Dokl. Akad. Nauk USSR* 218(1978), 268-271.

[9] W.M. Gentleman - G. Sande, Fast Fourier transforms for fun and profit, *Proc. AFIPS Fall Joint Computer Conference* 29(1966), 563-578.

[10] I.J. Good, The interaction algorithm and practical Fourier analysis, *J. Roy. Stat. Soc., Ser. B* 20(1958), 361-372.

[11] I.J. Good, The relationship between two fast Fourier transforms, *IEEE Trans. Comput.* C-20(1971), 310-317.

[12] P. Henrici, *Einige Anwendungen der schnellen Fourier transformation.* Moderne Methoden der Numerischen Mathematik, Birkhäuser Verlag, 1976, 111-124.

[13] J. Neveu, *Discrete-Parameter Martingales,* North--Holland/American Elsevier, 1975.

[14] F. Schipp, On a generalization of the concept of orthogonality. *Acta Sci. Math.*, 37(1975), 279-285.

[15] N.Ja. Vilenkin, On a class of complete orthonormed systems, *Izv. Akad. Nauk USSR* 11(1977), 363-400. (in Russian).

F. Schipp
Department of Mathematics
Eötvös Loránd University
Muzeum krt. 6-8.
1088 Budapest, Hungary

SETS OF COMPLEX MATRICES WHICH CAN BE TRANSFORMED TO TRIANGULAR FORMS

OLGA TAUSSKY

The matrices considered are square, $n \times n$, and have complex numbers as entries. The report given here does not deal with single matrices, but with pairs and with the pencil or even with the algebra generated by a pair. It has been observed that more insight into a pair can be gained in this way. In particular, the consideration of pencils of symmetric matrices is a classical problem. The report is to some degree a brief survey, but includes new treatments and work from two theses. Some of the survey was included in lectures at Keszthely and two Gatlinburg lectures (at Los Alamos and at Asilomar, California) but was not published previously.

When a pair A, B of matrices is given, the first consideration is usually to ask whether they commute or do not commute. There is much classical work done on such pairs, mainly by F r o b e n i u s and by S c h u r . Two important facts had been discovered concerning commuting pairs. On one side, such a pair can be transformed to triangular form simultaneously by a

similarity.[1] On the other side the eigen values of any
polynomial $P(A,B)$ can be obtained easily if the eigen
values of A and of B are known. Let $\alpha_1,\ldots,\alpha_n$ be
the eigen values of A and $\beta_1,\ldots,\beta_n$ the eigen values
of B. Then there exists an ordering α_i, β_i, independent
of p, such that $p(A,B)$ has eigen values $p(\alpha_i,\beta_i)$. By
polynomial $p(x,y)$ is meant a polynomial in noncommuting
variables x, y with scalar coefficients.

Triangular matrices play a big role in practical.
and in theoretical work. In practical work because they
display the eigen values of the matrix and because every
matrix can be transformed to triangular form, even by a
unitary similarity; in theoretical work, e.g. because
they form an algebra, a subalgebra of the full matrix
algebra.

However, commuting matrices are not characterized
by the property of simultaneous triangularization.

There is actually no canonical form for pairs of
commuting matrices, as was pointed out by I.M. G e l -
f a n d and V.A. P o n o m a r e v. They explain that
it makes no sense to look for one.

I had observed that the known conditions for commutativity are all necessary, not sufficient.

Pairs of matrices with either of the two properties
mentioned earlier, enjoyed by commuting matrices, are
said to have the P-property. Examples showing that
pairs with the P-property do not necessarily commute
are easily obtained, e.g.

$$A = \begin{bmatrix} 1 & a \\ 0 & -1 \end{bmatrix}, \quad B = \begin{bmatrix} 1 & b \\ 0 & 1 \end{bmatrix}, \quad AB = \begin{bmatrix} 1 & a+b \\ 0 & 1 \end{bmatrix},$$

---

[1] In what follows transforming a matrix will alwys
mean transforming by a similarity.

$$BA = \begin{bmatrix} 1 & a-b \\ 0 & -1 \end{bmatrix}$$

$$AB-BA = \begin{bmatrix} 0 & 2b \\ 0 & 0 \end{bmatrix} \neq 0.$$

The last fact verifies that the polynomial $xy-yx$ has the eigen values $\alpha_i\beta_i - \beta_i\alpha_i = 0$.

While sets of commuting matrices cannot be characterized, sets of simultaneously triangularizable, called s.t. matrices here, can. This was done by M c C o y :

THEOREM OF McCOY. *The matrices A, B have the P-property if and only if one of the two following (equivalent) conditions holds:*

(1) All matrices $p(A,B)(AB-BA)$ are nilpotent when p is an arbitrary polynomial in A and B.

(2) All polynomials $p(x,y)$ with scalar coefficients in the non commuting variables x, y have as eigen values $p(\alpha_i,\beta_i)$ where α_i are the eigen values of A, and β_i the eigen values of B, under a fixed ordering, independent of p.

There is much connection between these theorems and Lie algebra, as also observed by F l a n d e r s . M c C o y himself had studied this fact in one of his approaches. Later new proofs were found by D r a z i n, D u n g e y and G r u e n b e r g in a much cited paper. However, their proofs are rather complicated, though the proof for (1) has a computational flavor, starting with an arbitrary vector v and by some kind of iteration process constructing a polynomial $p(A,B)$ such that $p(A,B)v$ is a common eigen vector of A and B. Once that step is achieved Property P follows by induction.

Briefer proofs can be obtained via the theory of algebras and their radicals. Condition (1) implies that the commutators lie in the radical of the algebra formed by the polynomials in the matrices under consideration. But, since the commutators lie in the radical the quotient algebra with respect to the radical is commutative. These facts can be exploited to arrive at a briefer proof. Theorem 5 and 6 in a paper by F l a n d e r s concerns this. These proofs aim at a reduction of the veftor spaces under the linear transformations given by the algebra of matrices. Once such a reduction is proved induction on the dimension completes the argument. This idea of proof can be condensed even more by using a theorem of B u r n s i d e and the fact that a full matrix algebra admits no radical: The Burnside Theorem states that an algebra of matrices which is not a full matrix algebra, over an algebraically closed field can be transformed to matrices of the form $\begin{bmatrix} P & Q \\ 0 & R \end{bmatrix}$ where P, Q are square and 0 is the zero matrix. We then have the following proof for condition (1) of M c C o y :

There are two cases, the case where the algebra consisting of all matrices has a radical and the case where it does not. The first case is settled by Burnside's Theorem and the fact that a full matrix algebra has no radical. The second case implies that $AB-BA$ is the zero matrix, thus the algebra is commutative and this case is classical.

In a recent paper by L a f f e y it is shown that the P-property follows under the mere condition that all commutators $q(A)B-Bq(A)$ are nilpotent for all polynomials q, provided all eigen values of A are different.

The relation $AB = BA$ can be generalized to studying $AB = \alpha BA$, α a scalar, or a polynomial in A, B. However, replacing $AB = BA$ by $AB-BA = 0$ and weakening this condition has influenced a good bit of

research. There is, of course the Lie product $(A,B) =$
$= AB-BA$ and its use in Lie algebra, the operator $\Delta_A X =$
$= AX-XA$ and its powers, higher commutativity of matrices, i.e. $(A,(A,...,(A,B))) = 0$ (see R o t h) and
generalized commutativity, where scalar polynomials in
the commutator operator are considered (see T a u s s -
k y, G a i n e s , M a r c u s and A l i).

McCoy's Theorem is an outcome of generalizing the
relation $AB-BA = 0$ instead of $AB = BA$.

A weakening of condition (2) of McCoy's Theorem is
in the L-property, a concept suggested by M. K a c : A
pair of $n\times n$ matrices A, B with eigen values $\alpha_1,...$
$...,\alpha_n$, respectively $\beta_1,...,\beta_n$ is said to have the
L-property if every linear combination $\lambda A+\mu B$ has as
eigen values $\lambda\alpha_i+\mu\beta_i$, for a suitable ordering of the
eigen values independent of λ, μ. While the L-property
implies the P-property for $n = 2$, it ceases to do so
for $n = 3$. If however, the additional assumption is
made that AB is to have eigen values $\alpha_i\beta_i$, under the
same ordering, then L implies P for $n = 3$ in the
case of complex entries of the matrices. This was shown
by T a u s s k y and was greatly extended by
L a f f e y and by Z a s s e n h a u s .

Over the complex numbers the L-property is equivalent with the fact that the characteristic polynomial of
the pencil $\lambda A+\mu B$ is of the form $\Pi(\lambda\alpha_i+\mu\beta_i+\nu)$. This
fact leads itself to immediate generalization when one
simply postulates that the characteristic polynomial is
reducible. This play a big role when pairs of simultal-
taneously block triangularizable matrices are studied.

The L-property is a rather difficult concept. It
was studied by a number of researchers, starting with
joint work by M o t z k i n and T a u s s k y , later
Taussky alone, K a p l a n s k y, W i e l a n d t ,

W i e g m a n n, R.C. T h o m p s o n, H. S c h n e i-
d e r, S m i l e y and N.F. R o b i n s o n ,
W a l e s and Z a s s e n h a u s, Zassenhaus alone,
G u r a l n i c k, G u r a l n i c k, and T a u s s-
k y, G e r s t e n h a b e r, A. L a x, K a t o ,
D r a z i n, P. B r e n n e r, W i d l u n d , K o p a-
c e k, M a r i k.

If one assumes one member of the pair, say A, in
triangular canonical form, then B has a structure
which "tries" to be triangular. Under further conditions
triangular can be forced on B as well, i.e. A, B
have the P-property as well. Interesting combinatorial
and graph theoretic situations can arise in this connection.

In the paper by D r a z i n, D u n g e y,
G r u e n b e r g the case of diagonable sets of matrices is treated too, as well as their simultaneous
transformation to diagonals.

Another way of generalizing Frobenius' Theorem for
the case $AB = BA$ was found by D r a z i n:

*If all commutators of A and B of a fixed order
k vanish then A, B have the P-property.*

In computational work complete commutativity is not
really a meaningful term. There one talks about "almost
commuting" matrices and also about "near commuting
pairs"; a number of people have worked on these concepts.
G a i n e s and the author studied also almost s.t.
matrices.

I come now to the subject of simultaneous block
triangularization (s.b.t.) of a pair of $n{\times}n$ matrices
A, B. By this is meant that the matrices can be transformed simultaneously to upper block triangular matrices

$$
\begin{bmatrix} A_1 & & & * \\ & A_2 & & \\ 0 & & \cdot & \\ & & & A_t \end{bmatrix}, \qquad
\begin{bmatrix} B_1 & & & * \\ & B_2 & & \\ & & \cdot & \\ 0 & & & B_t \end{bmatrix},
$$

where A_i, B_i are square matrices and A_i has the same
dimension as B_i.

McCoy's Theorem goes over if we assume that $A_i B_i - B_i A_i =$
$= 0$ for all i. Further, the transformation can again
be carried out unitarily. This was treated in a joint
paper by F. G a i n e s and R.C. T h o m p s o n. But
clearly, it is desirable to do something for the case
where not all A_i, B_i commute. In this case $AB - BA$
does not have 0-blocks in the diagonal and is not
necessarily nilpotent. Some basic ideas are studied in
this joint paper.

Two further investigations are to be found in the
Caltech thesis of G a i n e s, and in a Caltech thesis
under preparation by H. S h a p i r o, part of which
has already appeared and in work of L a f f e y. There
is also the problem of simultaneous block diagonalization which is studied by W a t t e r s, L a f f e y
and by S h a p i r o. This leads to the concept of
generalized normality, i.e. the property of a complex
matrix to be unitarily similar to a block diagonal.

The work of G a i n e s uses the technique of
generalized commutativity, mentioned earlier. This arose
out of a commutator identity obtained by T a u s s k y
and W i e l a n d t. It can be described simplest via
the characteristic polynomial of the commutator operator,
coming from, say, a matrix A. If this polynomial of
degree $n(n+1)/2$ is applied to any $n \times n$ matrix the
result is the zero matrix. If, however, the matrix is of

a special nature a polynomial of smaller degree may have
the same property. (Using generalized commutativity
T a u s s k y obtained a generalization of a well known
fact concerning simultaneous block diagonals: Let $AB =$
$= BA$ with A block diagonal $\begin{bmatrix} A_1 & \\ & A_2 \end{bmatrix}$ and eigen values of
A_1 differing from eigen values of A_2. Then B is
block diagonal $\begin{bmatrix} B_1 & \\ & B_2 \end{bmatrix}$ with matching dimensions of
blocks.)

The application of this by G a i n e s lies in
studying the case where polynomials in the commutator
operator, applied to the algebra generated by A and
B, lie in the radical, instead of the commutator $AB-BA$.

In the work by H. S h a p i r o a different approach is used. Here polynomial identities for matrices
are being used, like the Amitsur-Levitzki identity or
an identity arising from central polynomials. These are
polynomials in non commuting variables, which for a
fixed n have as value over the ring of $n\times n$ matrices
only scalar matrices. Taking then the additive commutator of such a polynomial and a matrix variable leads to
an identity.

Polynomial identities are developing into a very
important subject and tool in various connections. In
this respect the ring of matrices of a fixed dimension
can contribute to this subject greatly and hence can
contribute to abstract ring theory by studying rings in
which the same identities hold that turn up for the
matrices (see K a p l a n s k y).

SELECTED REFERENCES

(References included in O. Taussky's article in
Amer. Math. Monthly (1951) are, in general not
repeated here)

[1]   S.A. Amitsur - J. Levitzki, Minimal identities for
algebras, *Proc. Amer. Math. Soc.* 1(1950), 449-463.

[2]   G.P. Barker - L.Q. Eifler - T.P. Kezlan, A non commutative spectral theorem, *Lin. Alg. and Appl.*
20(1978), 95-100.

[3]   H. Flanders, Methods of proof in linear algebra,
Amer. Math. Monthly 63(1956), 1-15.

[4]   E. Formanek, Central polynomials for matrix rings,
Journ. of Algebra 23(1972), 129-132.

[5]   F. Gaines, Simultaneous block triangularization and
block diagonalization of sets of matrices, *Calif.
Inst. Tech. Thesis* 1966.

[6]   F. Gaines, Kato-Taussky-Wielandt commutator relations, *Lin. Alg. and Appl.* 1(1968), 127-138.

[7]   F. Gaines, Kato-Taussky-Wielandt commutator relations and characteristic curves, *Pacific Journ.
Math.* 61(1975), 121-128.

[8]   F. Gaines - R.C. Thompson, Sets of nearly triangular
matrices, *Duke Math. Journ.* 35(1968), 441-453.

[9]   I.M. Gel'fand - V.A. Ponomarev, Remarks on the
classification of a pair of commuting linear
transformations in a finite-dimensional space,
Funksional. Anal. i Prilozen. 3(1969), 81-82.
(In Russian)

[10] T. Kato, *Perturbation Theory for Linear Operators*,
Springer 1966.

[11] T.J. Laffey, Simultaneous quasidiagonalizations of
complex matrices, *Lin. Alg. and Appl.* 16(1977),
189-201.

[12] T. Laffey, Simultaneous triangulaization of a pair
of matrices, *Journ. Alg.* 44(1977), 551-557.

[13] T.J. Laffey, Simultaneous triangularization of
matrices - low rank cases and the non derogatory
case, *Lin. and Multilin. Alg.* 6(1978), 269-305.

[14] M. Marcus, The minimal polynomial of a commutator,
Portugaliae Mathematica 23(1964), 73-76.

[15] M. Marcus - M. Shafqat Ali, Minimal polynomials of
additive commutators and Jordan products, *Journ.
Algebra* 22(1972), 12-33.

[16] M. Marcus - Nisar A. Khan, On matrix commutators,
Can. J. Math. 12(1960), 269-277.

[17] T.S. Motzkin - O. Taussky, Pairs of matrices with
property L, *Trans. Amer. Math. Soc.* 73(1952), 108-
-114.

[18] T.S. Motzkin - O. Taussky, Pairs of matrices with
property L II, *Trans. Amer. Math. Soc.* 80(1955),
961-963.

[19] T.S. Motzkin, On $L(S)$-tuples and ℓ pairs of
matrices, *Pac. Journ. of Math.* 40(1972), 639-645.

[20] Ju.P. Razmyclov, A certain problem of Kaplansky,
Izv. Akad. Nauk SSSR 37(1973), 483-501.

[21] D.W. Robinson, A note on k-commutative matrices,
Journ. Math. Physics 2(1961), 776-777.

[22] S. Rosset, A new proof of the Amitsur-Levitzki
theorem, *Isr. J. Math.* 23(1976), 187-188.

[23] W.E. Roth, On k-commutative matrices, *Trans. Amer. Math. Soc.* 39(1936), 483-495.

[24] H. Shapiro, Simultaneous block triangularization and block diagonalization of sets of matrices, *Lin. Alg. and Appl.* 25(1979), 129-137.

[25] D.A. Suprenenko - R.I. Tyškeviç, *Commutative Matrices*, Acad. Press, 1968.

[26] O. Taussky, Commutativity in finite matrices, *Amer. Math. Monthly* 64(1957), 229-235.

[27] O. Taussky, A weak property L for pairs of matrices, *Math. Zeitschr.* 71(1959), 463-465.

[28] O. Taussky, On a matrix theorem of A.T. Craig and H. Hotelling, *Proc. Nederl. Akad. Wet. Amsterdam* 20(1958), 139-141.

[29] O. Taussky, A characterization of property L, *Journ. de Mathématiques* 41(1962), 291-293.

[30] O. Taussky, A generalization of matrix commutativity, *Linear Alg. and Appl.* 2(1969), 349-353.

[31] O. Taussky - H. Wielandt, Linear relations between higher additive commutators, *Proc. Amer. Math. Soc.* 13(1962), 732-735.

[32] O. Taussky, Some results concerning the transition from the L- to the P-property for pairs of finite matrices, *J. Alg.* 20(1972), 271-283.

[33] O. Taussky, Some results concerning the transition from the L- to the P-property for pairs of finite matrices II, *Lin. and Multilin. Algebra* (1974), 195-202.

[34] W. Watkins, Polynomial identities for matrices, *Amer. Math. Monthly* 82(1975), 364-368.

[35] J.F. Watters, Simultaneous quasi-diagonalization
     of normal matrices, *Linear Alg. and Appl.* 9(1974),
     103-117.

[36] W. Wielandt, Pairs of matrices with property L,
     J. Res. Nat. Bur. Stand. 51(1953), 89-90.

[37] J. Williamson, The simultaneous reductions of two
     matrices to triangular forms, *Amer. J. Math.*
     57(1935), 281-293.

O. Taussky
Department of Mathematics
California Institute of Technology
Pasadena, California 91125, U.S.A.

SOME APPLICATIONS OF ELLIPTIC FUNCTIONS AND INTEGRALS

J. TODD

Among the recent applications are the following:

1. DESIGN OF FILTER CIRCUITS.

This is due to *W. Cauer* [5] and *S.D. Darlington* [10]. See, e.g., *F. Oberhettinger* and *W. Magnus* [19], *Z. A. Melzak* [21], *E. Stiefel* [34], *R. A.-R. Amer* and *H. R. Schwarz* [2].

2. OPTIMAL PARAMETERS IN THE ADI METHOD.

This is due to *E. L. Wachsspress* and *W. B. Jordan* [37]. See, e.g., *J. Todd* [35, (1967)].

3. OPTIMAL STARTING VALUES FOR THE NEWTONIAN PROCESS FOR SQUARE ROOTS.

See below.

4. RAPID COMPUTATION OF THE ELEMENTARY FUNCTIONS

See *R. P. Brent* [5], *E. P. Salamin* [33].

5. MANY VALUES OF MIXED MEANS

This derives from a problem of Gauss. See *J. Todd* [35,(1978)].

The first three of these applications can be regarded as special cases of work of *E. I. Zolotareff* [39, (1877)]. In this paper we confine our attention to the third topic; the list of references, however, covers all the applications mentioned. A comprehensive survey is in preparation.

The Newtonian process for the evaluation of $\sqrt{x}$ is a very efficient one. We discuss the optimal choice of the initial approximation, as a rational function of x, when x is restricted to an interval: $(\frac{1}{16},1)$ or $(\frac{1}{10^2},1)$, are important practical cases. Whereas the existence of such an optimal choice follows from the general Chebyshev theory and an infinite constructive process, following the ideas of *Remez*, can be given, we intend to show how to express the coefficients of the rational function explicitly in terms of elliptic functions whose parameters depend on the interval and on the type of the rational function. This type of result was first obtained by *Zolotareff* [39,(1877),(1878)]. The specific problem to which we come was handled by *Achieser* [1,(1953/6)] and by *Ninomiya* [25]; we give a somewhat different approach, similar to one given for the ADI problem (*Todd* [35,(1967)]). This problem was already considered by *Chebyshev* [7,(1884)], who gave the explicit form in the case of functions of type $(1,1)$ and gave a functional equation for the general (n,n) type. By type (m,n) we understand a rational function with numerator of degree at most m, and denominator of degree at most n.)

§.1. ORIENTATION

Consider, for orientation, the following problem. We are dealing with x restricted to an interval $[a,b]$.

We first ask what is the best constant estimate q for $\sqrt{x}$, in the sense of minimizing the maximum absolute relative error, i.e., we want to choose q so that

$$\max_{a \le x \le b} \left| \frac{q-\sqrt{x}}{\sqrt{x}} \right| \text{ is minimum.}$$

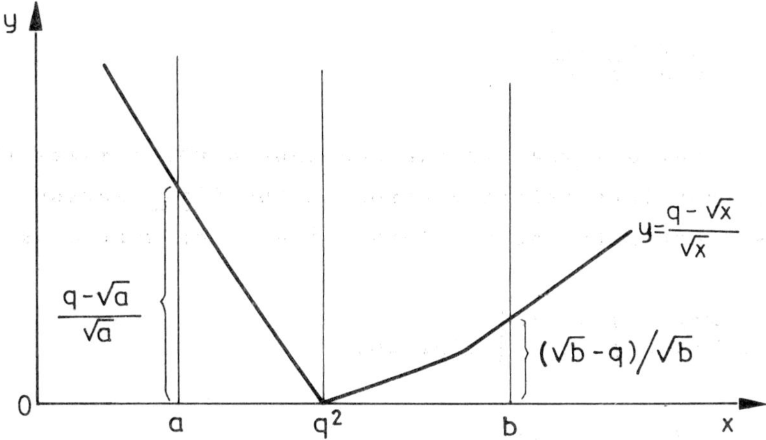

It is clear that this occurs when

$$\frac{q - \sqrt{a}}{\sqrt{a}} = \frac{\sqrt{b} - q}{\sqrt{b}}$$

which gives $q = 2\sqrt{(ab)}/(\sqrt{a}+\sqrt{b})$. The minimum of error for this q is:

$$\frac{\sqrt{b} - \sqrt{a}}{\sqrt{b} + \sqrt{a}} \; .$$

If we now apply the Newton–Raphson iteration once starting with this q we get

$$\frac{1}{2}\left(\frac{2\sqrt{ab}}{(\sqrt{a}+\sqrt{b})\sqrt{x}} + \frac{\sqrt{x}(\sqrt{a}+\sqrt{b})}{2\sqrt{ab}} \right)$$

and the corresponding relative error is maximum in absolute value at $x=a$ where it is

$$\frac{(\sqrt{b} - \sqrt{a})^2}{4\sqrt{a}(\sqrt{a}+\sqrt{b})} \; .$$

We now ask what is the constant q which makes the maximum absolute relative error in the <u>first</u> Newton-–Raphson estimate for $\sqrt{x}$ least, i.e., for what q is

$$\max_{a \le x \le b} \left| \frac{q_1 - \sqrt{x}}{\sqrt{x}} \right| \text{ minimum,}$$

where $q_1 = \frac{1}{2}(q + (x/q))$?

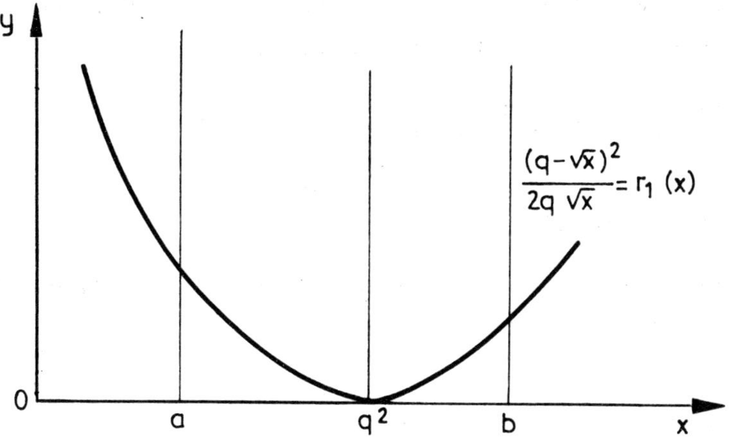

It is clear that this occurs when

$$\frac{(\sqrt{b}-q)^2}{2q\sqrt{b}} = \frac{(\sqrt{a}-q)^2}{2q\sqrt{a}}$$

which gives

$$\frac{\sqrt{b}-q}{q-\sqrt{a}} = \left(\frac{b}{a}\right)^{1/4}, \quad q = (ab)^{1/4}.$$

This is different from the value obtained above, since $a < b$. The min max in this case is

$$\frac{(b^{1/4} - a^{1/4})^2}{2(ab)^{1/4}}$$

which is smaller than

$$\frac{\sqrt{b}-\sqrt{a}}{(\sqrt{a}+\sqrt{b})} \quad .$$

We now ask what is the constant q which makes the maximum absolute relative error in the <u>second</u> Newton-Raphson estimate for $\sqrt{x}$ least, i.e., for what q is

$$\max_{a \le x \le b} \left[\frac{q_2 - \sqrt{x}}{\sqrt{x}} \right] \quad \text{minimum,}$$

where $q_2 = \frac{1}{2}(q_1 + (x/q_1))$?

We can answer this question without more computation by first observing that if

$$r_0 = \frac{q - \sqrt{x}}{\sqrt{x}}, \quad r_1 = \frac{q_1 - \sqrt{x}}{\sqrt{x}}, \quad r_2 = \frac{q_2 - \sqrt{x}}{\sqrt{x}},$$

then

$$r_2 = \frac{\frac{1}{2}(q_1 + (x/q_1)) - \sqrt{x}}{\sqrt{x}}$$

$$= \frac{q_1^2 - 2q_1\sqrt{x} + x}{2q_1\sqrt{x}}$$

$$= \left[\frac{q_1 - \sqrt{x}}{\sqrt{x}} \right]^2 \frac{1}{(2q_1/\sqrt{x})}$$

$$= \frac{r_1^2}{2(1 + r_1)} \ .$$

The behavior of r_2 as a function of r_1 is indicated by the graph:

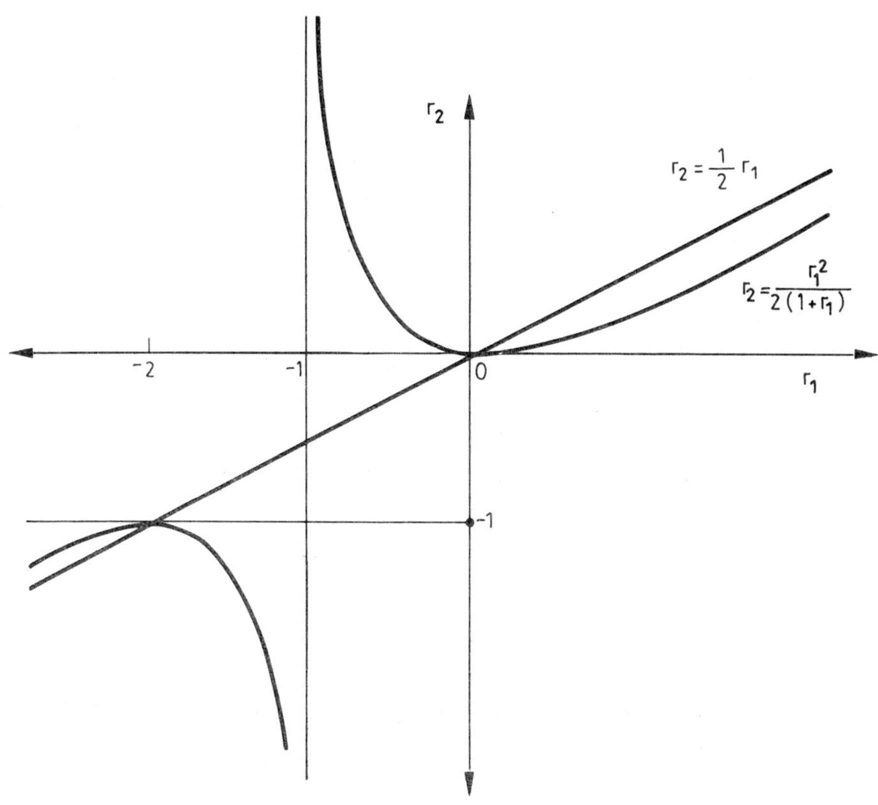

Since $r_1 \geqq 0$, it is clear that r_2 is a strictly increasing function of r_1 and so the graph of r_2 is of
the same shape as that for r_1 and the optimal value of
q is that which makes $r_2(a)=r_2(b)$, which is equivalent
to $r_1(a)=r(b)$, which gives $q=(ab)^{1/4}$ as before.

This remark is due to *D.G. M o u r s u n d* $((3)[8])^*$.
Wide generalizations of this are possible: The optimal

---

f.n.* (3) 1967 indicates a reference to the special
bibliography for this part.

q for all later Newton-Raphson estimates is the same as that for the first and these results remain true for a wide class of initial guesses q, in particular for the case when q is a rational function, and for functions other that $\sqrt{x}$. Moreover, in the square root case, the optimal starting functions $N(x)/D(X)$ for error criteria are scalar multiples of one another, specifically

$$\left(\frac{N(x)}{D(x)}\right)_L = \left(\frac{N(x)}{D(x)}\right)_{N-R} = \frac{1}{\sqrt{(1-e^2)}}\left(\frac{N(x)}{D(x)}\right)_C ,$$

where e is the optimal maximum absolute relative error in the Chebyshev case. L indicates that following $C\ h\ e\ b\ y\ s\ h\ e\ v$ [7,(1894)] a logarithmic criterion has been used, i.e., the maximum for $a \leq x \leq b$ of

$$|\ \log N(x)/D(x)\sqrt{x}\ |$$

is minimized over all relevant N,D ; similarly $N-R$ indicates that

$$\left|\left[\frac{1}{2}\left\{\frac{N(x)}{D(x)} + \frac{\sqrt{x}D(x)}{N(x)}\right\} - \sqrt{x}\right]/\sqrt{x}\right|$$

is used while C indicates that

$$\left|\left(\frac{N(x)}{D(x)} - \sqrt{x}\right)/\sqrt{x}\right|$$

is used.

The second equality, in the case when $N(x)$, $D(x)$ are constant, can be verified by using the calculation already given.

Some of these results are rather recent, some have been given by $A\ c\ h\ i\ e\ s\ e\ r$ (1930) and some are

rather delicate. See, for instance, the papers cited
below.

§.2. FORMULATION OF THE PROBLEM

The relative error committed in approximating $\sqrt{x}$
by a rational function $N(x)/D(x)$ is

$$e(x) = \frac{\sqrt{x}-N(x)/D(x)}{\sqrt{x}} = 1 - \frac{1}{\sqrt{x}}\frac{N(x)}{D(x)} \ .$$

Consider therefore the approximation of $f(x)\equiv 1$ by
functions of the form

$$\frac{1}{\sqrt{x}}\frac{N(x)}{D(x)} \ ,$$

where degree $N(x)\leqq n$, degree $D(x)\leqq d$. Suppose $N(x) =$
$= a_1 + a_2 x + \ldots + a_n x^n$ and $D(x) = b_1 + b_2 x + \ldots + b_d x^d$.

From the general theorem of A c h i e s e r
[1,p.55] it follows that in non-degenerate cases $e(x)$
assumes its maximum absolute value, L, say with alternating signs at $n+d+2$ points.

We apply this to the case when $a = k^2$, $0 < k < 1$, $b = 1$.
For clarity in our diagrams we assume $n = d = 1$; in non-
-degenerate cases we will have $|e| = L$ at 4 points.
Clearly $0 < L < 1$.

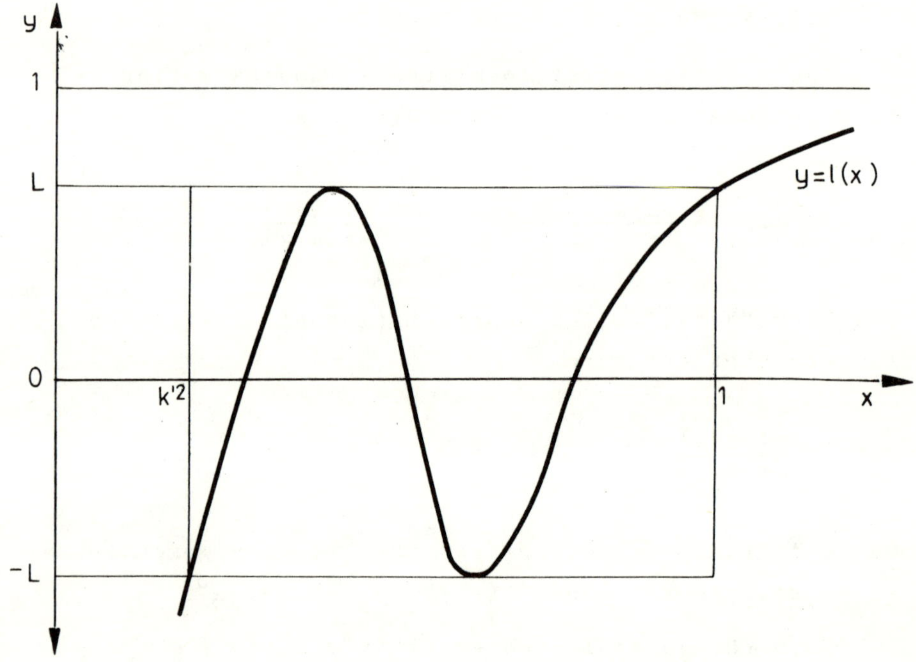

It will be convenient to change the variables from x to t^2, and from e to $r=1-e$. The new diagram will be of the form:

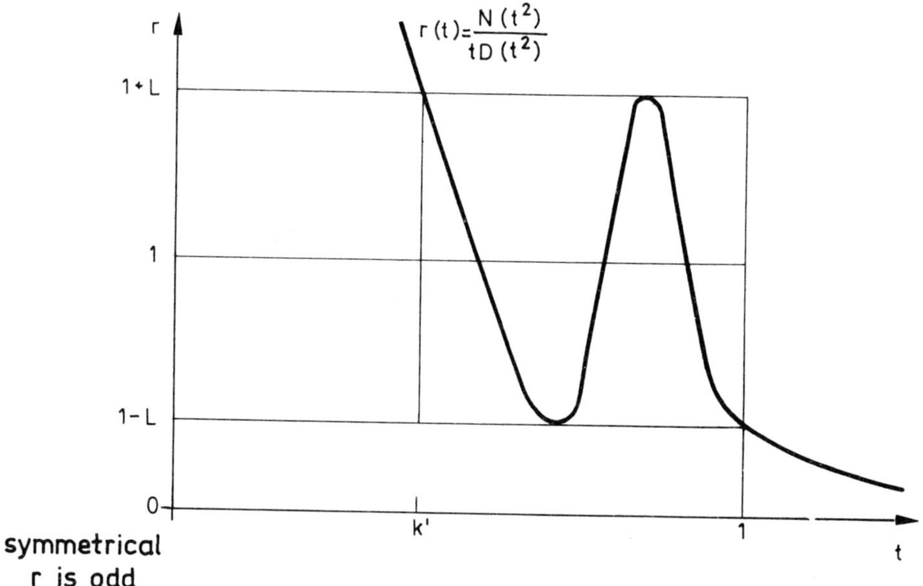

The characterization theorem assures us of the behavior in the interval $[k',1]$. Since the leading coefficient of $D \neq 0$ it follows that $r(t) \to 0$ as $t \to +\infty$; as $t \to 0$, $r(t) \sim b_1/(a_1 t) \to \infty$. The graph must be as indicated for if $r(t) \to -\infty$ as $t \to +0$, we would have too many intersections, e.g., with $r = 1$, for these are given by the cubic

$$N(t^2) = tD(t^2) \ .$$

Alternately, r' has a numerator of degree 4 and so four

601

turning points, which are used up in the interior of $(k',1)$ and $(-1,-k')$.

Consider now the behavior of $r(t)$ in the general case.

Clearly

$$f = ((1+L)^2 - r^2)(r^2 - (1-L)^2)$$

vanishes (twice) at each of the zeros of $r'(t)$ in the interiors of $(-1,-k')$, $(k',1)$ and also vanishes at ± 1, $\pm k'$. We shall show that f is a constant multiple of g, where

$$g = (r'(t))^2(1-t^2)(t^2-k'^2),$$

if $n=d$ or $n=d+1$.

We compute the degrees of the numerators and denominators of f and g:

$$f = \frac{((1+L)^2 t^2 D^2(t^2)-N^2(t^2))(-(1-L)^2 t^2 D^2(t^2)+N^2(t^2))}{t^4 D^4(t)}$$

$$= \frac{\text{polynomial of degree} \quad \max(4+8d,8n)}{\text{polynomial of degree} \quad 4+8d};$$

$$g = \left[\frac{2 N'(t^2)D(t^2)t^2-2N(t^2 D'(t^2)t^2-N(t^2 D(t^2))}{t\, D(t^2)}\right]^2 (1-t^2)(t^2-k'^2)$$

$$= \frac{\text{polynomial of degree } 4n+4d+4}{\text{polynomial of degree } 4+8d}.$$

Hence if $\max(4+8d,8n)=4n+4d+4$, i.e., if $n=d$ or $n=d+1$, f and g are rational functions of the same form; the denominators are identical and the numerators will have the same zeros.

§.3. THE (n,n) CASE.

We shall deal with the case $n=d$ first. The graph of N/D will be of the same general shape as that in the case $n=d=1$. The relation between f and g can be written in the form

$$\frac{dr}{[\{(1+L)^2-r^2\}\{r^2-(1-L)^2\}]^{\frac{1}{2}}} = \frac{dt}{\underline{+}M\{(1-t^2)(t^2-k'^2)\}^{\frac{1}{2}}}$$

(The ambiguous sign alternates between extrema; we shall omit this for the time being.)

Integrating this relation, remembering that $r(1)=$ $=1-L$ we have

$$(1)\qquad \int_r^{1-L}\frac{dR}{[\{(1+L)^2-R^2\}\{R^2-(1-L)^2\}]^{\frac{1}{2}}} = \frac{1}{M}\int_t^1\frac{dT}{[(1-T^2)(T^2-k'^2)]^{\frac{1}{2}}}$$

which we write as

$$M\,v = u\ .$$

From tables of elliptic integrals (e.g., NBS Handbook [23]), we find

$$(17.4.43)\qquad (1+L)v = \mathrm{nd}^{-1}(R/(1-L),\ 4L/(1+L)^2)$$

$$(17.4.44)\qquad u = \mathrm{dn}^{-1}(t,k^2)$$

so that

$$(2)\qquad \begin{aligned} r &= (1-L)\,\mathrm{dn}((1+L)v,\ 4L/(1+L)^2)\\[1mm] t &= \mathrm{dn}(u,k^2)\ . \end{aligned}$$

Let Λ, Λ' be the quarter periods corresponding to $\lambda^2=4L/(1+L)^2$. Then since

$$dn(u+K,k^2) = k'nd(u,k^2)$$

and since $\lambda'=(1-L)/(1+L)$ we have

$$(3) \qquad \begin{aligned} r &= (1+L)dn(1+L)v+\Lambda,\lambda^2) \\ &= (1+L)dn((1+L)uM^{-1}+\Lambda,\lambda^2) \, . \end{aligned}$$

The relations (2) and (3) give a parametric representation of the extremal curve. What we have yet to do is to determine M,L (which gives Λ) in terms of the data k^2,n.

In order that t describes the real axis from right to left, u must describe the perimeter of a rectangle:

$$u: \ -iK'\to 0\to K\to K-iK'\to K-2iK'\to -2iK'\to -iK'$$

$$t = dn(u,k^2): \infty\to 1\to k'\to 0\to k'\to -1\to -\infty.$$

Now the corresponding values of v must be, in the (n,n) case;

$$0; \ 1-L, \ 1+L, \ \ldots, \ 1-L, \ 1+L; \ \infty;$$

$$-(1+L), \ -(1-L), \ \ldots, \ -(1+L), \ -(1-L); \ 0 \, .$$

This means that the corresponding arguments of the dn function in (3) must be:

$$\Lambda-i\Lambda';\Lambda,2\Lambda,\ldots,(2n-1)\Lambda,2n\Lambda;2n\Lambda-i\Lambda';$$

$$2n\Lambda-2i\Lambda',(2n-1)\Lambda-2i\Lambda',\ldots2\Lambda-2i\Lambda',\Lambda-2i\Lambda';\Lambda-i\Lambda'.$$

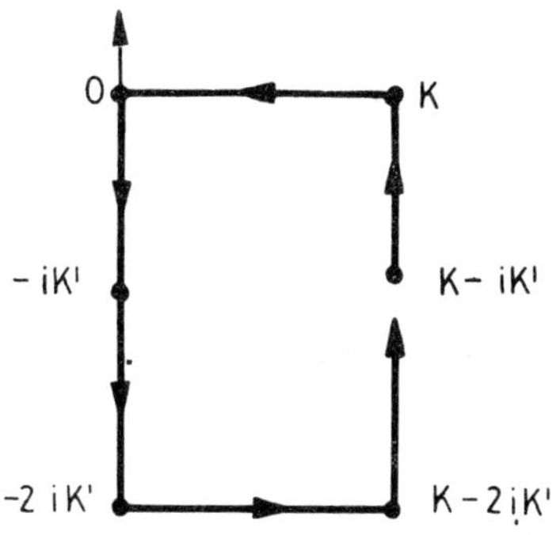

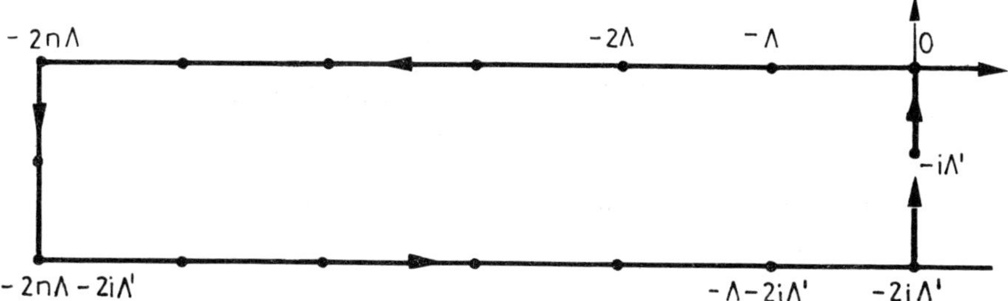

To get agreement we must have

$$(4) \qquad (1+L)M^{-1}K' = \Lambda', \quad (1+L)M^{-1}K = 2n\Lambda.$$

This means that $(K/K')=2n(\Lambda/\Lambda')$ and so the parameter q for the elliptic functions with quarter periods Λ, Λ' is q^{2n}. Since q is determined by k^2 it follows that we can find $\dot{q}=q^{2n}$ and hence Λ,Λ' and M. Explicit formulas are available, in particular we have, from $A\ c\ h\ i\ e\ s\ e\ r$ ([1,(1970),p.284]):

$$(5) \qquad \mathrm{dn}(uM^{-1}+\Lambda;\lambda^2) = \frac{\lambda'}{\mathrm{dn}(u,k)} \prod_{r=1}^{n} \frac{1-k^2 c_{2r}\mathrm{sn}^2(u,k^2)}{1-k^2 c_{2r-1}\mathrm{sn}^2(u,k^2)}$$

605

where

$$C_r = \mathrm{sn}^2(rK/2n, k^2),$$

$$\lambda = k^{2n}\prod_1^n C_{2r-1}^2,$$

$$M = \prod_{r=1}^n (C_{2r-1}/C_{2r}), \quad L = (1-\lambda')/(1+\lambda').$$

Returning to our original notation, since $(1+L)\lambda'=1-L$,

$$r(x) = \frac{1}{\sqrt{x}} \cdot \frac{2}{1+\lambda'} \prod_{r=1}^n \frac{(1-C_{2r})+C_{2r}x}{(1-C_{2r-1})+C_{2r-1}x}$$

and the optimal starting value sought is

$$(7) \qquad \frac{2}{1+\lambda'} \prod_{r=1}^n \frac{(1-C_{2r})+C_{2r}x}{(1-C_{2r-1})+C_{2r-1}x}$$

and the maximum absolute relative error is:

$$(8) \qquad L = \frac{1-\lambda'}{1+\lambda'} \ .$$

This checks in the case $n=0$, with the results obtained earlier.

This completes the discussion of the (n,n) case.

§.4. THE $(n,n-1)$ CASE.

We shall now briefly discuss the $(n,n-1)$ case. For simplicity take the $(1,0)$ case. Then the graph of the

error, $e(x)$, will be

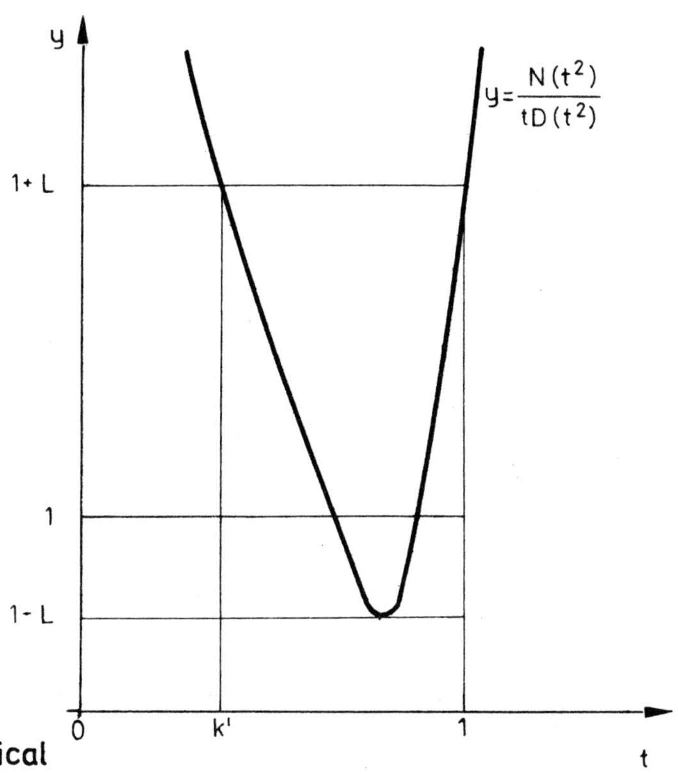

symmetrical

and that of $r(t^2)=1-e(t^2)$ will be

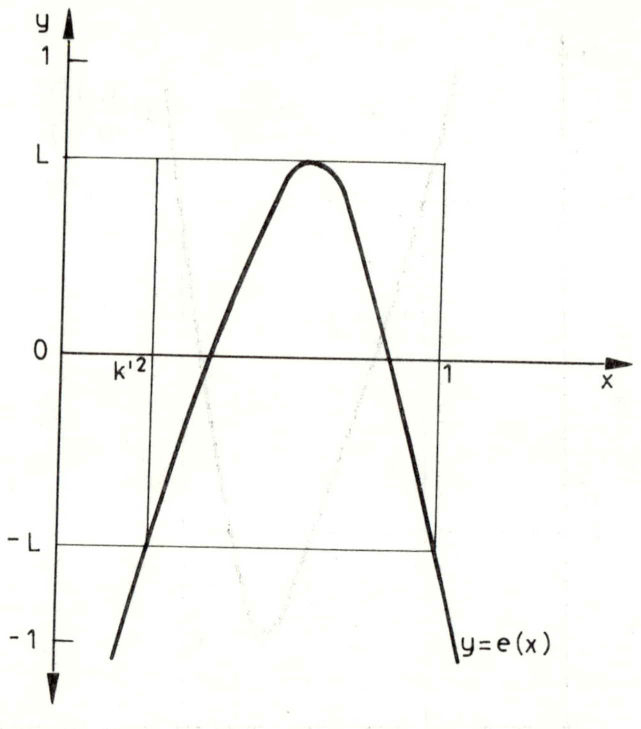

In place of (1) we have

$$(9) \qquad \int_r^{1+L} \frac{dr}{[\ldots]^{1/2}} = \frac{1}{M} \int_t^{k^1} \frac{dt}{[\ldots]^{1/2}} \ , \quad \text{say } v=u/M \ ,$$

which integrates to

$$(17.4.44) \quad (1+L)v = dn^{-1}(R/(1+L), \lambda^2),$$

$$(17.4.44) \quad u+K = dn^{-1}(t, k^2),$$

which give

$$r = (1+L)dn((1+L)uM^{-1}, \lambda^2),$$

(10)

$$t = dn(u+K, k^2) = k' nd(u, k^2).$$

The corresponding diagrams are:

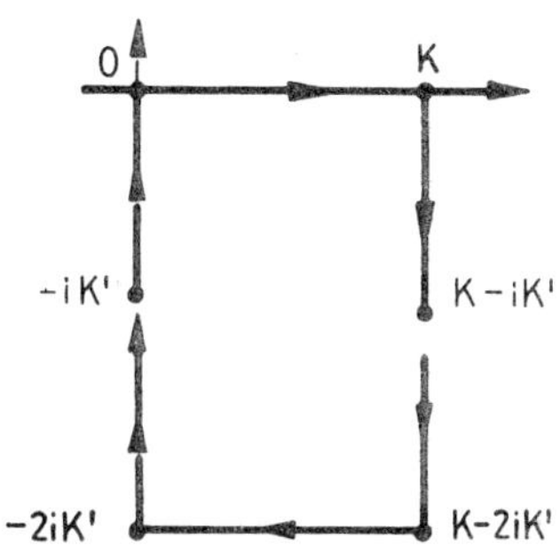

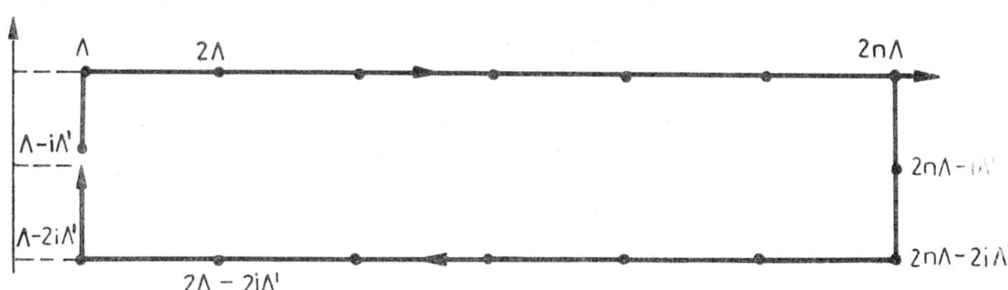

It follows that we must have

(11) $\quad (1+L)M^{-1}K' = \Lambda'$, $(1+L)M^{-1}K = 2n\Lambda$,

as before. From (4), using the relation $dn(uM^{-1}+\Lambda, \lambda^2) =$

$=\lambda' nd(uM^{-1},\lambda^2)$ we find

$$(12)\quad dn(uM^{-1},\lambda^2) = dn(u,k^2)\prod_{r=1}^{n}\frac{1-k^2 C_{2r-1}sn^2(u,k^2)}{1-k^2 C_{2r}sn^2(u,k^2)}\quad.$$

Returning to our original notation, since now $t=k'nd(u,k^2)$ we have

$$r(x) = \frac{k'}{\sqrt{x}}\cdot\frac{2}{1+\lambda'}\prod_{r=1}^{n}\frac{(1-C_{2r-1})x+C_{2r-1}k'^2}{(1-C_{2r})x+C_{2r}k'^2}\quad.$$

This product is a rational function of type $(n,n-1)$ because

$$C_{2n} = sn^2(K,k^2) = 1$$

and so the denominator in the last factor is a constant. Hence the optimal starting value is

$$(13)\quad \frac{2}{1+\lambda'}\prod_{r=1}^{n}\frac{(1-C_{2r-1})x+k'^2 C_{2r-1}}{(1-C_{2r})x+k'^2 C_{2r}}$$

and the maximum absolute relative error is

$$(14)\quad L = \frac{1-\lambda'}{1+\lambda'}$$

as before.

We note here that, in view of a remark on p. 611 the optimal Newton-Raphson estimate, e.g. in the (n,n) case, is just

$$(15) \qquad \frac{1}{\lambda'} \prod_{r=1}^{n} \frac{(1-c_{2r})+c_{2r}x}{(1-c_{2r-1})+c_{2r-1}x}$$

and the maximum absolute relative error is

$$(16) \qquad \frac{1}{\sqrt{\lambda'}} - 1 \ .$$

REMARKS

It is clearly desirable to see how the errors (15), (16) depend on n, k being fixed. For small n the results can be obtained from the numerical values given by I. N i n o m i y a [25] and the asymptotic behavior can be readily obtained from the corresponding investigation in the ADI case by D. G a i e r and J. T o d d [13]. However this is not of great interest since modifications of this Newton–Raphson algorithm lead to a process with much better convergence.

The latter development arises from the observation that the error in the Newton–Raphson case is always positive: for instance we have seen that

$$r_1(x) = (q_1 - \sqrt{x})^2 / 2q_1\sqrt{x} \ .$$

This suggests the modification of the formula to get a balanced error; this can be accomplished by multiplying the optimal estimate by a suitable constant. Indeed I. N i n o m i y a [25] has shown that given n and k'^2, constants c_i, $i=1,2,\ldots$ can be easily computed, once for all, such that if $y_o(x)$ is the optimal Newton–Raphson approximation of type (n,n), say, and if $c_o=1$ and

$$y_i(x) = \frac{1}{2} c_i \left(y_{i-1}(x) + x/y_{i-1}(x) \right), \quad i=1,2,\ldots$$

then y_i will be the optimal Newton-Raphson approximation of the type $(2^i n, 2^i n)-$ the proof depends on the Landen transformation. The corresponding error will be reduced from

$$[1/\sqrt{\lambda'}-1] = \frac{\lambda^2}{\sqrt{\lambda'}(1+\sqrt{\lambda'})(1+\lambda')} \sim \frac{\lambda^2}{4} \sim \frac{1}{2}(kq)^n$$

to

$$[1/\sqrt{\mu_i'} -1] \sim \frac{1}{2}(kq)^{2^i n},$$

where, as before, λ' is the complementary modulus for q^{2^n}, and μ_i' that for $q^{2^i n}$.

The optimality of this improvement and the extension to the case of $x^{1/m}$ is discussed by G. D. T a y l o r ((3)[13]).

We have included a special bibliography for papers which discuss the square root case, and also the case of $x^{1/m}$.

REFERENCES FOR (3)

[1] W.J. Cody, Double-precision square root for the CDC-300, *Comm. ACM* 7(1964) 715-718.

[2] J. Eve, Starting approximations for the iteractive calculation of the square roots, *Computer J.* 6(1963), 274-276.

[3] C.T. Fike, Starting approximation for square root calculations on IBM System/360. *Comm ACM* 9(1966), 297-298.

[4]   J.B. Gibson, Optimal rational starting approximations, *J. Approx. Theory* 12(1974), 182-198.

[5]   R.F. King, Improved Newton iteration for integral roots, *Math. Comp.* 25(1971), 299-304.

[6]   R.F. King - D.L. Phillips, The logarithmic error and Newton's method for the square root, *Comm. ACM* 12 (1969), 87-88.

[7]   G. Meinardus - G.D. Taylor, Optimal starting approximations for iterative schemes, *J. Approx. Theory* 9(1973), 1-19.

[8]   D.G. Moursund, Optimal starting values for Newton--Raphson calculation of $\sqrt{x}$, *Comm. ACM* 10(1967), 430-432.

[9]   D.G. Moursund - G.D. Taylor, Optimal starting values for the Newton-Raphson calculation of inverses of certain functions, *SIAM J. Numer. Anal.* 5(1968), 138-150.

[10] D.L. Phillips, Generalized logarithmic error and Newton's method for the mth root, *Math. Comp.* 24, (1970), 383-389.

[11] H. Rutishauser, Betrachtungen zur Quadratwurzel-iteration, *Monatsh. Math.* 67(1963), 452-464.

[12] P.H. Sterbenz - C.T. Fike, Optimal starting approximation for Newton's method, *Math. Comp.* 23 (1969), 313-318.

[13] G.D. Taylor, Optimal starting approximation for Newton's method, *J. Approx. Theory* 3(1970), 156-163.

[14] G.D. Taylor, An improved Newton iteration for calculating roots which is optimal, *Numerische Methoden der Approximationtheorie I.ISNM* 16(1972), 209-227.

[15] L. Wuytack, Some remarks on a paper of D.G.Moursund, *SIAM J. Numer. Anal.* 7 (1970), 233-237.

[16] K. Zeller, Newton-Čebyšev-Approximation, *Iterations-verfahren. Numerische Mathematik. Approximations-theorie,* ISNM 15 (1970), 101-104.

APPLICATIONS OF ELLIPTIC FUNCTIONS

[1] N.I. Achieser, *Vorlesungen über Approximations-theorie,* Akademie Verlag, Berlin, 1953., translated as,
On the Theory of Approximation, Ungar, New York, 1956.
Elements of the Theory of Elliptic Functions, 2nd edition (in Russian), 1970.
For complete bibliographies of Achieser see the articles on him in *Uspehi Mat. Nauk* $\equiv$ *Russian Math. Surveys* 16, 6 (102) (1961), p.14, 26, 6 (162) (1971), 257-261.

[2] R.A.-R. Amer - H.R. Schwarz, Contributions to the approximation problem of electrical filters, *Mitt. Inst. ang. Math., ETH,* Zürich,#9, 1964.

[3] Bateman Project, *Higher Transcendental Functions,* II. McGraw Hill, New York, 1953.

[4] E.K. Blum, *Numerical Analysis and Computation, Theory and Practice,* Addison-Wesley, 1972.

[5] R.P. Brent, Fast multiple precision evaluation of elementary functions, *J. ACM* 23 (1976), 212-251.

[6] W. Cauer, *Synthesis of Linear Communications Networks,* McGraw Hill, New York, 1958., German edition 1934. Bemerkung über eine Extremalaufgabe von E.Zolotareff *ZAMM* 20 (1940), 358.

[7] P.L. Chebyshev, *Oeuvres, I, II,* St.Petersburg 1899. Chelsea, New York.

Sur les fractions algébriques ..., *Bull. Soc.Math. France* 12(1884), 167-168 (≡II, 725).

[8]   A.R.Curtis, Tables of Jacobian elliptic functions whose arguments are rational fractions of the quarter period, *NPL Math. Tables,* v. 7, H. M. Stationery Office, London, 1964.

[9]   S. Dumas, Sur le développement des functions elliptiques en fractions continues, *Thesis, Zürich,* 1908.

[10] S.D. Darlington, Synthesis of resistance 4-poles which produce prescribed insertion loss characteristics including special application to filter design, *J. Math. Phys.* 18(1939), 257-353.

[11] A. Enneper, rev. F. Müller, *Elliptische Functionen, Theorie und Geschichte,* Halle, 1890.

[12] R. Fricke, *Lehrbuch der Algebra,* 1924.

[13] D. Gaier - J. Todd, On the rate of convergence of optimal ADI processes, *Numer. Math.* 9(1967), 452-459.

[14] N. Gastinel, Sur le meilleur choix des paramètres de sur-relaxation, *Chiffres* 5(1962), 109-126.

[15] W. Gautschi, Computational methods in special functions, pp. 1-98 in *Theory and Application of Special Functions,* Academic Press, 1975.

[16] E. Glowatski, Sechstellige Tafel der Cauer-Parameter, *Abh. Bayer. Akad. Wiss., Math. Nat. Klasse,* 67(1955).

[17] G. Greenhill, *The Applications of Elliptic Functions,* London, 1892.

[18] L.V. King, *On the Direct Numerical Calculation of Elliptic Functions and Integrals,* Cambridge, 1924.

[19] W. Magnus - F. Oberhettinger - R.P. Soni, *Formulas and Theorems for the Special Functions of Mathematical Physics*, {3}, Springer, Heidelberg, 1966.

[20] G. Meinardus, *Approximation von Funktionen und ihre numerische Behandlung*, Springer, 1964, translated by L.L. Schumaker as *Approximation of Functions, Theory and Numerical Methods*, Springer, 1967.

[21] Z.A. Melzak, *Mathematical Ideas, Modeling and Applications*, Vol. II of Companion to concrete mathematics, Wiley, New York, 1976.

[22] L.M. Milne-Thomson, *Jacobian Elliptic Function Tables*, Dover, New York, 1950.

[23] *NBS Handbook of Mathematical Functions*, ed. M. Abramowitz - I. A. Stegun, U.S. Government Printing Office, Washington, D.C. 1964.

[24] E.H. Neville, *Jacobian Elliptic Functions*, Oxford, 1946.

[25] I. Ninomiya, Generalized rational Chebyshev approximation, *Math. Comp.* 24(1970), 159-169. Best rational starting approximations and improved Newton iteration for the square roots, *Math. Comp.* 24(1970), 391-404.

[26] F. Oberhettinger - W. Magnus, *Anwendung der elliptischen Funktionen in Physik und Technik*, Springer-Verlag, Heidelberg, 1949.

[27] S. Paszkowski, *The Theory of Uniform Approximation*, I. Rozprawy Matem., 26, Warsaw, 1962.

[28] H. Piloty, Zolotareffsche rationale Funktionen, *ZAMM* 34(1954), 175-189.

[29] E.P. Ozigova - E.I. Zolotarev, 1847-1878. *Nauka*, Moscow, 1966. (in Russian).

[30] H.E. Rauch - A. Leibowitz, *Elliptic Functions, Theta-Functions and Riemann Surfaces*, Baltimore, 1973.

[31] J.R. Rice, *The Approximation of Functions*, 2 vols. Addison-Wesley, 1964, 1969.

[32] T.J. Rivlin, *An Introduction to the Approximation of Functions*, 1969.

[33] E. Salamin, Computation of π using arithmetic-geometric mean, *Math. Comp.* 30(1976), 565-570.

[34] E.L. Stiefel, Le problème d'approximation dans la théorie des filtres électriques, p.81-87. *Belgian Colloquium*, 1961.

[35] J. Todd, Optimal ADI-parameters, in *Functional Analysis, Approximationstheorie, Numerische Mathematik*, ISNM, 7, Birkhäuser, Basel, 1967.
Inequalities of Chebyshev, Zolotarev, Cauer and W.B. Jordan, pp. 321-328 in *Inequalities*, II., ed. O. Shisha, Academic Press, 1967.
Optimal parameters in two programs, *Bull. Inst. Math. Appl.* 6(1970), 31-35.
The lemniscate constants, *Comm. ACM* 18(1975),16-19.
The many values of mixed means I, pp. 5-22 in *General Inequalities*, ed. E.F. Beckenbach, ISNM 41, Birkhäuser, Basel, 1978.
Introduction to the Constructive Theory of Functions, Birkhäuser, Academic Press, 1963.

[36] R.S. Varga, *Matrix Iterative Analysis*, Prentice-Hall, Englewood Cliffs, 1962.

[37]   E.L. Wachspress, Extended application of ADI
       iteration model problem theory, *J. SIAM* 11(1963),
       994-1016.
       Optimum ADI iteration parameters for a model
       problem, *J. SIAM* 10(1962), 339-350.
       Iterative Solution of Elliptic Systems, Prentice-
       -Hall, Englewood Cliffs, 1966.

[38]   E.T. Whittaker- G.N. Watson, *Modern Analysis,*
       Cambridge Univ. Press, 1927.

[39]   E.I. Zolotarev, *Complete Collected Works,* I(1931),
       II (1932), Leningrad.
       Sur l'application des functions elliptiques aux
       questions de maxima et minima, *Bull. Acad. Sc.*
       St. Petersburg (3), 24(1878), 305-310, Melanges,
       5, 419-426 ($\equiv$ I, 369-376).
       Anwendungen der elliptischen Functionen auf Probleme
       über Functionen die von Null am wenigsten oder am
       meisten abweichen (in Russian), *Abh. St. Petersburg*
       30 (1877) 5. ($\equiv$ II, 1-59).

J. Todd
California Institute of Technology
Pasadena,
California 91125

NONZERO EIGENVALUES OF SINGULAR MATRICES, JORDAN NORMAL FORM OF MATRICES

GY. VARGA

1. INTRODUCTION

It is a well-known fact that difficulties may arise when calculating nonzero eigenvalues of nonsymmetric singular matrices by the usually applied methods. The commonly used iterative methods reqire the factorization of the matrix in question in each step, but this generally cannot be performed for singular matrices. For those of a Hessenberg form Wilkinson [1] gives a method to construct a Hessenberg matrix of lower order, from which the nonzero eigenvalues of the initial matrix can be calculated. However, if the initial matrix is not of a Hessenberg form, then it seems to be unnecessary and impractical to transform the matrix into Hessenberg form instead of directly reducing its order.

In this paper a numerical method is given that reduces the multiplicity of multiple zero eigenvalues of complex matrices and, as a special case, determines a regular matrix whose eigenvalues are the nonzero ones of the initial matrix. Furthermore it will be shown how the

method can be applied for determining the Jordan blocks
of a known multiple eigenvalue of a complex matrix that
may be nonsingular, too. Finally an example will be given.

2. DESCRIPTION OF THE METHOD AND A THEOREM CONNECTED
WITH IT

In this section an algorithm is given and a theorem
is proved in connection with the determination of the
Jordan blocks of matrices having zero or multiple eigenvalues.

Let $A \in \mathbb{C}^{n \times n}$ be a complex singular matrix. The
algorithm consists of the following steps (a) to (f):

(a) Perform Gaussian elimination for A with complete pivoting. It is the equivalent of post-
multiplying matrix A by a permutation matrix
F and premultiplying by a matrix G which is
the product of permutation matrices and lower
triangular matrices having ones in their diagonal. In notation the result is $C = GAF$. The
number of the successfully performed elimination
steps gives the rank r of matrix A.

(b) Bring matrix C into an Hermite normal form

$$(2.1) \qquad H = \begin{bmatrix} I & W \\ 0 & 0 \end{bmatrix}$$

by backward elimination steps and appropriate
normalizations of row. It is the equivalent of
premultiplying C by an upper triangular matrix V having ones in its diagonal and by a
nonsingular diagonal matrix D: $H = DVGAF$.

(c) In line with the Gaussian elimination, form matrix $B = F^{-1}AF$. Matrix B came about from A by changes of columns when performing Gaussian elimination and by the same changes of rows, so it is similar to A.

(d) Solve the matrix equation $BX = 0$ for determining the columns of X that span the null space of B. From the construction of H and B, it can be seen that equation $BX = 0$ is equivalent to equation $HX = 0$. Instead of solving $BX = 0$, it is simpler to solve equation $HX = 0$. From the equality

$$(2.2) \qquad \begin{bmatrix} I & W \\ 0 & 0 \end{bmatrix} \cdot \begin{bmatrix} -W \\ I \end{bmatrix} = 0$$

it is seen that

$$(2.3) \qquad X = \begin{bmatrix} -W \\ I \end{bmatrix}$$

is a full system of solutions of equation $HX = 0$.

(e) Substitute the above solution into equation $BX = 0$. Partition matrix B into blocks comparably to H as

$$(2.4) \qquad B = \begin{bmatrix} P & Q \\ R & S \end{bmatrix}.$$

After substitution one gets

$$(2.5) \qquad \begin{bmatrix} P & Q \\ R & S \end{bmatrix} \cdot \begin{bmatrix} -W \\ I \end{bmatrix} = \begin{bmatrix} -PW+Q \\ -RW+S \end{bmatrix} = 0.$$

(f) Finally perform a similarity transformation for B that is constructed by means of X:

$$(2.6) \quad (A\sim) B \sim \begin{bmatrix} I & W \\ O & I \end{bmatrix} \cdot \begin{bmatrix} P & Q \\ R & S \end{bmatrix} \cdot \begin{bmatrix} I & -W \\ O & I \end{bmatrix} = \begin{bmatrix} I & W \\ O & I \end{bmatrix} \begin{bmatrix} P & O \\ R & O \end{bmatrix} =$$

$$= \begin{bmatrix} P+WR & O \\ R & O \end{bmatrix}.$$

For constructing the matrix $P+WR$, P and R can be found in (2.4) and W in (2.1).

Let $Z \in C^{n \times n}$ that has a multiple eigenvalue λ. One has the following

THEOREM. *By one application of the above described algorithm for $A=Z-\lambda I$ the order of each Jordan block of λ is reduced by 1.*

The theorem can be proved by applying the algorithm for matrix $A=Z-\lambda I$ that is singular and has the same block structure as Z [2].

3. COROLLARIES

1) If applying the algorithm repeatedly for A_i $(i=0,1,\ldots)$ (where $A_0=A=Z-\lambda I$ and A_i is the upper left hand block in the matrix resulting from the algorithm when applied for A_{i-1}) as many times until the order of A_i equals its rank for some i, then one gets the order of each Jordan block belonging to λ by tabulating the reductions of the order of A_{i-1} after each application of the algorithm.

2) If Z is singular, then apply the algorithm for matrix Z. According to Corollary 1, one obtains the number and the order of the Jordan blocks belonging to eigenvalue O and, as a final

result, a regular matrix is obtained whose eigenvalues are the nonzero ones of the initial matrix.

Note that in order to determine the block structure
of a given eigenvalue, one has to apply the algorithm
exactly as many times as the order of the largest Jordan
block belonging to that eigenvalue.

4. EXAMPLE

For a matrix of order 17 that has a multiple eigenvalue λ, one can get the following table of numbers:

order	reduction of order
17	4
13	3
10	2
8	1
7	1
6	0

Then putting as many ones in the rows of the following
table as the reductions of order one gets:

4			
1	1	1	1
1	1	1	
1	1		
1			
1			
5	3	2	1

It is seen from the Theorem that the number of blocks
belonging to λ is 4, and their orders are 5,3,2 and
1.

REFERENCES

[1]   J.H. Wilkinson, *The Algebraic Eigenvalue Problem*,
      Oxford University Press, London, 1965.

[2]   Gy. Varga, Reduction of an eigenvalue problem, *Al-
      kalmazott Matematikai Lapok*, 2(1976), (in Hungarian).

Gy. Varga
Computer and Automation Institute
of the Hungarian Academy of Sciences,
H-1111 Budapest
Kende u. 13-17.

ON ERROR ESTIMATION FOR APPROXIMATE INTEGRATION ON THE CLASS $D_S^{\alpha,\beta,\gamma}(A,B)$

I.M.ZHILEIKIN - I.KÖRNYEI

It is well known, that if $f(x)$ is a continouos function of s variables and their partial derivatives exist up to the αs-th order and are continouos functions on the unit cube, then for all natural N there exists an N-point integration rule, for which the error term is $O(N^{-\alpha})$.

Precisely, let E_s be the s-dimensional unite cube, then with suitable points $x_i^{(N)}$ and suitable weights $A_i^{(N)}$

$$(1) \qquad \left| \int_{E_s} f(x)\,dx - \sum_{i=1}^{N} A_i^{(N)} f\left(x_i^{(N)}\right) \right| \le \frac{c_1 M}{N^\alpha},$$

where c_1 depends on α,s, but it is independent of N, and

$$(2) \qquad M = \max_{\alpha_i \ge 0} \ \max_{x \in E_s} \left| \frac{\partial^{\alpha_s} f(x)}{\partial x_1^{\alpha_1} \ldots \partial x_s^{\alpha_s}} \right|.$$

$$\alpha_1 + \ldots + \alpha_s = \alpha s$$

N. S. Bakhvalov [1] has proved the existence
of a constant c_2, independent of N, such that for arbitrary points $x_i^{(N)}$ and weights $A_i^{(N)}$ a function
exists, with continouos derivatives up to the αs-th
order, for which

$$(3) \qquad \left| \int_{E_s} f(x)dx - \sum_{i=1}^{N} A_i^{(N)} f(x_i^{(N)}) \right| \geq \frac{c_2 M}{N^\alpha} \; ,$$

where M is as before.

In this paper, the error estimation of approximate
integration over the entire s-dimensional space will be
examined.

For natural numbers α, s, for positive real
numbers A, B and for real numbers β, γ ($\beta > s$) let
$D_s^{\alpha, \beta, \gamma}(A, B)$ be the class of functions, which are continuous on the s-dimensional Euclidean space and let

$$(4) \qquad |f(x)| \leq \frac{A}{(1+|x|)^\beta} \; , \qquad x \in R_s$$

which have continuous derivates up to αs-th order and
let

$$(5) \qquad \left| \frac{\partial^{\alpha s} f(x)}{\partial x_1^{\alpha_1} \ldots \partial x_s^{\alpha_s}} \right| \leq \frac{B}{(1+|x|)^\gamma} \; , \qquad x \in R_s$$

$$\alpha_i \geq 0; \quad \alpha_1 + \ldots + \alpha_s = \alpha s$$

One may suppose, that $\gamma \leq \alpha s + \beta$. If this is not the case,
then f is an element of $D_s^{\alpha, \beta, \gamma'}(A', B')$ with suitable
γ', $\gamma' < \gamma$.

Let ℓ_N be an N-point integration rule
$$\ell_N(f) = \sum_{i=1}^{N} A_i f(x_i) \quad \text{and } R_s' \text{ a domain in } R_s, \text{ then}$$

$$\rho(\ell_N, f, R'_s) = |\int_{R'_s} f(x)dx - \ell_N(f)|.$$

The following theorem will be proved:

THEOREM. *For an arbitrary number N there exists an N-point integration rule ℓ_N, such that the error term of the formula for all elements of $D_s^{\alpha,\beta,\gamma}(A,B)$ is less than*

$$c_3 \cdot N^\alpha, \qquad c_4 \frac{(\ell nN)^{\alpha+1}}{N^\alpha} \quad or \quad c_5 \cdot N^{\frac{\alpha(s-\beta)}{\alpha s+\beta-\gamma}}$$

depending on whether γ is greater, equal or less than $(\alpha+1)s$, where c_3, c_4 and c_5 are dependent on the class $D_s^{\alpha,\beta,\gamma}(A,B)$ but independent of N.

The construction of the N-point integration rule ℓ_N is as follows.

Let $G(T)$ be the s-dimensional cube

$$G(T) = \{x=(x_1,\ldots,x_s); \quad |x_i| \le T, \quad i=1,\ldots,s\}$$

If T is large, then $\int_{R_s} f(x)dx$ and $\int_{G(T)} f(x)dx$ are equal approximately. Integers K_0 and I will be chosen later. Let $K_i = K_0 \cdot 3^i$ and $T = K_I$.

$G(K_i) \backslash G(K_{i-1})$ is a union of cubes, each of which is congruent to $G(K_{i-1})$. The number of these cubes is $3^s - 1$. Let $G_i^{(j)}$ be the j-th of these cubes. Let N_i be the same number for all $G_i^{(j)}$, $j=1,2,\ldots,3^s-1$. In each cube one can choose an N_i-point rule $\ell_i^{(j)}$ such that

$$(6) \qquad \rho(\ell_i^{(j)}, f, G_i^{(j)}) \le \frac{c_1 (2K_i)^{(\alpha+1)s} B}{(1+K_i)^\gamma} \frac{1}{N_i^\alpha}$$

for all $f(x) \in D_s^{\alpha,\beta,\gamma}(A,B)$

as a consequence of (1) and (4).

Naturally $G(k+1)\backslash G(k)$ is a union of unit cubes $G_k^{(j)}$, the sides of which are parallel to the coordinate-planes. Their number is $(2(k+1))^s - (2k)^s$. Let M_k be the same number of points chosen for all cubes and one can choose M_k-point formulae $\tilde{\ell}_k^{(j)}$, such that the following inequality is valid

(7)
$$\rho(\tilde{\ell}_k^{(j)}, f, G_k^{(j)}) \leq \frac{c_1 B}{(1+k)^\gamma} \frac{1}{M_k^\alpha},$$

$$f \in D_s^{\alpha,\beta,\gamma}(A,B).$$

Let

(8)
$$\ell(f) = \sum_{i=1}^{I} \sum_{j=1}^{3^s-1} \ell_i^{(j)} + \sum_{k=0}^{K_o-1} \sum_{j=1}^{(2(k+1))^s-(2k)^s} \tilde{\ell}_k^{(j)},$$

then

$$\rho(\ell_N, f, R_s) \leq c_1 B (3^s-1) \sum_{i=1}^{I} \frac{(2K_i)^{(\alpha+1)s}}{(1+K_i)^\gamma} \frac{1}{N_i^\alpha} +$$

(9)
$$+ c_1 B \sum_{k=0}^{K_o-1} \frac{(2(k+1))^s - (2k)^s}{(1+k)^\gamma} \frac{1}{M_k^\alpha} +$$

$$+ c_7 T^{s-\beta}.$$

Namely

$$\rho(\ell_N, f, R_s) \leq c_8 \sum_{i=1}^{I} 3^{((\alpha+1)s-\gamma)i} \frac{1}{N_i^\alpha} +$$

$$+ c_9 \sum_{k=0}^{K_o-1} \frac{(2(k+1))^s - (2k)^s}{(1+k)^\gamma} \; \frac{1}{M_k^\alpha} + c_7 T^{s-\beta}$$

The number of the points, employed in the integration formula, is

$$N = (3^s-1) \sum_{i=1}^{I} N_i + \sum_{k=0}^{K_o-1} \left(((2(k+1))^s - (2k)^s\right) M_k$$

The functions of the numbers N_i and M_k

$$\sum_{i=1}^{I} 3^{((\alpha+1)s-\gamma)i} \; \frac{1}{N_i^\alpha} \qquad \text{and}$$

$$\sum_{k=0}^{K_o-1} \frac{(2(k+1))^s - (2k)^s}{(1+k)^\gamma} \; \frac{1}{M_k^\alpha}$$

have constrained global minimum subject to

$$\sum_{i=1}^{I} N_i = c_{10} \cdot N \qquad \text{and} \qquad \sum_{k=0}^{K_o-1} ((2(k+1))^s - (2k)^s) M_k = c_{11} \cdot N,$$

where c_{10} and c_{11} are constants.

The coordinates in the point of minimum are not necessarily integers, but one can choose integers near to them.

Let N_i or M_k be chosen according to the inequalities

$$\gamma > (\alpha+1)s \qquad\qquad \gamma = (\alpha+1)s \qquad\qquad \gamma < (\alpha+1)s$$

$$N_i \qquad \left[c_{12} N \cdot 3^{-\varepsilon i} \right] \qquad\qquad \left[c_{13} \frac{N}{\ln N} \right] \qquad\qquad \left[\frac{c_{14} \cdot 3^{\left(s-\frac{\gamma}{\alpha+1}\right)i}}{3^{\left(s-\frac{\gamma}{\alpha+1}\right)I}} \right]$$

$$M_k \qquad \left[c_{15} \frac{N}{(k+1)^{\gamma/\alpha+1}} \right] \qquad \left[\frac{c_{16}}{(k+1)^s} \frac{N}{\ln N} \right] \qquad \left[c_{17} \frac{N}{K_o^{\,s-\frac{\gamma}{\alpha+1}} (k+1)^{\frac{\gamma}{\alpha+1}}} \right]$$

where $\varepsilon, c_{12}, \ldots, c_{17}$ are positive constants.

If T and K_o are chosen as

$$\gamma > (\alpha+1)s \qquad\qquad \gamma = (\alpha+1)s \qquad\qquad \gamma < (\alpha+1)s$$

$$T \qquad \left[c_{18} N^{\frac{\alpha}{\beta-s}} \right] \qquad\qquad \left[c_{19} N^{\frac{\alpha}{\beta-s}} \right] \qquad\qquad \left[c_{10} N^{\frac{\alpha}{\alpha s+\beta-\gamma}} \right]$$

$$K_o \qquad \left[c_{21} N^{\frac{\alpha+1}{\gamma}} \right] \qquad\qquad \left[c_{22} \left(\frac{N}{\ln N}\right)^{1/s} \right] \qquad \left[c_{23} N^{1/s} \right]$$

then one can find constants $c_{18}, \ldots, c_{23}$ such that the following inequality is true

$$(3^s-1) \sum_{i=1}^{I} N_i + \sum_{k=o}^{K_o-1} \left((2(k+1))^s - (2k)^s \right) M_k \leq N$$

For sufficiently large N, the arguments of the expressions for N_i and M_k are greater than 1. We get the statement of the theorem by using the inequality $[x] \geq \frac{1}{2}x$.

A stronger statement as stated in the theorem is not

true. There are constants c_{24}, c_{25}, c_{26} such that for
all integration rule ℓ_N there exists a function in the
class $D_s^{\alpha,\beta,\gamma}(A,B)$, for which the statement of the theorem
does not hold if c_{24}, c_{25}, c_{26} are substituted for
c_3, c_4 and c_5.

REFERENCES

[1] N.S. Bakhvalov, On optimal convergence of quadrature
    rules and Monte-Carlo integration methods on classes
    of functions, in *Numerical Methods for Solving
    Differential and Integral Equations and Quadrature
    Formulae*, Nauka, Moscow, 1964, pp. 42-48 (in Russian).

I.M. Zhileikin
Computing Centre, Moscow State University
117234 Moscow
USSR
I. Környei
Computing Centre for Universities
Budapest, Dimitrov tér 8.
H-1098